T0144484

PREDICTION VERSUS PERFORMANCE IN GEOTECHNICAL ENGINEERING

PROCEEDINGS OF THE SYMPOSIUM ON PREDICTION VERSUS PERFORMANCE
IN GEOTECHNICAL ENGINEERING / BANGKOK / 30 NOVEMBER - 4 DECEMBER 1992

Prediction versus Performance in Geotechnical Engineering

Edited by
A.S. BALASUBRAMANIAM
YUDBHIR, D.T. BERGADO, N. PHIEN-WEJ
T.H. SEAH & P. NUTALAYA
Asian Institute of Technology, Bangkok, Thailand

A.A. BALKEMA / ROTTERDAM / BROOKFIELD / 1994

The texts of the various papers in this volume were set individually by typists under the supervision of either each of the authors concerned or the editor.

Authorization to photocopy items for internal or personal use, or the internal or personal use of specific clients, is granted by A.A. Balkema, Rotterdam, provided that the base fee of US$1.00 per copy, plus US$0.10 per page is paid directly to Copyright Clearance Center, 222 Rosewood Drive, Danvers, MA 01923, USA. For those organizations that have been granted a photocopy license by CCC, a separate system of payment has been arranged. The fee code for users of the Transactional Reporting Service is: 90 5410 355 8/94 US$1.00 + US$0.10.

Published by
A.A. Balkema, P.O. Box 1675, 3000 BR Rotterdam, Netherlands
A.A. Balkema Publishers, Old Post Road, Brookfield, VT 05036, USA

ISBN 90 5410 355 8
© 1994 A.A. Balkema, Rotterdam
Printed in the Netherlands

Prediction versus Performance in Geotechnical Engineering, Balasubramaniam et al. (eds)
© 1994 Balkema, Rotterdam, ISBN 90 5410 355 8

Table of contents

3 Embankments, excavations and buried structures

4 Earth structures, mines and slopes

5 *Dynamic behaviour of soil and earthquake*

Preface

This volume is the ninth in a series of Balkema Publications on the Annual Geotechnical Symposium sponsored by the Asian Institute of Technology and the Southeast Asian Geotechnical Society. The previous volumes are devoted to Geotechnical Aspects of Restoration Works on Infrastructures and Monuments (1990), Computer and Physical Modeling in Geotechnical Engineering (1989), Environmental Geotechnics and Problematic Soils and Rocks (1988), Geotechnical Aspects of Mass and Material Transportation (1987), Recent Developments in Laboratory and Field Tests and Analysis of Geotechnical Problems (1986), Recent Developments in Ground Improvement Techniques (1985), Geotechnical Aspects of Coastal and Offshore Structures (1983), and Geotechnical Problems and Practices of Dam Engineering (1982).

The new volume deals with the Prediction versus Performance in Geotechnical Engineering and includes thirty two papers of international significance categorized into five sections for ease of reference. The first section on Foundation Engineering and Field and Laboratory Testing contains six valuable contributions from Japan, Italy, Malaysia, Korea and India related to modeling and/or field investigation of flexible sluiceway, spliced piles, levee in the Po Delta, Italy, and geogrid reinforcements in various kinds of soft and compressible soils.

Section two is on Ground Improvement Techniques and Reinforced Earth containing seven papers dealing with such techniques as geomembranes, vertical drains, micropiling, coal-ash treatment of asphalt subgrade, and computer grouting. The third section is on Embankments, Excavations and Buried Structures with eight papers while section four is on Earth Structures, Mines and Slopes with nine papers. Finally section five is on Dynamic Behaviour of Soil and Earthquake with two papers which investigated the ground failures in Japan and Iran during earthquakes.

The release of this volume was made possible through the efforts, help and support received from many individuals and organizations or institutions, and are therefore gratefully acknowledged. First of all, acknowledgement is due to the following members of the General Committee of the Southeast Asian Geotechnical Society: Dr Ooi Teik Aun (President), Dr C.D.Ou, Dr Za-Chieh Moh, Dr S.B.Tan, Dr E.W.Brand, Dr W.H.Ting, Prof. S.L.Lee, Dr A.W.Malone, Prof. Sambhadharaksa, Dr K.Y.Yong, Dr J.C.Li, Dr S.F.Chan and Dr Clive Franks.

At the Asian Institute of Technology, the editors would also like to express their sincere thanks among others to Prof. Alastair North, President and to Dr Yordphol Tanaboriboon for their wholehearted support. The meticulous and untiring work and the patience of Elmer Bandalan in checking and proofreading the manuscripts have made possible the making of this volume in its present form.

A.S. Balasubramaniam

1 Foundation engineering and field and laboratory testing

Prediction versus Performance in Geotechnical Engineering, Balasubramaniam et al. (eds)
© *1994 Balkema, Rotterdam, ISBN 90 5410 355 8*

Performance of flexible sluiceway in compressible ground

Norihiko Miura & Madhira R. Madhav
Institute of Lowland Technology, Saga University, Japan

Yoshiyuki Kawakami
Ministry of Construction, Takeo Office, Saga, Japan

ABSTRACT: Sluiceway across an embankment in Japan has been constructed on end bearing piles. When this method is adopted in a compressible soil, large differential settlements occur between the sluiceway and the ground, leading to a harmful cavity under the sluiceway. To avoid this problem, a flexible system of sluiceway with floating type foundation is investigated. This is a report of the proposed system, describing the performance of test embankments on prototype of flexible sluiceways with skin resistance piles and improved soil columns.

INTRODUCTION

Chikugo-Saga plain is located along the northern coast of Ariake Sea in Kyushu. The sea has a very large (6m) tidal range and an extremely soft deposit beneath. Rokkaku river, one of the big rivers flowing into the Ariake bay, is significantly affected by this tidal action. In fact, the slope of the river is so gentle that sea water comes 29km inland from the estuary at high tide. Thus, along the Rokkaku river, a highly compressible marine clay, Ariake clay, is deposited in a thickness of 15 to 20 meters (Nakamura et al. 1985). Because of its high compressibility, high sensitivity and low strength, public works adjacent to or on the river are met with serious geotechnical difficulties.

One of these problems is in the foundation of a sluiceway across an embankment on soft ground. Presently, sluiceway is constructed on end bearing piles according to an officially recommended design method. When the sluiceway is constructed on soft clay the following problems are met with: a) the surrounding ground settles due to embankment weight resulting in a cavity under the sluiceway; and b)

large differential settlements occur near the shoulder of the structure causing cracks in the embankment.

As a countermeasure for this, a pilot project was carried out by the Takeo Office, Ministry of Construction, where a rigid raft on a floating foundation system was adopted using skin resistance piles. As described in an earlier report (Yoshioka 1988), the sluiceway suffered many cracks and this alternative was not successful either.

The above experience suggests that a flexible type of structure should be provided for the sluiceway with a floating type foundation. To investigate a flexible system for a sluiceway in soft ground, Ministry of Construction, Japan, started field tests in 1990 at three places - Ishikari plain in Hokkaido where extremely compressible peat deposits, and Kanto plain near Tokyo and Takeo in Saga plain where soft clay deposits exist. The authors have taken part in the investigation and are responsible for the Takeo project. The performances of the prototype models of sluiceway constructed in 1991 are being monitored. This is a report based on the data obtained so far.

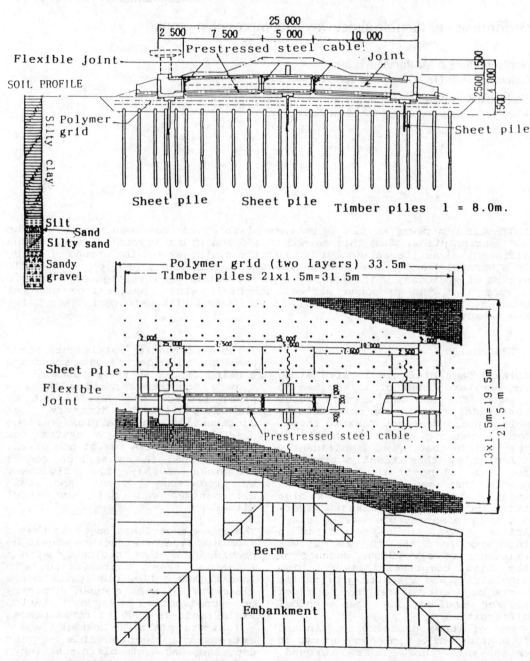

Fig.1 Configuration of timber pile - grid foundation Type A

OUTLINE OF FLEXIBLE FOUNDATION SYSTEM

The basic idea of the flexible foundation system is that the sluiceway deformations follow the controlled differential settlement of the ground, which is reinforced by a floating foundation. The maximum settlement should be smaller than the allowable value for the

4

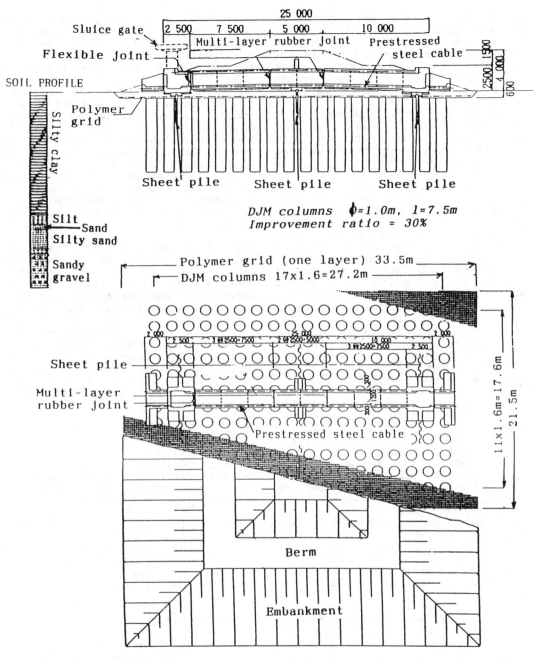

Fig.2 Configuration of DJM column - grid foundation Type B

sluiceway for its proper functioning. Tentative code recommends 30cm for this value, in Japan. Among the different types of floating foundations, a) Timber pile-grid method (Type- A), and b)Improved column-grid method (Type-B) were investigated at the Takeo test field.

Type-A foundation uses timber pile as skin resistance pile in soft clay in a configuration as shown in Fig.

5

1. On top of the pile group, a polymer grid reinforced granular course was placed to confine the lateral displacements of the piles. On this reinforced course, camber was provided to minimise excess differential settlement of the sluiceway.

Present code for the design of skin resistance of a pile does not permit the consideration of the skin resistance when the N value of soft ground is less than 2. This means floating type piles cannot be used at this site, since N value of the clay is actually zero. However, Miura et al.(1992) have established that a certain amount of skin resistance is available for timber piles in clay even if N value is less than two. Timber pile is advantageous because: a) timber absorbs water from the film formed at the pile-soil interface; b) since the timber pile tapers naturally, it leads to a closer contact with clay; c) timber pile is lighter compared to a concrete pile. Unit shaft resistance of a timber pile can be taken as equal to the undrained shear strength of clay.

In Type-B test, improved columns of clay of 1 m diameter are formed by the DJM method, a mechanical mixing method of quicklime with clay in-situ (Chida, 1982, Miura, et al., 1987). The DJM method has been widely used for the foundation of a high embankment on soft clay, with the columns reaching a bearing stratum under the clay layer. To use the DJM method as a floating founda-tion, the lengths of the DJM columns are made shorter than the depth of the clay layer. Fig. 2 shows the outline of the Type-B foundation, where the ratio of improved area to the original ground is 30%. This improvement ratio is smaller than the usual value of 50%. A reinforced granular base course with a single layer of polymer grid and camber was provided as in Type-A foundation.

FLEXIBLE SLUICEWAY

To make the sluiceway flexible, its body is separated into several blocks which are then connected by special type of joints. The sluiceway blocks were placed along an estimated settlement curve with each joint lying exactly on the settlement curve. The maximum differential settlement was designed to be smaller than the allowable value. An analysis based on the theory of a beam on elastic foundation was made, and the stress induced in the reinforced concrete body of the sluiceway estimated. Iterative calculations were made for several cases until the stresses and the differential settlement were within the allowable values. Based on the calculations, two types of sluiceways were selected and constructed by connecting the four blocks as shown in Figs. 1 and 2.

Fig. 3 illustrates two types of joints. A flexible joint, which is used in Type-A foundation construction, is capable of 5cm extension, 1cm compression and 30cm differential settlement. Multi-layer rubber joint, used in Type-B foundation, is attached to the concrete body in a compressed state to permit 6cm expansion. Each joint works for differential settlement up to 30cm. The details of each block of the two sluiceways are depicted in Fig.4.

FIELD TEST

The main purpose of the field tests is (a)to verify the effectiveness of the foundation for controlling the differential settlement; b) estimate the magnitudes of deformation and stresses induced in the concrete body of the sluiceway by the differential settlement of the ground; c) Check the functioning of the joints; and d) provide a practical design method.

Ground Condition

The field test site is located 1.3 km from the estuary of Rokkaku river, where the thickness of the Ariake clay layer is 12.5m. The soil properties are: Natural water content = 90 - 140%; Liquid limit = 80 - 130; Plastic limit = 30 - 45; Unit weight = 1.35 - 1.55t/m^3; Unconfined compressive strength = 2 - 7 t/m^2. The consolidation yield stress values scatter around the

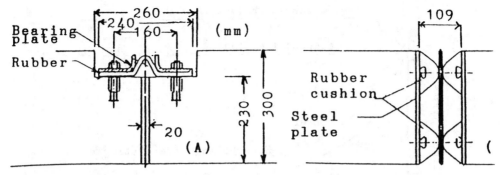

Fig.3 Flexible joint A and Multi - layer rubber joint B

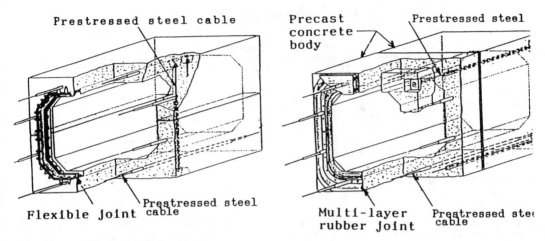

Fig.4 Details of sluiceways Type A and Type B

line of effective overburden stress indicating that the clay is normally consolidated. The coefficient of consolidation = $1 \times 10^2 \text{cm}^2$ - $4 \times 10^2 \text{cm}^2$/d, and the compression index = 0.49 to 1.82.

Construction and Predictions

Calculations indicate that the final settlement due to the weight of the 4m high embankment would be 190 cm and the safety factor for slip failure is 0.9 when the sluiceway and embankment are placed on the ground without any foundation. In order to decrease the settlement and to increase the bearing capacity, Type-A foundation uses timber piles 8m long, in a square configuration

with 1.5m spacing as shown in Fig. 1. With this type of foundation, the stresses from the superstructure are assumed to be distributed uniformly at a depth equal to two-third the length of the piles so that the 7.2m thickness of the compressible layer below would be one third of the length of the pile (=2.7m) plus the depth of clay below the pile tip (13.5m - 8.0m- 4.5m). Final settlement due to consolidation of this layer is estimated to be 80 cm. The safety factor for slip failure increases to 1.3. Polymer-grid reinforced base course was constructed, on which a camber with 30cm height at the center was also provided. The final level of sluiceway at the center after consolidation is estimated to be

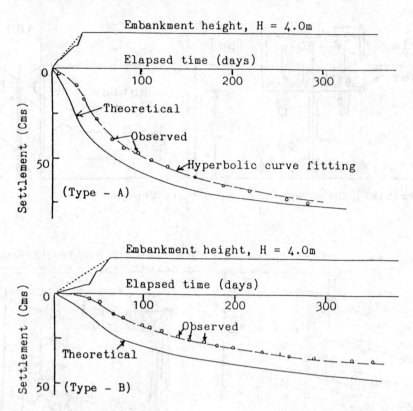

Fig.5 Time - settlement curves

50cm below the horizontal base line. The four blocks of the sluiceway were connected by flexible joints, and the prestressing steel wires were installed through the sluiceway to prevent opening of the connections, as shown in Fig. 4. The direct cost for the Type-A foundation was 7 million yen for the pile foundation and 6 million yen for the granular base with two layers of polymer grid.

Type-B foundation was constructed with DJM columns of 7.5m length in 1.6m pitch configuration as shown in Fig.2. The thickness of the compressible layer beneath the DJM columns is assumed to be 12.5m - 7.5m = 5m, and the final settlement of the layer is estimated as 50 cm. For this foundation the safety factor of 2.2 is expected. The camber provided is with 20 cm height at the center. Hence, the final level of the sluiceway at the center is expected to be 30cm below the original ground level. For joining

the concrete blocks multi-layered rubber joints were used. To ensure the connections of the four blocks, steel wires were installed in tension as in the case of Type-A foundation design. Direct cost for this foundation was 12 million yen for DJM column construction and 3 million yen for base course with a single layer of polymer grid.

OBSERVATION

Settlement of Ground

The observed time - settlement relations are plotted in Fig.5 and a hyperbolic curve fitted. The final settlements are estimated as 96cm for Type-A, and 61cm for Type-B foundation alternatives. These values are about 20% larger than the predicted ones.
Fig. 6 shows the horizontal displacements of the test embankments at their toes. Dotted line in

8

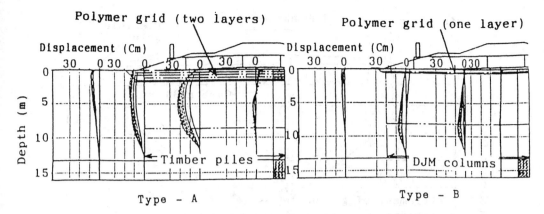

Fig.6 Horizontal displacement at the toe of the embankment

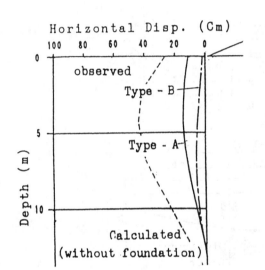

Fig.7 Horizontal displacements

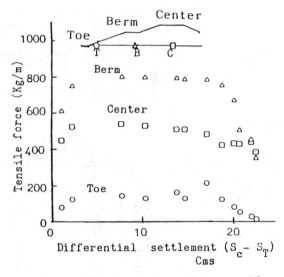

Fig.8 Tensile force in polymer grid

Fig. 7 indicates the horizontal displacements of untreated ground calculated by FEM, and the other two lines show the observed displacements. Comparing the observed values with those predicted, the horizontal displacements at the surface are 40% and 10% of the value for the untreated case for Type-A and Type B foundations, respectively, suggesting that the latter alternative is working very effectively.

The reductions in the horizontal displacements are considered to result partly from the action of the polymer grid. In Fig. 8, one can note the mobilisation of the tensile force in the polymer grid as the ground settles. The magnitude of the tensile force is largest below the berm. This indicates that the polymer grid is actually effective in confining the soil and in reducing the horizontal displacements of the piles. In the region where the differential settlement exceeds 20cm, the value of the tensile force drops rapidly suggesting that slip between the soil and the polymer

9

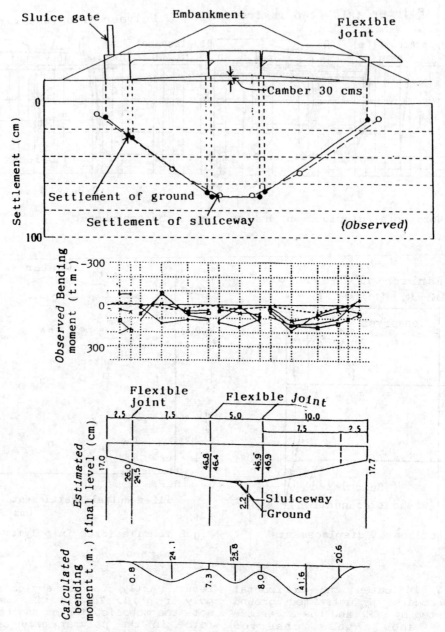

Fig.9 Settlement distribution of Type A sluiceway

grid, and/or breaking of the polymer grid might have occured.

Settlement of Sluiceway

Fig. 9 shows the observed settlement distribution curves of both the ground surface and the bottom of the sluiceway. The two curves overlap each other, indicating that the sluiceway settled along with the ground. It also depicts the final settlement distribution and the

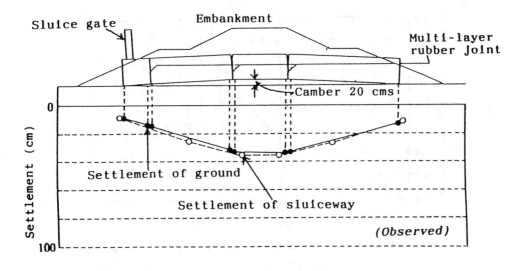

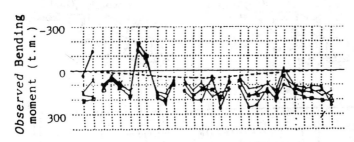

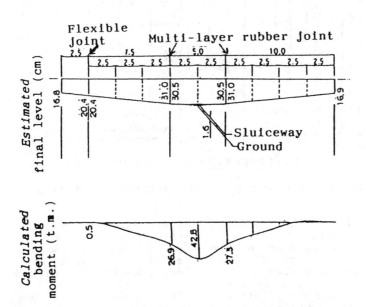

Fig.10 Settlement distribution of Type B sluiceway

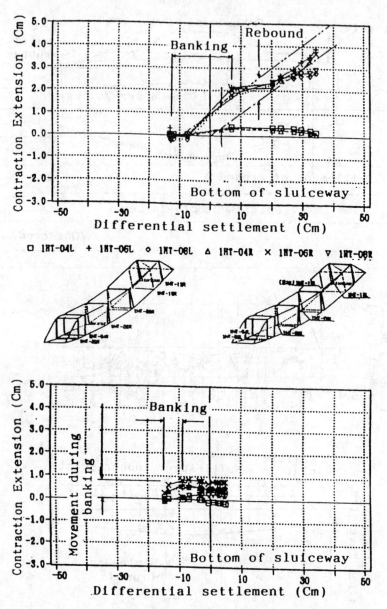

Fig.11 Movements of flexible joint A and multilayer rubber joint B

corresponding values of bending moments of the Type-A sluiceway. Observed value of bending moment scatter between 100 to 200 t.m. These values are less than 30% of the allowable ones. The observed results indicate that the joints are working well as anticipated and there is no leakage of water. The behavior of the Type-B foundation design is shown in Fig.10. As in Type-A foundation, the sluiceway settled along with the adjacent ground. The observed maximum moment was 260 t.m, which is considerably less than the allowable value.

Fig. 11 shows the magnitudes of extension and compression of the flexible joints at the bottom of the sluiceway in Type-A foundation test, and also those of the multi-layer rubber joint used in Type-B foundation test. The allowable extension of the former joint is 5 cm, and of the latter 6cm. Thus the observed values are smaller than the allowable ones. At the roof of the sluiceway, Type-A Design showed 0.5 cm of compression, and Type-B 1cm of compression, which are less than the allowable values.

CONCLUDING REMARKS

Observations at the test fields are continuing and the final evaluation on this project will be made six months later. At this stage, however, a positive perspective can be noted on the new system of flexible sluiceway with floating foundation constructed on compressible ground. Based on the available results, the following remarks are made:
1) Two types of floating foundations, timber pile - grid system and DJM column-grid system, are investigated. Both of them are working well and as anticipated, for controlling the differential settlement of ground under the load of sluiceway and embankment.
2) Lateral displacements of timber piles and DJM columns were suppressed and are much smaller than those predicted for the ground without floating foundation, suggesting that the polymer-grid reinforced course contributed to the confining effect.
3) Prototype models of sluiceway deformed appropriately following the settlement profile of the ground. The stresses and distortions induced in the concrete body are considerably lower than the allowable values. No cracks in the concrete body and no cavity under the sluiceway are detected till now.
4) Flexible joint in Type-A Design showed an extension of 4cm which is smaller than the allowable value of 5 cm. Multi-layer rubber joint in Type-B Design extended by only 1cm which is smaller than the allowable value of 6 cm. No water leaked from the joints. Thus, they seem to be working very efficiently.
5) The new system of flexible sluiceway with floating foundation is a good alternative for practical use provided that the individual blocks are joined properly.

REFERENCES

Chida, S. 1982. Dry Jet Mixing Method. State of the art on improvement methods for soft ground," JSSMFE:69-76.
KKC (Kokudo Kaihatsu Center) 1992. Report on flexible foundation with flexible sluiceway on soft ground. Research Report:1- 215. (in Japanese).
Miura, N., Bergado, D.T., Sakai, A. and Nakamura, R. 1987. Improvement of soft marine clay by special admixtures using dry and wet jet mixing methods. 9th Southeast Asian Geotechnical Conference:8/35 - 8/46.
Miura, N., Wu, Wenjing, and Nagayasu, T. 1992. Behaviour of piled-raft foundation in soft clay. Report of Faculty of Science and Engineering, Saga University, 21:1:65-75.
Nakamura, R., Onitsuka, K., Aramaki, G., and Miura, N. 1985. Geotechnical properties of the Ariake clay. Proc. Symposium on Problematic Soils and Rocks, AIT, Bangkok:533-544.
Yoshioka, H. 1988. Problems with foundations for structures on soft ground. Proc. of the Int. Symposium on Shallow Sea and Lowland, Saga:211-229.

Prediction versus Performance in Geotechnical Engineering, Balasubramaniam et al. (eds)
© *1994 Balkema, Rotterdam, ISBN 90 5410 355 8*

Foundation behaviour of a levee in the Po Delta

F.Colleselli & G.Cortellazzo
Istituto di Costruzioni Marittime e di Geotecnica, Università di Padova, Italy

ABSTRACT: This paper reports studies on the behaviour in time of the foundation soils of a levée constructed in the Po Delta on soft compressible clayey soils. Measurements of their behaviour, taken during and after construction, were compared both with the results of predictions made during the design phase, carried out using ordinary methods for settlement assessment, and with numerical simulation of back-analysis with finite-element programs. This paper also considers the main factors influencing the calculation methods used.

1 INTRODUCTION

Study of structures on very thick, soft, compressible soils subject to considerable settlement in time is of great interest and involves large volumes of soil in sufficiently well-defined boundary conditions. In these cases of behaviour prediction and simulation, hardening elasto-plastic constitutive laws (critical-state models) must be used for the soils and consolidation phenomena must be analysed with models often differing from Terzaghi's simple traditional model (Britto & Gunn 1987, Lepidas & Magnan 1990, Magnan 1986, Wroth 1977).

In the case in question, in situ measurements carried out using complex geotechnical instrumentation to verify the foundation soil behaviour of the levée of Volta Vaccari, built between 1980 and 1985 in the Po Delta, were compared with results obtained by traditional design prediction methods available at the time of construction and with more recent ones deriving from numerical simulation using finite-element programs. The study reveals the degree of approximation which may be obtained with the various methods and the difficulties inherent in the various types of analysis.

2 CHARACTERISTICS OF LEVEES AND FOUNDATION SOILS

The Volta Vaccari levée was built further back from and replacing an already existing levée, due to the need to widen and modify the bed of the terminal stretch of the river Po, the main Italian river (Colleselli & Soranzo 1987). The levée is about 1000 m long and is substantially symmetrical. Its foundation plane is 55.0 m wide at the base and 8.0 m at the top; it has two platforms at +0.50 and +2.50 m a.s.l., and is 6.0 m high (Fig. 1).

Boreholes, penetrometric tests and laboratory tests showed the following stratigraphy for the foundation soils, starting from ground level at -2.0 (Fig. 1):

1. a 2.0 m thick low-consistency layer of silty clay and clayey silt (tip resistance $q_c = 0.4 \div 1.0$ MPa; natural water content w at about the liquid limit w_L);

2. from -4.0 to about -10.0, a layer of quite loose sandy silt and fine silty sand ($q_c = 0.5 \div 1.5$ MPa; $\phi = 27°$ for silts, $\phi = 37°$ for sands);

3. from -10.0 to -30.0 ÷ -32.0, a very thick, normal-consolidated layer of silty clay ($c_u/\sigma'_{vo} = 0.23$), whose characteristics improve with increasing depth (undrained shear strength

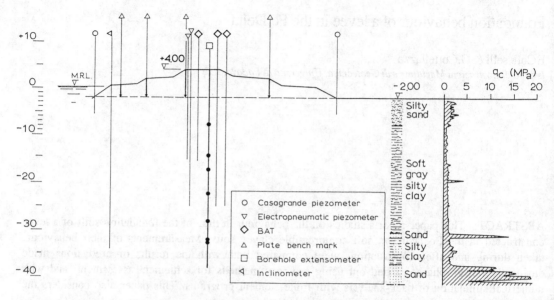

Fig. 1: Soil foundation characteristics and geotechnical instrumentation.

$c_u = 20\div30$ kPa, $w_L = 43\div46$, $I_p = 14\div17$, $C_c = 0.32\div0.39$ in upper layers; $c_u = 25\div33$ kPa, $w_L = 51\div55$, $I_p = 25\div38$, $C_c = 0.32\div0.39$ in central layers; $c_u = 40\div80$ kPa, $C_c = 0.44\div0.66$ in deepest layers);

4. from -32.0 to -45.0, a very dense sandy layer ($q_c \geq 10.0$ MPa).

The values of shear strength and strain parameters obtained from in situ and laboratory tests agreed well with data from many researches carried out by Italian workers on these types of clay (Jamiolkowski, Lancellotta & Tordella 1980, Bilotta & Viggiani 1975).

The number of data available was such that sufficiently good and significant characterization of the soils could be obtained, and the features necessary for processing in order to use Cam-clay and modified Cam-clay constitutive laws were thus determined. These determinations are generally the most difficult aspect in applying elasto-plastic constitutive laws, as the number of available data is usually too low.

The behaviour of the levée and foundation soil was followed in time by complete geotechnical instrumentation, composed of plate bench marks, borehole extensometers, inclinometers, electropneumatic piezometers

and BATs, and Casagrande piezometers (Fig.1).

The properties of the soils used in the various analyses (Fig. 1, Tables 1,2) were obtained from in situ and laboratory test results, in some cases resorting to empirical correlations (Ladd, Foott, Ishihara, Schlosser & Poulos 1977).

Table 1 : Soil foundation parameters of F.E.M. analysis

Layer	κ	λ	Γ	M	E [MPa]
1					1.4
2	0.030	0.165	2.56	1.30	
3	0.035	0.170	2.79	1.07	
4	0.052	0.269	3.39	1.02	
5					40.0

κ = swelling line slope;
λ = normal consolidation line slope;
Γ = specific volume value corresponding to $p' = 1.0$ kNm^{-2};
M = critical state line gradient;
E = elastic modulus.

Table 2 : Soil foundation parameters of elastic F.E.M. and piece-wise linear consolidation method of analysis.

Layer	E' [MPa]	Eu [MPa]	Cc
1	1.4		
2		12.0	0.38
3		21.0	0.39
4		22.0	0.62
5	40.0		

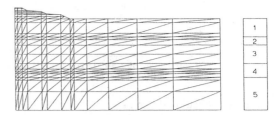

Fig. 2 : Mesh and soil layers.

Measurements using Casagrande piezometers showed that pore pressure in the various layers from the hydrometric level of the river was independent in time.

3 CONVENTIONAL CALCULATION METHOD AND NUMERICAL ANALYSES

During the design phase, the total settlement was obtained as the sum of immediate and consolidation settlements (Skempton & Bjerrum 1957).

For the best assessment both of the extent of theoretical settlements and pore overpressures and their time trends, back-analyses were carried out using two different methods of calculation.

In the first, the stress state induced by overloads in the soil and the immediate settlement were calculated using a finite-element program, hypothesizing elastic behaviour by the soil. The consolidation settlement was assessed by the piece-wise linear consolidation method, adapted for analysis of consolidation phenomena in very thick layers (Yong, Siu & Sheeran 1983).

The second method consisted of another finite-element analysis in which, however, hardening elasto-plastic (Cam-clay, modified Cam-clay) laws were adopted for the clayey layers; elastic constitutive laws were deemed to be valid for the sandy layers. The behaviour of

the structure in time was also simulated using coupled analysis. The element used was the six-noded linear strain triangle with additional degrees of freedom for assessment of pore overpressures. The mesh was composed of 469 elements and 257 nodes (Fig. 2).

In order to simulate levée construction realistically and thus take into account the effective load diagram, the elements representing the structure were added in various time-steps and their weight was applied, subdivided inside each time-step.

4. ANALYSIS OF RESULTS

The design analyses, which were developed on the simple basis of the elasticity theory for evaluation of pressure increments and initial settlements and on the results of oedometric tests for consolidation settlements, identified a maximum final theoretical settlement in the centre of the levée of about 1.15 m. Times required to reach 50% and 90% of the average degree of consolidation were assessed using Terzaghi's classic one-dimensional theory, and turned out to be respectively 920 and 3960 days.

Settlements and pore overpressures were also calculated using two more complex methods.

The first, the piece-wise linear consolidation method, was two-phase. Elastic analysis with a finite-element program was first used to assess initial settlements and stresses in the foundation soil. A finite-differences program followed, accounting for variability in void ratio e and the coefficient of permeability k according to the effective pressures acting in the clayey layers,

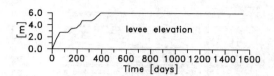

Fig. 3: Loading diagram.

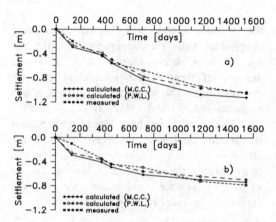

Fig. 4 : Settlement-time relationship according to F.E.M. (modified Cam-clay), piece-wise linear consolidation method, measured: a) point C; b) point D.

thus determining their settlement in time and identifying the time diagram of superficial settlements of two significant points (Figs. 3,4). In the case in question, according to laboratory and in situ tests, the following equation for the variation of k according to effective stress was adopted (Juarez-Badillo 1986):

$$k(\sigma) = 0.49 * 10^{-9} + (\sigma_v'/ 300)^{-0.9367} \ [m/s]$$

Reference to the oedometric tests carried out at various depths was made for void ratio versus vertical effective stress relationships. The settlements thus assessed turned out to vary between 0.33 and 1.03 m along the foundation plane, and the degree of average consolidation was about 43% when the last measurement was read (1550 days).

Further analyses were then carried out, again using finite-element programs but with constitutive laws of the Cam-clay and modified Cam-clay type, in order to assess the overall behaviour of the volume of soil under the levée. The results of these two analyses did not differ greatly: settlements assessed using the Cam-clay model were only slightly greater than those assessed by the modified Cam-clay model.

Theoretical settlements along the whole foundation plane, ranging between 0.35 and 1.10 m, and theoretical settlements along the borehole extensometers at various depths agreed well with effectively measured real values and those calculated using the piece-wise linear consolidation method (Figs. 4,5,6).

Pore pressures at two points in the clayey layer, although different in the two analyses (lower values being found in the uncoupled analysis) fitted experimental values quite well (Fig. 7).

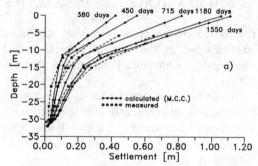

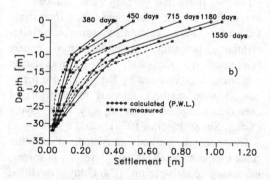

Fig. 5: Settlements versus depth: a) F.E.M. (modified Cam-clay) - measured; b) P.W.L. consolidation method - measured.

18

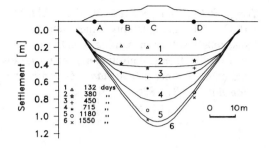

Fig. 6: Calculated (F.E.M., modified Cam-clay) and measured foundation settlements.

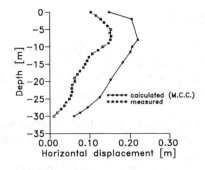

Fig. 8: Calculated (F.E.M., modified Cam-clay) and measured horizontal displacements.

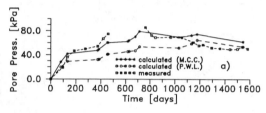

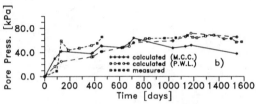

Fig. 7: Excess pore pressure: a) PEP1(-15.0); b) PEP2(-25.0).

Simulations made with both methods may thus be considered sufficiently reliable.

Simulation of the elasto-plastic program also allowed assessment of horizontal displacements, which showed trends similar to the effectively measured ones but differed as regarded absolute values. The maximum measured displacement of 0.15 m did not correspond to the maximum calculated displacement of 0.22 m (Fig. 8). The difficulties of using Cam-clay models for correct simulation of horizontal displacements, as already noted in previous reports (Lepidas & Magnan 1990, Magnan 1986), were thus highlighted.

5. CONCLUSIONS

Analyses using the piece-wise linear consolidation and elasto-plastic finite-element methods simulated quite well the behaviour of soft, compressible clayey foundation soils in levées subject to considerable settlements. They also demonstrated that they can correctly identify pore overpressure trends. As regards settlements and overpressures, only slight differences were found between the various types of analysis.

This study also confirmed that the reliability of results depends on accurate characterization of soil stratigraphy and structure geometry, on suitable simulation of its construction, and on the choice of appropriate boundary conditions.

The elasto-plastic program allows determination of the behaviour of the whole soil volume in question, although processing takes considerably longer than uncoupled analyses and assessment of horizontal displacements does not seem to be sufficiently accurate.

The less accurate calculation methods used during the design phase only gave useful information on some aspects of the problem. Currently available design methods now allow more complete and satisfactory simulations of the behaviour of levée foundation soils in time, although more numerous in situ and laboratory tests than those normally available are required.

REFERENCES

Bilotta, E. & C. Viggiani 1975. A full-scale field investigation on the behaviour of a thick layer of normally consolidated clays. *XII It. Nat. Conf. on Soil Mechanics, Cosenza*: 223-240.

Britto, A.M. & M.J. Gunn 1987. *Critical State Soil Mechanics via Finite Elements*. Chichester, Ellis Horwood Ltd..

Colleselli, F. & M. Soranzo 1987. The seepage control in excavations and river levees. *Proc. IX ECSMFE Groundwater Effects in Geotechnical Engineering, Dublin:* 129-132. Rotterdam: Balkema.

Jamiolkowski, M., Lancellotta, R. & M.L. Tordella 1980. Geotechnical properties of Porto Tolle NC Silty Clay. *Proc. VI Danube Europ. Conf. Soil Mech., Varna Bulgaria:* 151-181.

Juarez-Badillo, E. 1986. General Theory of Consolidation for Clays. In R. N. Young & F.C. Towsend (eds.), *Consolidation of Soils: Testing and Evaluation*, ASTM STP 892: 137-153. Philadelphia.

Ladd, C.C., Foott, R., Ishihara, K., Schlosser, F. & H.G. Poulos 1977. Stress-Deformation and Strength Characteristics. *Proc. 9th ICSMFE State-of-Art Report, Vol. 2, Tokio*: 421-494.

Lepidas, I. & J.-P. Magnan 1990. *Fluage et consolidation des sols argileux: modélisation numérique*. Laboratoire Central des Ponts et Chaussées, Paris, Rapport de Recherche LPC N°157.

Magnan, J.-P. 1986. *Modélisation numérique du comportement des argiles molles naturelles*. Laboratoire Central des Ponts et Chaussées, Paris, Rapport de Recherche LPC N°141, 1986.

Skempton, A.W. & L. Bjerrum 1957. A Contribution to the Settlement Analysis of Foundations on Clay. *Geotechnique*, 7: 168-178.

Wroth, C.P. 1977. The predicted performance of soft clay under a trial embankment loading base on the Cam-Clay model. In Gudehus (ed.), *Finite Elements in Geomechanics*, 191-208. New York, Wiley.

Yong, R.N., Siu, S.K.H. & D.E. Sheeran 1983. On the stability and settling of suspended solids in settling ponds. Part I. Piece-wise linear consolidation analysis of sediment layer. *Canadian Geotechnical Journal*, 20: 817-826.

Prediction versus Performance in Geotechnical Engineering, Balasubramaniam et al. (eds)
© 1994 Balkema, Rotterdam, ISBN 90 5410 355 8

Predictive ability of strain rate effect on soft clay behavior

S.R.Lee & S.B.Oh
Korea Advanced Institute of Science and Technology, Daejon, Korea

ABSTRACT: Using a spacing ratio of critical state, an elasto–plastic constitutive model is proposed in order to flexibly represent undrained behavior of various normally consolidated clays. The spacing ratio of critical state can be simply evaluated from the normalized undrained strength, and the proposed model has precisely predicted the stress paths and stress–strain relationships, compared with the modified Cam–clay model, with respect to undrained triaxial test results. Furthermore, the proposed model has an ability to predict the effects of strain rate on the constitutive relationship for the interested range of strain rates. It might be said that strain rate effect of normally consolidated clay is related to the secondary consolidation.

1 SPACING RATIO OF CRITICAL STATE

In original or modified Cam–clay constitutive models, a critical state line(CSL) could be fully defined on a volumetric stress versus void ratio($p-e$) plane, once a normally consolidated line(NCL) is determined from an isotropic compression test. According to the assumed equations of yield loci in the original and modified Cam–clay models, the ratio of volumetric stress at critical state to the isotropically preconsolidated pressure(p_c) must be constant irrespective of various soil properties; i.e., $0.368\ p_c$[1] and $0.5\ p_c$[2] in the original and modified Cam–clay models, respectively.

However, experimental data obtained for various types of clay materials generally show different values of volumetric stress at critical state. If the CSL and NCL are assumed to have a unique slope(λ) in ln $p-e$ relationship, a volumetric stress ratio, p_{cr}/p_c, between CSL and NCL could be regarded as a constant for a specific soil. Thus, a spacing ratio of critical state (termed s) can be defined as the ratio of critical state volumetric stress to the preconsolidation stress along the swelling line(Fig.1).

$$s = p_{cr}/p_c \qquad (1)$$

It seems that the spacing ratio of critical state could be more reasonably selected in order to suitably characterize the material behavior. In the original and modified Cam–clay models, the spacing ratios of critical state are fixed as 0.368 and 0.5, respectively, for all materials.

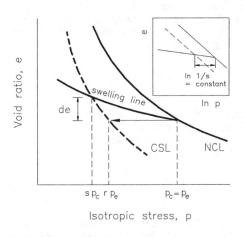

Fig. 1 Spacing ratio of critical state

Using the spacing ratio, a yield locus may be proposed as a following equation with the assumption of elliptic shape.

$$F = q^2/M^2 + \frac{s^2}{(1-s)^2} \times \{p-(2s-1)p_c\}(p-p_c) \qquad (2)$$

To evaluate the spacing ratio, s, it is required to determine the CSL, NCL and swelling line, as shown in Fig.1. For the more simplified procedure of evaluating s without relying on the CSL, a volumetric stress on the CSL can be represented as rp_e along the state line of no void ratio change as shown in Fig.1. The r value would represent the spacing ratio of the volumetric stress along undrained stress paths, and is constant for any confining stress of a normally consolidated soil. According to Fig.1, the change of void ratio(de) can be formulated as follows:

$$de = \kappa\{ \ln p_c - \ln (sp_c)\} \\ \lambda\{ \ln (rp_c) - \ln (sp_c)\} \qquad (3)$$

Based on eq.(3), the relationship between spacing ratios of critical state and undrained path(s and r) can be determined

$$s = r^{1/\Lambda} \qquad (4)$$

,where $\Lambda = 1 - \kappa/\lambda$, and r can be evaluated from an isotropically consolidated undrained(ICU) triaxial test. By virtue of the normalized undrained strength concept (SHENSAP) for normally consolidated(NC) clays[3], the spacing ratio, r, can be evaluated as follows:

$$r = \frac{2}{M}(S_u/\sigma_{v0})_{NC} \qquad (5)$$

,where $(S_u/\sigma_{vo})_{NC}$ is the normalized undrained strength of NC clay. From the equations (4) and (5), the spacing ratio, s, is obtained.

$$s = \left[\frac{2}{M}(S_u/\sigma_{v0})_{NC}\right]^{1/\Lambda} \qquad (6)$$

From the practical point of view, especially in soft clay, it is more convenient to evaluate the spacing ratios from undrained strengths without resorting to the CSL. The spacing ratios of critical state are presented in Table 1 for various types of clays. It seems to be very much affected by

strain rate and secondary consolidation time. As shown in Fig.2, different yield loci can be defined for the various clays using the obtained spacing ratios.

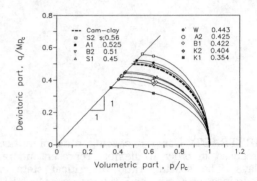

Fig. 2 Yield loci of proposed model

2 ANALYSIS OF UNDRAINED TRIAXIAL (ICU) TEST RESULTS

According to the theory of elasticity for isotropic materials, elastic strains can be evaluated as follows[7]:

$$d\varepsilon_v^e = \frac{1}{K}dp, \quad d\varepsilon_d^e = \frac{1}{3G}dq \qquad (7)$$

,where the bulk modulus, $K=(1+e)p/\kappa$ and the shear modulus $G=3K(1-2\nu)/2(1+\nu)$. The elastic strain occurs in such a small amount in contrast to plastic strains that the Poisson's ratio can be assumed to be a constant value, $\nu=0.25$. Furthermore, in an undrained case, total volumetric strain is not allowed to occur.

The undrained behavior is predicted in comparison with the triaxial test results of remolded Weald clay(Fig.3) and undisturbed San Francisco bay mud(Fig.4), using the parameters shown in Table 1(W and B1). For both clays, it has been found that the modified Cam-clay model overpredicts undrained strength and underpredicts axial strains. In contrast, the predictions of the proposed model is in good agreement with the test results. It can also be observed from the measured behavior that after arrowed points, deviatoric strain is increased suddenly and stress paths sloped in more horizontal direction at the same time. Since the proposed yield locus would be appropriate for the stress paths after the

Table 1. Spacing ratios for various clays

data	M	λ	κ	e_0	S_u/σ_{v0}	s	remark	reference
A1	1.50	0.10	0.02	0.720	0.446	0.525	different	Adachi
A2					0.378	0.425	strain rate	& Oka[4]
S1	1.49	0.32	0.08	1.651	0.420	0.450	different	Shen et al
S2				1.662	0.490	0.560	consolidation	[5]
B1	1.40	0.37	0.054	2.200	0.251	0.422	time	Bornaparte
B2				2.215	0.395	0.510		[6]
W	0.95	0.093	0.035	0.592	0.285	0.443		Weald clay[1]
K1	1.40	0.0714	0.0125	0.619	0.297	0.354	different	remolded Ildo
K2					0.331	0.404	strain rate	Ildo(Korean) clay

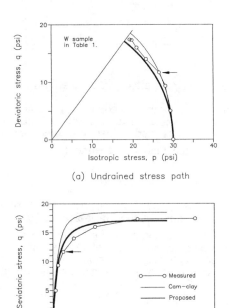

(a) Undrained stress path

(b) Stress-strain relationship

Fig. 3 Undrained behavior of Weald clay

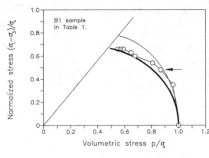

(a) Undrained stress path

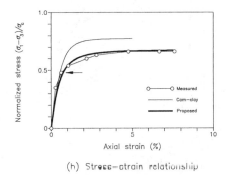

(h) Stress-strain relationship

Fig. 4 Undrained behavior of undisturbed San Francisco bay mud

arrowed points, the measured behavior may
well be predicted.

3 STRAIN RATE EFFECT

The strain and/or loading rate effect on
undrained strength and constitutive relation-

ship are important aspects which must be considered in the analyses of a number of geotechnical problems. Especially, a constitutive model which is able to represent in-situ strain rate is necessary to simulate the complex time-dependent behavior of geothechnical problems related with soft clay deposit.

According to Vaid and Campanella[8], variations in the rate of strain could affect the stress-strain relationship and hence the undrained strength of soft clay. In addition, there is essentially a linear increase in undrained strength with the log of strain rate in the region of higher rates.

Fig.5 shows the measured and predicted strain rate effects on undrained constitutive behavior of Ildo remolded clay using the parameters shown in Table 1(K1,K2). The proposed model has an ability to predict the different stress paths induced by different strain rates, since various **shapes** of yield locus might be selected in contrast to the modified Cam-clay model. The spacing ratio was much affected by the strain rate, and the undrained strength and constitutive relationship are reasonably predicted by the proposed model. Only considered here is the variation of spacing ratio with strain rate.

Fig.6 shows linear relationships between log of strain rate and spacing ratio of undrained stress path, r, for Ildo and Kunsan remolded clays. Those relationships are consistent with Vaid and Campanella's results by assuming that a critical strength parameter, M, is not changed. This means that, for almost full range of interested strain rates, the undrained behavior of normally consolidated clay can be modeled by considering exclusively the s value estimated from the value of r.

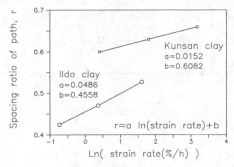

Fig. 6 Strain rate effect on stress path

It has been hypothesized that a deviatoric stress, q, is a function of current strain level, ε, and instantaneous strain rate, $\dot{\varepsilon}$, in the following manner:

$$q = q(\varepsilon, \dot{\varepsilon}) \qquad (8)$$

For various strain rates, the predictive relationship among the variables of q, ε, and $\dot{\varepsilon}$ is represented in Fig.7 for Ildo clay. To evaluate spacing ratio of critical state, r can be determined using the linear relationship as shown in Fig.6 for the given rate.

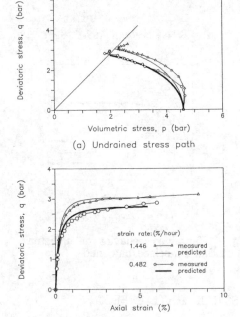

(a) Undrained stress path

(b) Stress-strain relationship

Fig. 5 Undrained behavior of Ildo clay

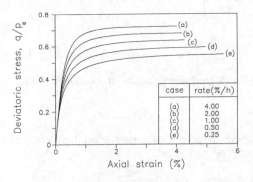

Fig. 7 Predictive constitutive relationships

At critical state, the volumetric stresses develop differently with respect to the corresponding strain rate and then the CSL's for various rates may be paralleled as shown in Fig.8. For two arbitrary critical volumetric stresses,$(p_{cr})_n$ and $(p_{cr})_{n+1}$, there is a difference of void ratio, Δe, along the swelling line. The time-dependent response induced by different strain rates is assumed to occur due to a creep phenomenon including secondary consolidation, and it is also assumed that the strain rate effect is caused exclusively by secondary consolidation. Then, Δe can be considered as void ratio change due to secondary consolidation during an elapsed time from t_n to t_{n+1}.

$$\Delta e = \psi \ln (t_{n+1}/t_n) \qquad (9)$$

,where ψ is a secondary consolidation index, defined as $de/d \ln t$, and, from Fig.8, Δe could also be evaluated as

$$\Delta e = \kappa \ln \frac{(p_{cr})_n}{(p_{cr})_{n+1}} \qquad (10)$$
$$= \kappa \ln r_n/r_{n+1}$$

From equations (9) and (10), the following relationship can be obtained:

$$r_n/r_{n+1} = (t_{n+1}/t_n)^{\psi/\kappa} \qquad (11)$$

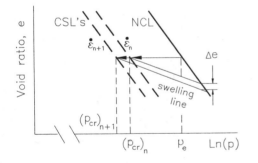

Fig. 8 Critical volumetric stresses

Furthermore, it is assumed that strains at failure are not varied with strain rates, and thus the ratio of elapsed time could be determined from strain rates,

$$t_{n+1}/t_n = \frac{\dot{\varepsilon}_n}{\dot{\varepsilon}_{n+1}} \qquad (12)$$

By substituting eq.(12) into eq.(11), $d\ln r/d\ln \dot{\varepsilon}$ can be evaluated as ψ/κ, and the slope of relationship between strain rate and r can be determined as follows.

$$\frac{dr}{d \ln \dot{\varepsilon}} = \frac{\partial r}{\partial \ln r} \frac{\partial \ln r}{\partial \ln \dot{\varepsilon}} \qquad (13)$$
$$= r \psi/\kappa$$

The calculated value of $dr/d\ln \dot{\varepsilon}$ (by eq.13) is compared with the measured value in Table 2, and it is shown that the calculated slope is varied within a small range and similar to the measured value.

Table 2 Comparison of measured and calculated slope

λ	κ	ψ	r	calculated	mesured
0.0714	0.0125	0.0016	0.42– 0.54	0.054– 0.069	0.0486

4 SUMMARY

From the result of this study, the following may be summarized.

(1) Using an spacing ratio of critical state, an elasto-plastic constitutive model is proposed in order to flexibly represent undrained behavior of various normally consolidated clays.
(2) The spacing ratio of critical state can be simply evaluated from the normalized undrained strength, and the proposed model has precisely predicted the stress paths and stress-strain relationships, compared with the modified Cam-clay model, with respect to undrained triaxial test results.
(3) The proposed model has an ability to predict the effects of strain rate on the constitutive relationship for almost the full range of interested strain rates.
(4) It might be said that the strain rate effect of normally consolidated clay is related to secondary consolidation.

REFERENCES

1.Scofield, A.N., and Wroth, C.P. 1968. *Critical state soil mechanics*: McGraw Hill
2.Roscoe, K.H., & Burland, J.B. 1968. On the generalized stress-strain behavior of 'wet'

clay. in *Engineering plasticity*, J. Heyman and Leckie(Eds.) Cambridge Univ. Press, Cambridge: 535-609

3.Ladd, C.C., & Foott, R. 1974. New design procedure for stability of soft clays. *J. Geotech. Engrg., ASCE* 100(7): 763-786

4.Adachi, T., & Oka, F. 1982. Constitutive equations for normally consolidated clay based on elasto-viscoplasticity. *Soils and Foundation* 22(4): 57-70

5.Shen, C.K., Arulanandan, K., & Smith, W.S. 1973. Secondary consolidation and strength of clay. *J. Geotech. Engrg.,ASCE* 99(1): 95-110

6.Bonaparte, R. 1981. *A time-dependent constitutive model for cohesive soils.*: Thesis presented to the Univ. of Califonia, at Berkeley, Calif. in partial fulfilment of requirements for the degree of Doctor of Philosophy.

7.Atkinson, J.H., and Bransby, P.L. 1978. *The mechanics of soils*: McGraw Hill

8.Vaid, Y.P., and Campanella, R.G. 1977. Time-dependent behavior of undisturbed clay. *J. Geotech. Engrg.,ASCE* 103(7): 693-709

Behaviour of spliced piles – A field study

B.G. Rao & M.P. Jain
Central Building Research Institute, Roorkee, India

Gopal Ranjan
University of Roorkee, India

ABSTRACT: In several situations use of long piles becomes necessary. The inconvenience of use of long piles can be off setted by using "Spliced Piles" provided the splices/joints between segments do not introduce weakness in piles. Since splice pile technology has not picked up in India yet, efforts have been made to develop joints for piles and thus to introduce splice technology for use in field. Four types of joints have been developed, and their performance evaluated through use of these joints in RCC precast piles installed in loose to medium dense cohesionless, and soft clay deposits. The results obtained from their field behaviour both during driving and in situ load testing have demonstrated their suitability with respect to performance under compressive, pullout and lateral loads.

Further with a view to eliminate/reduce the negative drag on piles, the ambient soft clay around the piles was transformed into composite mass of low compressibility by resorting to ground treatment techniques utilizing Plain Granular Piles (PGP), Mini Grouted Piles (MGP) and Self Setting Soil Slurry Piles (SSSSP). Adoption of these techniques resulted in significant reduction in settlement of ambient ground and consequently the negative drag on piles. The introduction of bitumen slip layer also resulted in significant reduction of negative drag on piles.

It is concluded that the performance ofsplices developed and used with RCC segment of precast piles during driving, load testing under vertical compressive, lateral and pull out loads have been found satisfactory and in no way inferior to intact piles. This is true both for cohesionless and soft clay deposits. For reducing/eliminating negative drag all the three methods of ground treatments are effective in transforming the ambient soil mass into an improved composite soil mass.

INTRODUCTION

Deep deposits of soft saturated clays are usually met with coastal area, around lakes, ports and harbours. The conventional practices of providing foundation in such deep deposits are either to use driven cast-in-situ or bored piles. Bored concrete piles, many a times, suffer from several defects (Mohan et al 1978). These are not always on account of materials and mixes or laxity in quality control, but at times

selection of a piling system which may not be well suited to highly compressible deep deposits. In such a situation precast driven piles provide a better alternative. In case of precast driven piles also driving of long concrete piles, in single length becomes inconvenient and costly. Precast piles in smaller segments driven and jointed together with a suitable joint may provide a cost effective and efficient alternative, provided the joint between the two segments does not introduce weakness.

Though the joints in timber piles are in use for over a century and are common in steel piles also, yet the economical and structurally safe joints in concrete piles are yet to be developed, tested and standardised for use in the field in India.

Failure of pile foundations, in general are attributed to pile defects, caused during installation in poor subsoil conditions. The common pile defects are honey combing, necking, dislocation of toe and absence of concrete altogether in pile toe (Mohan et al, 1978.)

However, many foundation failures or their unsatisfactory performances are often explained in terms of either unexpected or under expected added axial load on the piles, on account of the negative drag. The additional loads thus result in over stressing of the pile which may lead to unacceptable settlements or even failure of the underlaying supporting soil.

In a classical situation, negative drag on a point bearing pile develops when the ground in which it is installed settles relative to the pile. Such settlements may take place due to (a)the under consolidated foundation soil (b) remoulding of the soil that occurs during pile driving (c) a freshly placed fill around a pile and (d) ground water lowering.

The Indian sub-continent has in abundance soft saturated clay deposits varying in depth from 6 m to 30 m and even more, along the coastal belt, shores of Gulf of Cambay and Bay of Bengal. In such areas deep piles provide an efficient and cost effective foundations for medium to high rise structures. Therefore, it gains favour from designers. However, in such situation negative friction on piles in soft saturated clay deposits, leads to reduction in factor of safety and ultimately failure of pile foundation.

The state-of-the art reveals that the negative drag is reduced or eliminated by using (a) protection piles; (b) bentonite slurry, (c) bitumen coating; (d) electro-osmosis; (e) casing and predrilling and (g) reducing the point resistance. Bitumen coating, though used widely all over the world, yet, its usefulness is restricted to precast driven piles only. However, chances of damage of coating during driving cannot be over ruled and evaluated. Further, the negative drags on piles can also be significantly reduced by reducing point resistance in piles by adopting small diameter mini grouted piles (Rao et al. 1992).

Still another approach to overcome the problem of negative drag is to load the ambient ground before the installation of piles. Though the time required for complete consolidation may be long which can be significantly reduced by providing drains.

It is therefore clearly brought out that the settlement in the ambient ground is the main cause of negative drag on piles which is generated with time. It may therefore be appropriate that if the compressible ambient ground around the pile is transformed into a strata of stiff composite mass having reduced compressibility and improved strength by treating the virgin ambient ground, which will result into significant reduction in settlement, thus reducing the main cause of negative drag.

The chapter describes four types of joints designed and developed at the Central Building Research Institute, Roorkee (India) along with their performance evaluation during driving and testing under different loading conditions, both in the laboratory and field in cohesionless soil deposits (Rao et al. 1992, 1993, Jain 1993).

The performance of these spliced piles vis a vis intact piles have also been compared in soft marine clay deposits.

Further, the ambient soft clay around the piles has been transformed into a composite mass of low compressibility by resorting to ground treatment techniques, which resulted in significant reduction in the settlement of ambient ground and under load and subsequently negative drag on piles (Jain 1993, Rao et al 1993).

Sub-Soil Conditions

Field studies were carried out at two different sites viz. (i) Site (I) having predominantly, cohesionless soil deposits (ii) Site (II) having soft marine clay deposits.

Detailed sub-soil investigations were carried out at both the sites. Soil exploration consisted of boring, standard penetration test, and dynamic cone penetration tests. These tests were carried out in accordance with IS:2131-1981 and IS: 4968 (Part I) 1976. The field test results at SITE I showing bore log, standard penetration test and also dynamic cone penetration test are presented in Fig.1.

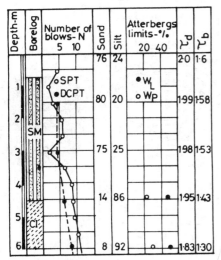

Fig.1- Bore log and other sub soil characteristics (Site 1)

A study of bore log indicates the presence of 0.78 m thick hard desicated sandy silt soil followed by a 4.5 m thick layer of sandy silt (SM) deposit. Further, extension of the bore hole revealed the presence of a layer of clay

deposit with medium plasticity, (CI) upto 6 m, the depth at which the bore hole was terminated. The water table at the time of investigation during November was found to be at 4 m below the ground level.

The average standard penetration test values (N-value) upto 4 m depth was observed to be 4 beyond which it increased practically at a constant rate. The value was noted 8 at 6 m depth. Similar trend was indicated from dynamic cone penetration tests (Fig.1). In general, the sub-soil strata was thus found to be as loose silty sand deposit (SM) having angle of shearing resistance of 29° from triaxial tests on samples collected from different depths.

Similarly boring, standard and dynamic cone penetration tests and vane shear tests, as per IS 2131 * 1981), IS: 4968 Part I (1976) and IS:4434(1978) at site II were conducted. The test results are presented in Fig.2. A study of the bore log indicated that the sub-soil from 0.5 m to 4 m depth is soft marine clay underlain by a filled up layer of 0.5 m thickness. The soft marine clay was followed by a poorly graded sand deposit (SP) from 4 to 15 m.

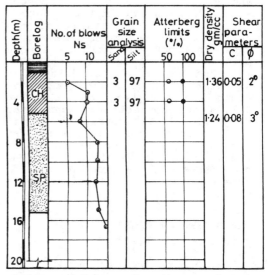

Fig.2- Borelog and other subsoil characteristics-Site II

The number of blows, N_C of dynamic
cone tests from 0.3 m to 3.3 m were
noted to vary between 4 to 8,
whereas below this and upto 5 m,
the value of N_C increased to 18.
The natural moisture content varied
from 49 to 64 percent and bulk
density from 1.04 gm/cm^3 to 1.38
gm/cm^3. The liquid limit ranged
from 94% to 96% and plasticity
index from 60% to 61%. The value of
cohesion was found to vary between
0.05 kg/cm^2 to 0.08 kg/cm^2 and the
angle of shearing resistance,
$\emptyset= 2^o$ However, the value of
cohesion was found to range between
0.15 kg/cm^2 to 0.23 kg/cm^2 and0.043
kg/cm^2 to 0.086 kg/cm^2 as obtained
from vane shear tests from the
insitu tests conducted on
undisturbed and disturbed state
respectively. As a matter of fact
high moisture content clearly
indicates the reasons for low shear
value.

DEVELOPMENT OF PILE JOINTS

Four types of joints analogous to
Hercules, ABB, Coupler-ring, and
square (Bruce and David 1974, and
Bredbergard, Broms, 1979) were
selected for the study. These
joints hereafter shall be nominated
as Splice A, Splice B, Splice C,
and Splice D, respectively. Splices
A to D are shown in Fig.3 thro 5
respectively. The joining of the
splice A with the reinforcement is
shown in Fig.6. The male and female
part of joint B is identical. The
male and female part of joint C is
circular in shape and has a groove
along the periphery and the two
halves are connected by connector
ring (Fig.4). The connector ring is
a plate in two halves projected at
the ends equal to the dimensions of
the groove. The details of various
components of Splice D, male,
female parts and locking keyis
shown in Fig.5.

PRECASTING OF PILES
SITE I

Piles, square in section (200 mm x
200 mm) size were cast in two
segments, each 1.5 m long. When the

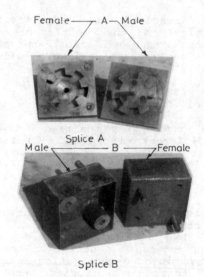

Fig.3- Two types of splices
developed

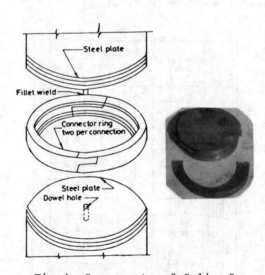

Fig.4- Components of Splice C

two segments are joined together, a
precast pile of 3 m length was
obtained. Five precast piles with
each splices A,B,C, and D (Fig.3 to
5) were cast. In addition, five
intact piles; without any joint
having same cross-section and
length were also cast. The details
of piles are presented in Table 1.

Table 1- Pile details (Site I)

Pile Designation		Remarks
Length (m)	No.of segment	
3	–	Intact Pile
3	Two	Splice A
3	Two	Splice B
3	Two	Splice C
3	Two	Splice D

* Cross-section of pile 200 mm x 200 mm
* Length of segment is 1.5 m.

Fig.6- Showing Splice after Welding with Reinforcement Cage

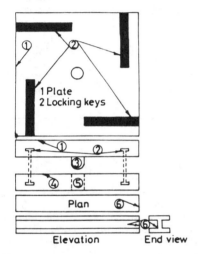

1 Plate
2 Locking keys

Plan

Elevation   End view

Fig.5- Various parts of the Splice D

Fig.7- Alignment with reinforcement cage Splice C before Casting

These splices after welding with reinforcement cage with male and female parts (Fig.6) were placed in alignments to recheck the proper fitting of joints, prior to the casting of piles, by placing and casting in the specially fabricated timber form work. The precast segments are shown in Fig.7 and 8.

PRECASTING OF PILES

SITE II

Piles, square in section, (200 mm x 200 mm) size were cast in two segments, each having 3 m length.

When two segments were joined together, a precast pile of 6 m length was obtained. In the present study five such piles were cast and out of four joints (A,B,C and D), only splice B was used (Fig.1). Reinforcement cage after welding with the splices, with male and female parts (Fig.9) were placed in alignment with a view to recheck proper fitting of joints, prior to casting of piles. The piles were then cast in specially prepared steel former with an inside lining of thin G.I. sheet. The full details of piles are presented in Table 2 and those of the treated ambient ground in Table 3. Further

Fig.8- Casted segments of Splice B

Fig.9- Reinforcement with Joints

Table 2- Pile details (Site II)

Pile designation		No.of segment	Remarks
Length pile (cm)	Length of segment (m)		
6.0	6.0	-	Intact Pile
6.0	3.0	Two	With one splice segment 3 m length
6.0	2.0	Three	With two splice each segment of 2 m length
6.0	1.5	Four	With three splices each segment of 1.5 m length
6.0	1.5	Four	Group of four piles each segment of 1.5 m length

* Cross-section of pile 200mmx200mm

details of precasting of piles are presented in Fig.10 thro Fig.13.

DRIVING OF PILES

Site I

The precast piles whether spliced or intact were driven in the field with a hammer (weight 250 kg and free fall 1 m). The hammer was operated through a tripod hoist and a mechanically operated winch. All the piles appeared to behave satisfactorily since chipping of concrete or failure of joints or any erratic behaviour was not observed during driving. It is further confirmed from driving record showing number of blows vs depth of pile penetration relationship for both, intact and spliced piles (Fig.14).

Site II

The precast spliced piles in this site were driven by a 800 kg weight hammer with a free fall of 100 cm, through a tripod hoist and a mechanically operated winch, (Fig.15). The joining of the segments during driving is shown in Fig.16. During driving all the piles behaved satisfactorily at this site also for joints no erratic behaviour was observed. It is further confirmed from driving record Fig.17 also. Thus from the driving records it may be concluded that the behaviour of spliced and intact piles during driving is found satisfactory.

At the same site, three piles each 200 mm x 200 mm in cross-section and 6 m long, were also driven to study the performance of precast spliced piles by loading the ambient ground under sustained vertical loads and measuring the settlements of precast piles as an indirect measure of negative drag. The details have been presented in Table 3.

Table 3- Details of Piles and Treatment of ambient clay subsoil

Pile Section mmxmm	Length (m)	Treatment of Ambient Subsoil and Testing
200x200	6	Pile coated with bitumen. The ambient compressible clay was loaded through a specially fabricated steel pile cap.
200x200	6	Pile coated with bitumen. The surrounding soil was loaded through the same pile cap and settlement recorded.
200x200	6	The ambient clay around the pile improved by installing granular piles, self setting soil slurry and minigrouted piles having 3 m depth and then load tested. Simultaneously, movement of intact pile was also monitored.

FLEXURAL TESTS

Flexural tests on two precast piles, 3 m long, having two segments joined together with a joint in the middle with Splice B and Splice C besides one 3 m long intact pile were performed in the

Fig.11- Casting of Piles (above)

Fig.12- Curing of Piles

Fig.10- Splices with Reinforcement

Fig.13- Bitumen Coated and Uncoated Piles

33

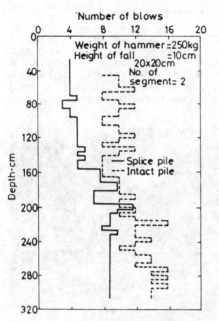

Fig.14- Driving record of intact and splice pile with splice C.

Fig.16- Joining of Segments during Driving

Fig.15- Driving of Bitumen Coated Pile

laboratory. The modulus of elasticity of the pile material was 2.8 kg/cm^2. The pile element was placed on two supports located 3 m apart, in such a way that the splice was located at the centre between the two supports (Fig.18). The load was applied by a remote controlled hydraulic jack placed on a steel girder. The girder in turn was supported on two edge support placed over pile, 1 m apart. The deflection of pile were measured at three locations a,b and c. Observation points a and c were 700 mm away from the centre of the splice, and the point b was at the centre of splice. The load deflection curves for piles with Splice B and Splice C along with intact piles have been presented together in Fig.18.

TEST RESULTS AND DISCUSSIONS

A glance of load-deflection curves of jointed piles (Fig.18) in bending indicates that the loads corresponding to 8 mm to 20 mm deflection (Table 4) varies between 2.6-10.8 tonnes for the spliced and intact piles. The curve for the other two splices A and D lie within the shaded portion (Fig.18). Thus the study Fig.18 and Table 4 indicates that the flexural loads for all types of splices, vary

34

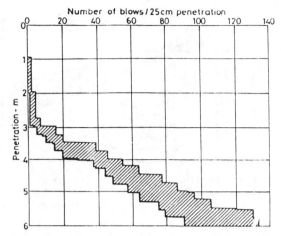

Fig.17- Pile Penetration Records

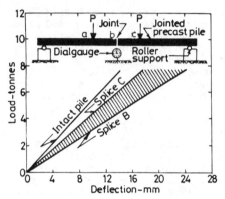

Fig.18- Test Setup and Load Deflection Curves of Piles in Bending Test (Mid Span)

Table 4 - Comparative Behaviour of Spliced in Bending Test

Splice Type	Load in tonnes at deflection at the mid span (Point B)				Percentage of load at deflection at mid span with reference to intact piles			
	8 mm	12 mm	16 mm	20 mm	8 mm	12 mm	16 mm	20 mm
Splice B	2.6	3.9	5.0	6.5	53.0	61.0	57.5	60.0
Splice C	3.5	5.5	7.0	8.7	71.0	86.0	89.6	80.0
Intact Pile	4.9	6.4	8.7	10.8	100.0	100.0	100.0	100.0

between 53-89%, of the loads taken by the intact piles when the deflection varied from 8 mm to 20 mm. It is further noted that the loads taken by splice B varied from 53% to 60% which is lowest from all

the splices, however, this load is sufficient to support actual bending stresses during driving and also in service period under actual live structures.

VERTICAL COMPRESSIVE LOAD CAPACITY OF SPLICED AND INTACT PILES

SITE I

Study of load settlement behaviour of spliced piles and intact piles (Fig.19) indicates that the load corresponding to (i) 12 mm settlement and (ii) 20 mm settlement (1/10 of the pile diameter) (IS:2911-Part IV)-1985) varies between (i) 8.0-9.3 tonnes and (ii) 9.5-10.5 tonnes respectively (Table 5). The corresponding safe loads (from both the criteria) vary between 4.75-6.0 tonnes. The ultimate load capacity was also computed using soil properties from the static formula for cohesionless soils as per IS: 2911-Part I/Sec 4 (1984) and the value obtained was 7.0 tonnes. The value of shearing resistance of soil assumed in the analysis was 29°. The corresponding settlement from the load settlement curve

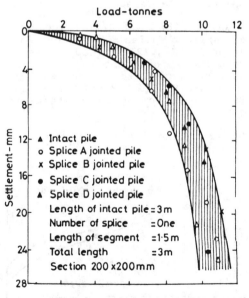

Fig.19- Load settlement curves under compressive loads

35

(Fig.19) at 7.0 tonnes load was found to vary from 4.0 mm - 7.5 mm.

It is therefore concluded that the behaviour of intact piles vis-a-vis spliced piles (with developed splices) on the basis of in-situ loading in compression is practically the same.

Site II

Study of load settlement curves of spliced piles (Fig.20 and Fig.21) indicate that the load corresponding to (i) 12 mm settlement and (ii) 20 mm settlement (1/10 of the pile diameter) varies between (i) 24.7-26.2 tonnes and (ii) 24.7 - 27.7 tonnes respectively for splice B including intact piles (IS:2911 Part IV-1985). The corresponding safe loads (from both the criteria) vary between 16.8-18.8 tonnes. The ultimate load capacity computed from soil properties from the static formula for cohesive soils as per IS:2911 (Part I/Sec4)-1984 and the value obtained was 17.5 ton The corresponding settlement from the load settlement curve (Figs.20 and 21) was found to vary from 5.0 mm - 7.9 mm.

Spliced Pile Group

Study of load settlement behaviour of spliced piles in a four pile group (Fig.22) (Cap resting condition) indicates that the load correspondsing to 6 mm was 120 tonnes. IS:2911 (Part IV) - 1985 lays the following two criteria for groups.

(a) Final load at which the total settlement attains a value of 25 mm.

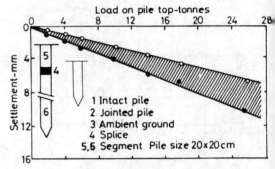

Fig.21- Load settlement curves for intact and four spliced piles with 2,3,4 segments

(b) Two thirds of the final load at which the total displacement attains a value of 40 mm.

None of the two criteria of the safe load on group of piles could be reached. In the absence of the two criteria the final load at 4.5 mm settlement is taken as 100 tonnes from which individual capacity of the single piles worked out to be 25 tonnes.

It is therefore concluded that the behaviour of intact piles visa-a-vis jointed piles with (splice B) on the basis of insitu load testing in compression is identical. The above finding corroborated well for all types of joints in cohesionless soils also. It is thus inferred that pile group behaviour of spliced piles is identical to that of intact piles.

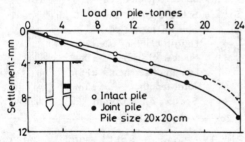

Fig.20- Load Settlement Curves of Intact and Spliced Piles

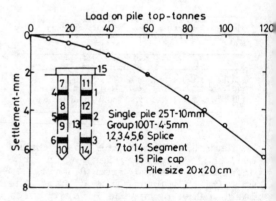

Fig.22- Load settlement curve of four pile groups

Table 5- Comparative Behaviour of Spliced and Intact Piles

Type of Splice	Load corresponding to 12 mm settlement (tonnes)		Load corresponding to settlement 1/10 of the diameter i.e. 20 mm - (tonnes)		Safe Load (tonnes)			
	Spliced Pile	Intact Pile	Spliced Pile	Intact Pile	Spliced pile*	Intact Pile*	Spliced pile**	Intact pile**
A	8.0	8.9	9.5	9.5	5.33	5.92	4.75	4.75
B	9.3	8.9	11.0	9.5	6.20	5.92	5.55	4.75
C	9.0	8.9	10.0	9.5	6.00	5.92	5.00	4.75
D	9.0	8.9	10.5	9.5	6.00	5.92	5.25	4.75

All intact piles were 200 mm x 200 mm square in section and 3 m long

*   The safe loads corresponds to 2/3rd of the load at 12 mm settlement

** The safe load corresponds to 50% of the load at 20 mm settlement i.e. (1/10 of the pile diameter). The safe loads were computed as per IS: 2911 (Part 4- 1985).

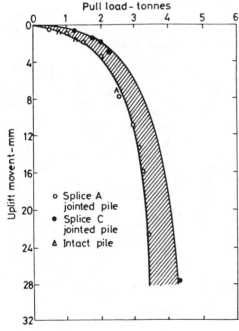

Fig.23- Full load-pull movement curve under pull load

PULL OUT TESTS

Pull load tests at site I were also conducted on spliced piles with splices A and C besides intact pile. The pull load versus pull movement curve for all the three tests are shown in Fig.23.

The study of load movement curve of spliced piles indicate that the load corresponding to 12 mm movement and the load at which the load-movement curve shows the clear break are 3.0 - 3.5 tonnes and 3.5 - 4 tonnes.

The corresponding safe loads from both the criteria vary between 1.75 - 2.3 tonnes. Conclusions similar to compression tests hold good in pull out also.

Lateral Load Tests

Lateral load tests were conducted on spliced piles using splices A, D, and C besides intact pile. The lateral load versus deflection (at ground level) curves are presented in Fig.24. On the basis of the criteria, (i) fifty percent of load at 12 mm displacement and (ii) load at which total displacement corresponding to 5 mm settlement the loads are found to lie between (1.25 - 1.75) tonnes and (1.6 - 2) tonnes. Thus, in general, there is not much difference between the safe loads

37

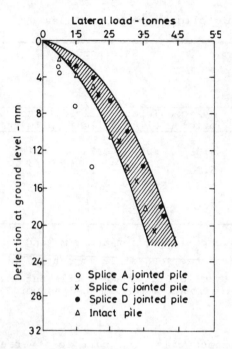

Fig.24- Load-deflection curve under
lateral loads

Fig.25- Closeup view of Precast
Pile with Bitumen Coating.

Fig.26- Bitumen Coated Pile

of spliced piles and intact piles
from both the criteria. It may be
concluded that the behaviour of
spliced piles is practically the
same as that of intact piles in
lateral loading.

It can therefore be summarised,
that the behaviour of spliced
piles as compared to intact piles
during load test upto their
ultimate and safe capacity under
vertical compression, lateral and
pull out loads is satisfactory and
in no way inferior to intact
piles. Thus these four splices may
be used with confidence in
practice.

REDUCTION OF NEGATIVE DRAG

It has been well established that
negative drag on piles is of great
significane as it results in
lowering of allowable load on
piles and/or increases
settlements. Efforts are on to
estimate this drag and to reduce
the same. Some of the studies on
this aspect are discussed in
following sections.

(a) Bitumen Coating

A precast pile was coated with hot
bitumen of 60-70 grade after the
pile was thoroughly cleaned and
dried. The bitumen compound was
poured at $190^{o}C$, on the surface of
the pile (Fig.25) and thereafter it
was brushed along the length of the
pile to have a smooth finished
surface (Fig.26). Total thickness
of 5 mm of the bitumen was
achieved by applying bitumen in two
stages. An interval of 6 to 8 hours
was maintained between the two
layers of bitumen coating.The
brushing was done in one direction
only. In this manner all the four
sides of the pile were coated. It
was then left for 48 hours before
it was taken for driving.

(b) Granular Piles

A group of 12 granular piles
(250 mm installed pile diameter) at
a spacing of 300 mm centre, 3 m
deep (Fig.27) were installed around
the spliced piles using simple
auger boring method (Rao 1983,
Ranjan and Rao 1990). After making
the bore hole upto 3 m, granular
piles were cast using 20-70 mm
stone aggregate with 15-20 percent

locally available sand. The stone aggregate was placed in the bore hole in layers of 30 cm followed by 5 to 6 cm of sand layer. A cast iron hammer having 65 kg weight and diameter less than the diameter of the bore hole was used to compact the stone aggregate and sand mixture.

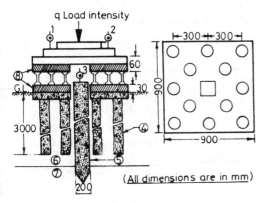

(a) Sectional elevation.   (b) Plan view

1,2,3 dial gauges, 4. Plain Granular Piles (PGP), or Mini Grouted Piles (MGP), or Self Settle Slurry Piles, (SSSSP), 5. Spliced RCC Pile, 6. Ambient Soft Saturated Clay, 7. Hard Strata, 8. Fabricated Steel Cap.

Fig.27- Set-up for Loading the Treated Ambient Ground and Measuring Pile Settlement.

(c) Mini-Grouted Piles

A group of 12 minigrouted pile at a spacing of 300 mm c/c, 3 m deep arranged in a staggered pattern (Fig.27) were cast around spliced piles using simple auger boring method. Subsequent to the installation of bore holes 3 m deep, minigrouted piles were cast by lowering a single tor steel bar reinforcement 22 mm in diameter with a centrally grooved 8 mm thick steel plate welded to the bottom of the reinforcement. The bore hole was then charged with M-20 cement:sand grout by Tremie method. The water cement ratio was kept as 0.5.

Suitable ingradients were added in

the predetermined quantity to maintain the workability of the grout for longer duration. The grouting pressure was maintained between 1.75 to 3.0 kg/cm^2. The homogeneous grout was continuously circulated into the bore hole until the grout attained the desired consistency to form a pile (Rao, 1993).

(d) Self-Setting Soil Slurry Piles

Similarly, self-setting slurry piles were cast around the spliced piles using simple auger boring method. After making the bore hole 3 m deep soil slurry from the soil having ingradient as clay (5-10%), sand (60-70%) and silt (20-30%) with 5% to 6% of cement was prepared. The soil mixture was thoroughly mixed dry in a mixer and water was added. Following the procedure soil slurry was prepared by thorough mixing. The prebored hole was then flushed with water before injecting the soil slurry. All the loose soil particles present in the bore hole were thus removed along with water during injection. The slurry was then injected at a pressure of 1-3 kg/cm^2, starting from the bottom of the bore hole by Tremie method. (Rao et al 1992). The lowering of the level of soil slurry in the bore hole after a lapse of 3-4 hours, was made good by charging the bore hole with more slurry. The soil slurry piles were cast only upto the cut off level followed by excavating a 30 cm deep trench and size in plan equal to the size of pile cap. A netlon polymer grid of the same size as the pre-fabricated steel cap and was placed in the bottom of the cap. The trench was filled with soil slurry upto the surface.

INDUCING NEGATIVE DRAG

The negative drag on the central precast spliced concrete pile was created by loading the treated/untreated ambient ground as demonstrated in Fig.27, through the specially fabricated steel cap. Thus, using the test set-up, for each load increments, the respective settlement of the steel cap and the central pile were recorded. Following the procedure, negative drag was created on the central pile.

INSITU LOAD TEST AND DISCUSSIONS

Insitu load test on spliced piles were conducted using maintained load method (IS:Part IV-1985). Load increments were applied on the treated ambient ground through the steel cap 900 mm x 900 mm in size allowing the central precast spliced pile to pass through centrally and freely. The corresponding settlement of the ambient ground were thus recorded (Fig.27(a) (1 and 2).

Simultaneously, the settlement of the central splice pile was also recorded for the same load increment (Fig.27(a) 3).

Following the procedure described above five full scale in-situ tests were carried out at the site, viz., (a) central splices RCC pile and virgin ambient ground around the pile (b) bitumen coated central spliced pile and virgin ambient ground around (c) spliced RCC pile and ambient ground treated with PGP (d) spliced RCC pile and ambient ground treated with MGP and (e) spliced rcc pile and ambient ground treated with SSSSP. As stated earlier in all the five cases the ambient ground whether treated or untreated were loaded through (900 mm x 900 mm fabricated) steel cap for creating indirect negative drag. The test results have been presented through Fig.28 thro 32. Study of Fig.28 indicates that the movement of the central RCC pile is found to be more than 6 mm under a load of 24 tonnes on the pile cap due to which the ambient ground has reached its ultimate capacity. This clearly establishes the significance of negative drag on the performance of central pile. On the other hand the movement of the ambient virgin ground to reach a settlement of 20 mm is not effective in inducing even a measurable settlement which is found to be less 2 mm in the case of bitumen coated pile (Fig.29).

Similar behaviour is noticed in cases where the ambient clay

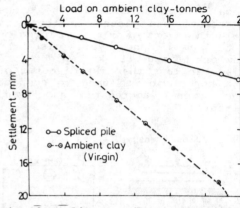

Fig.28- Spliced Pile Settlement due to Compression of Virgin Ambient Clay.

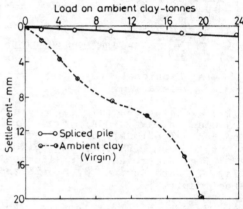

Fig.29- Settlement of Bitumen Coated Pile due to Compression of Virgin Ambient Clay.

treated with either PGP or MGP or SSSSP around the untreated central spliced pile (Fig.30 thro 32). It is further noted that a relative movement of even upto 20 mm, the negative drag is not created on central pile (Fig.30 thro 32) though it should happen at a relative movement of only 10 mm (Broms 1979).

The test results thus suggest that if ambient ground around the pile is transformed into a composite soil mass by reinforcing through PGP or MGP or SSSSP the negative drag on pile could be significantly reduced or eliminated altogether.

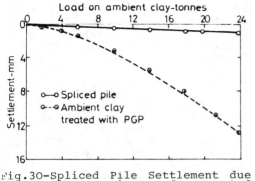

Fig.30-Spliced Pile Settlement due to Compression of Treated Ambient Clay.

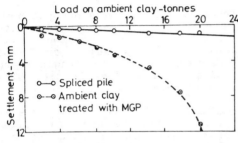

Fig.31-Settlement of Spliced Pile due to Compression of Improved Ambient Clay.

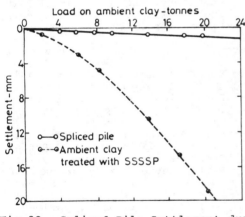

Fig.32- Spliced Pile Settlement due to Compression of Treated Ambient Clay

CONCLUSIONS

Based on the experimental investigations the following conclusions were drawn:

(i) Four types of splice namely A,B,C and D have been developed and their performance evaluated both in cohesionless and cohesive deposits vis-a-vis intacts.

(ii) The performance under flexural load of all the splices in the laboratory have been found satisfactory and the flexure loads varied from 53-89 percent as that of intact piles. These flexural loads are sufficient to sustain actual bending stresses during installation and also during service condition under actual structures as an alternative to intact piles.

(iii) The behaviour of spliced piles during driving as compared to intact piles is satisfactory and is in no way inferior to the intact piles. This is true for both in cohesionless and soft clays deposits.

(iv) The behaviour of spliced piles during load tests under vertical compressive, lateral and pull out loads have been found satisfactory and in no way inferior to intact piles. This is valid for all the four types of splices developed and used in cohesionless soils. It may therefore be concluded that all the four splices could be used in precast piles with confidence.

(v) The behaviour of spliced piles (with splice B) at their ultimate capacity under vertical compressive load when compared to intact piles is satisfactory. It is also revealed that the piles having splice varying from one to three and same length of pile in all the cases have almost the same capacity. It therefore indicates that the number of splices in a

41

pile does not have any effect on the carrying capacity. The study also concludes that the joint B, could be used with confidence.

(vi)   The behaviour of the splice piles in cluster during driving as well as during load testng in compression is same.

(vii)   Ambient clay treated with Plain Granular Piles (PGP) or Minigrouted piles (MGP) or Self Setting Soil Slurry Piles (SSSSP) is found to reduce negative drag significantly. So also the introduction of a bitumen slip layer by coating the pile with bitumen.

(viii)   All the three methods of ground treatments are effective in transforming the adjacent soil mass into improved composite soil mass. However, SSSSP and PGP are preferred.

(ix)   The proposed method is preferable even over bitumen slip layer since these may find their applicability on all types including bored cast in situ piles, where a bitumen slip layer method can be used only for prefabricated precast driven piles.

(x)   The introduction of spliced pile technology in India is likely to provide a cost effective, or hard strata in deep deposits of soft saturated clays where the installation, performance and the integrity of driven cast in situ piles and also bored piles was questionable. The technology has since been patented and is ready for field application.

ACKNOWLEDGEMENT

The paper is based on the study performed as a part of collaborative research programme between Central Building Research Institute, Roorkee, U.P. (India) and the Department of Civil Engineering, University of Roorkee, U.P. (India).The authors wish to express their gratitude to Dr. R.K. Bhandari, former Director, Central Building Research Institute, Roorkee for encouragement during the entire period of research programme. The financial assistance provided by the Department of Science and Technology, New Delhi for the field study is gratefully acknowledged. Thanks are also due to Shri A.K. Sharma and A.K. Mishra for their valuable help.

REFERENCES

Alphousus, I.M., Chaessen and Endre Horvat (1974), Reducing Negative Friction with Bitumen Slip Layers'. Journal of the Geotechnical Engineering Division No.10764, August, GT8, pp.925-974.

Bredenberg, H and Broms, B.B. (1979), 'Joints used in Sweden for precast concrete piles, recent development in the design and construction of piles', Proc. Conference held at Institute of Electrical Engineers, 21-22 March. The Institution of Civil Engineers, London, pp.11-22.

Broms, B.B. (1979), Negative Skin Friction, State-of-the-art report, Proc. Sixth Asian Regional Conf. Soil Mech. and Foundation Engineering, Singapore, Vol.2, pp.41-75.

Bruce, R. N. and David, C.H. (1974), Splicing of Precast Prestressed Concrete Piles', Part I-Review and Performance of Splices, Jr. of Prestressed Concrete Institute, Vol.19, No.1, Sept.-Oct.1974, pp.70-97.

Bruce, R.N. and David, C. (1974), Splicing of Prestressed Concrete Piles', Part 2- Tests and Analysis of Cement-dowel Splice', Journal of Prestressed Concrete Institute, Vol.19, No.6, Nov-Dec. 1974, pp.40-66.

Endo, M., Minov, A., Kawaski, T., and T.Shibata (1969), 'Negative

Skin Friction acting on Steel Pipe Piles in Clay', Proceedings 7th ICSMFE Mexico, Vol.2, pp.423-428.

Fellenius, B.H., and Broms, B.B. (1969), 'Negative Skin Friction on Long Piles Driven in Clay, Proceeding 7th ICSMFE, Mexico, Vol.2, pp.93-98.

Grant, E.V., Stephens, J.E., and L.K. Moulton (1958), 'Measurement in Produced in Piles Settlement of Adjacent Soil', HRB Bulletin 173: Analysis of Soil Foundation Studies, pp.22-37.

IS: 2131-1981: Indian Standard, Method for Standard Penetration Test for Soils, ISI, New Delhi.

IS: 4968 (Part I) 1976: Indian Standard Method for Subsurface Sounding for Soils- Part I Dynamic Method using Cone and without Bentonite Slurry, ISI, New Delhi.

IS: 2911 (Part IV) 1979, Code of Practice for 'Design and Construction of Pile Foundations, Part IV Load Test on Piles.

Jain, M.P., (1993), 'Behaviour of Spliced and Intact in Soft Clays', Ph.D.Thesis (under submission), University of Roorkee, Roorkee.

Kandasamy, M.K. (1979), 'Bituminous Coating to reduce Skin Friction in Piles", M.E. Thesis, I.I.T. Madras, pp.1-64.

Mohan, Dinesh, Jain, G.R.S., and Bhandari, R.K. (1978), 'Remedial Underpinning of Steel Tank Foundation', Journal of Geotechnical Engineering Division, ASCE, Vol.104, No.GTS, Proc. Paper 3756, May, pp.639-655.

Mohan, D. (1981), 'A Close Look at Problems of Research and its Applications to Pile Foundations', The IIIrd IGS, Annual Lecture Indian Geotechnical Journal, Vol.11, No.1, pp.1-44.

Mohan, D. et al. (1981), 'Negative Drag on an Instrumented Pile A Field Study', Proc. Xth ICSMFE, Stockholm, Vol.2, pp.767-790.

Rao, B.G. (1982), 'Behaviour of Granular Piles under Load', Ph.D. Thesis, Civil Engineering Department, University of Roorkee, Roorkee (U.P.).

Rao, B.G., Gopal Ranjan, Jain, M.P. (1992), 'Performance of Jointed Piles in Cohesionless Deposits', Symposium on Prediction versus Performance in Geotechnical Engineering, GEOTECH 92, Nov. 30 - Dec. 04, 1992, Bangkok, Thailand.

Rao, B.G., Gopal Ranjan, Jain, M.P. (1993), 'Foundations for Tall Buildings', Conference on Tall Buildings held at University of Roorkee, Roorkee in March.pp.1-3.

Rao, B.G., Gopal Ranjan, Jain, M.P. (1993), 'Negative Drag on spliced piles in soft clay-A Field Study', Eleventh South East Asian Geotechnical Conference (11th SEAGC), at Singapore, May 6-8, 1993.

Rao, B.G. and Jain, M.P. (1993), Innovative Foundations in Poor Third International Conference on Caste Histories in Geotech. Engg. to be held at Missouri Rolla, U.S.A. June 1-6, 1993, Vol.3.

Rao, B.G., Jain, S.K. and Dinesh (1992), 'Performance of Geofabric Reinforced Pad Foundation on Composite Ground', Geotechnique Today, Proc. IGC-18-20 Dec. pp.251-254.

Rao, B.G. (1993), 'Behavioural Prediction and Performance of Structures on Improved Ground and Search for New Technologies', 15th IGS-Annual Lecture, Indian Geotechnical Journal, Vol.23, No.1, pp.1-192.

SBN 1975, Rules for the Design Installation of Piles, Swedish Board of Regional Planning and Building-Forlag, Stockholm, Sweden.

Sawaguchi, M. (1971), 'Approximate Calculation of Negative Skin Friction of a Pile', Soils and Foundations, Vol.11, No.2, pp.31-49.

Victor, D.J. (1977), A new method Splicing Concrete Piles for Bridge Foundations', I.R.C. Journal, Vol.38-2, October 1977, pp.321-326.

Prediction versus Performance in Geotechnical Engineering, Balasubramaniam et al. (eds)
© *1994 Balkema, Rotterdam, ISBN 90 5410 355 8*

Cyclic load-induced settlement of shallow foundation on geogrid-reinforced sand

B. Yeo, S. C. Yen, V. K. Puri, B. M. Das & M. A. Wright
Southern Illinois University at Carbondale, Ill., USA

ABSTRACT: Laboratory model test results for the settlement of a shallow square foundation supported by geogrid-reinforced sand and subjected to cyclic loading are presented. During the application of the cyclic load, the foundation was subjected to a sustained static load. The ultimate cyclic load-induced settlement of the foundation is a function of the magnitude of the sustained static load and the amplitude of the cyclic load. The model tests were conducted at 70% relative density of compaction of sand. Only one type of geogrid was used.

1 INTRODUCTION

The results of several laboratory model tests are presently available in the literature which evaluate the beneficial effects of soil reinforcement in improving the ultimate and allowable bearing capacities of a shallow foundation supported by granular soil (Akinmusuru and Akinbolande, 1981; Binquet and Lee, 1975; Fragaszy and Lawton, 1984; Guido et al. 1985, 1986; Huang and Tatusoka, 1988, 1990). The materials used for soil reinforcement in the existing studies were metal strips, wire mesh, aluminum foil, rope fibers, geotextiles, and geogrids. The studies cited above have, in general, evaluated the optimum values of the following parameters (Fig. 1) which provide the maximum benefit from the soil reinforcement:

a. Extent of reinforcement, d, measured from the bottom of the foundation;

b. Width of reinforcement, b; and

c. Location of the first layer of reinforcement with respect to the bottom of the foundation, u.

The increase of the ultimate bearing capacity due to soil reinforcement has traditionally been expressed in a nondimensional form called the bearing capacity ratio (BCR) which is defined as

$$BCR = \frac{q_{u(R)}}{q_u} \qquad (1)$$

where $q_{u(R)}$ = ultimate bearing capacity with the inclusion of soil reinforcement, and q_u = ultimate bearing capacity in unreinforced soil.

Shallow foundations are sometimes subjected to cyclic loading. The cyclic loading may arise from various types of machinery or other devices supported by the foundation and will induce some settlement. A review of existing literature shows that practically no studies have been undertaken to evaluate the cyclic load-induced settlement of shallow foundations supported by reinforced soil. The purpose of this paper is to present some recent laboratory model test results related to cyclic load-induced settlement of shallow square foundations (B × B) supported by geogrid-reinforced sand.

2 PARAMETERS STUDIED IN LABORATORY MODEL TESTS

In order to evaluate the cyclic load-induced settlement of a foundation, it was decided that the sand supporting the square foundation

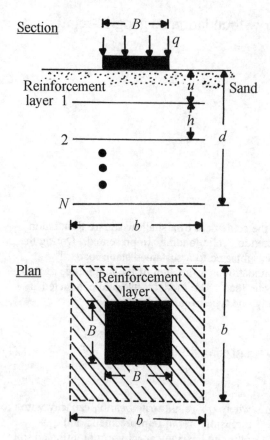

Fig. 1 Square foundation on sand with N
number of reinforcement layers

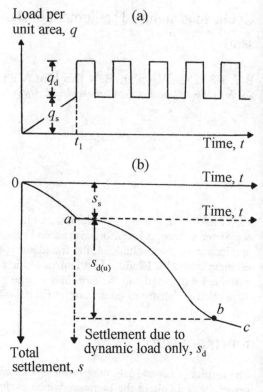

Fig. 2 Variation of load and foundation
settlement with time

needed to have optimum geogrid reinforcement.
Therefore, in the first phase of the study, it was
essential to determine the critical values of d/B
[= (d/B)$_{cr}$] and b/B [= (b/B)$_{cr}$] under static
loading conditions. For strip foundations,
Binquet and Lee (1975) observed that, in order
to obtain maximum benefit from soil reinforcement, it is desirable that u/B be less than about
0.67. Based on bearing capacity tests on a
model square foundation supported by geogrid-
reinforced sand, Guido et al. (1986) determined
that the critical value of u/B is about 0.75. In
most practical cases, however, u/B is kept
between 0.25 to 0.4. With the above factors in
mind, the magnitude of u/B for the present
study was kept at 0.333.

The second phase of the study relates to the
determination of settlement of the square foundation under cyclic loading. With geogrid rein-

forcement layers in place (with d = d$_{cr}$, b = b$_{cr}$,
and u = 0.333B), if a static load q$_s$ is applied on
the foundation (Fig. 2a), then the settlement of
the foundation with time will take the path Oa
as shown in Fig. 2b. The foundation will
undergo a settlement of s = s$_s$ at time t = t$_1$. It
is important to realize that

$$q_s = \frac{q_{u(R)}}{FS} \qquad (2)$$

where FS = factor of safety. For time t ≥ t$_1$, if
a cyclic load having a period T and an amplitude of q$_d$ is applied to the foundation (Fig. 2a),
then it will undergo further settlement which
can be represented by path abc (Fig. 2b). For
all practical purposes, the maximum settlement
due to dynamic loading only is equal to s$_{d(u)}$. In
this phase of laboratory testing, with various
combinations of q$_s$ and q$_d$, the nature of variation of s$_d$ and s$_{d(u)}$ has been evaluated.

3 LABORATORY MODEL TESTS

The laboratory bearing capacity tests were conducted using a model foundation, made from an aluminum plate, with dimensions (B × B) of 76.2 mm × 76.2 mm. A fine round silica sand was used for the model tests. The grain-size distribution of the sand is shown in Fig. 3. All model tests were conducted at an average relative density, D_r, of 70%. The average physical properties of the sand during the laboratory tests are given in Table 1. A biaxial geogrid was used for reinforcement, and its physical properties are given in Table 2.

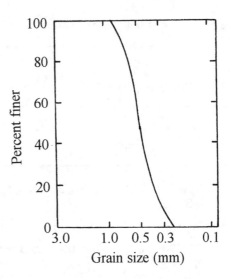

Fig. 3 Grain-size distribution of sand

Table 1. Average physical properties of the sand during the model tests

Parameter	Quantity
[a]Maximum dry unit weight (kN/m^3)	18.94
[a]Minimum dry unit weight (kN/m^3)	14.07
Dry unit weight during model tests (kN/m^3)	17.14
Relative density of compaction during model tests (%)	70
[b]Angle of friction, ϕ, during model tests(deg)	40.3

[a]ASTM test designation D-4253; [b]From direct shear test

Table 2. Physical properties of the geogrid

Parameter	Description/quantity
Structure	Punctured sheet drawn
Polymer	PP/HDPE co-polymer
Junction method	Unitized
Aperture size (MD/XMD)	25.4 mm/33.02 mm
Nominal rib thickness	0.762 mm
Nominal junction thickness	2.286 mm

Laboratory model tests were conducted in a Plexiglas box measuring 760 mm × 760 mm × 760 mm. Rough base condition of the model foundation was achieved by cementing a thin layer of sand onto its base with epoxy glue. In conducting the tests, sand was poured into the box in 25.4 mm high layers using a raining technique. The accuracy of sand placement and consistency of placement density were checked during raining by placing small cans with known volumes at different locations in the box. Geogrid layers were placed in the sand at desired values of u/B and h/B (h = center-to-center spacing of the geogrid layers as shown in Fig. 1). After completion of the sand placement, the model foundation was placed on the surface of the sand for starting the model test. As mentioned before, the laboratory study was divided into two phases. A brief description of each phase of the laboratory tests follows.

3.1 Phase I - Static Tests

The tests in this phase were conducted to determine $(d/B)_{cr}$ and $(b/B)_{cr}$. In conducting the tests, load to the model foundation was applied by a hydraulic jack. The load on the foundation and the corresponding settlement were measured by a proving ring and two dial gauges. Three series of tests were conducted, and the details are given in Table 3.

Table 3. Details of static tests--phase I

Test series	Constant parameters	Variable parameters	Purpose
I-A	D_r = 70%	-------	To determine q_u on unreinforced sand
I-B	D_r = 70%, u/B = h/B = 1/3 b/B = 6	N = 1, 2, 3, 4, 5, 6	To determine $(d/B)_{cr}$
I-C	D_r = 70%, u/B = h/b = 1/3 N = 4	b/B = 1, 2, 3, 4, 5, 6	To determine $(b/B)_{cr}$

3.2 *Phase II - Dynamic Tests*

It will be shown in the following section that the magnitudes of $(d/B)_{cr}$ and $(b/B)_{cr}$ obtained from tests conducted in Phase I were about 1.32 and 4, respectively. Hence in this phase, for all tests, the following constant parameters were adopted: u/B = h/B = 0.333; b/B ≈ $(b/B)_{cr}$ = 4; d/B ≈ $(d/B)_{cr}$ = 1.32; and average relative density of sand, D_r = 70%.

For conducting the dynamic tests in this series, a universal testing machine was used for application of load on the foundation. The magnitude of the load and the foundation

settlement were recorded by a data acquisition system. In order to start a test, a static loading (q_s) with a desired factor of safety (FS) was first applied to the foundation, followed by the application of the cyclic load. The nature of the dynamic loading pattern is shown in Fig. 4. The period, T, of the dynamic load for all tests was 1.0 sec. Details of the laboratory tests conducted in this phase of the study are given in Table 4.

Table 4. Details of dynamic tests--phase II

Test series	$q_s/q_{u(R)}$ (%)	$q_d/q_{u(R)}$ (%)
II-A	13.2	4.36, 10.67, 14.49, and 22.33
II-B	25.0	4.36, 10.67, 14.49, and 22.33
II-C	33.3	4.36, 10.67, 14.49, and 22.33

Dynamic load per unit area

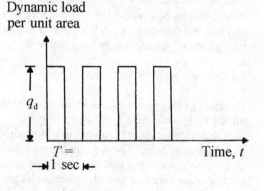

Fig. 4 Pattern of dynamic loading

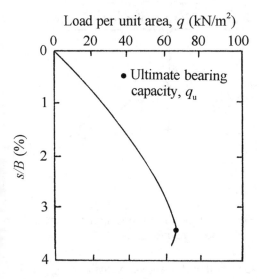

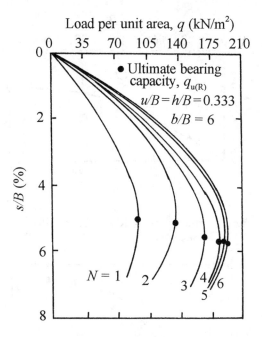

Fig. 5 Variation of load per unit area vs. s/B (Test Series I-A)

Fig. 6 Variation of load per unit area vs. s/B (Test Series I-B)

4 MODEL TEST RESULTS OF PHASE I - STATIC TESTS

4.1 Test Series I-A

The load per unit area (q) versus foundation settlement (s) obtained from the tests on unreinforced sand is shown in Fig. 5. For the present tests, the ultimate bearing capacity, q_u = 64.2 kN/m^2, was achieved at an s/B value of about 3.43%.

4.2 Test Series I-B

The tests in this series were conducted to determine the critical depth ratio, $(d/B)_{cr}$, of geogrid reinforcement beyond which the increase in the bearing capacity ratio (BCR) is practically negligible. For these tests, the magnitudes of u/B, h/B and b/B were kept constant at 0.333, 0.333, and 6, respectively. Figure 6 shows the plots of q versus s for various numbers of reinforcement layers, N. The depth ratio of reinforcement can be related to N as

$$\frac{d}{B} = \frac{u}{B} + (N - 1)\frac{h}{B} \qquad (3)$$

The ultimate bearing capacity, $q_{u(R)}$, for each

one of these q versus s plots is also shown in Fig. 6. In general, the ultimate bearing capacity increased with the increase of d/B accompanied by an increase of the settlement at ultimate load. Based on the values of q_u and $q_{u(R)}$ obtained from Figs. 5 and 6, the variation of the bearing capacity ratio with d/B is shown in Fig. 7. From the plot, it appears that the magnitude of BCR increases with d/B and reaches an approximate maximum value at d/B = $(d/B)_{cr} \approx$ 1.32. For d/B > $(d/B)_{cr}$, the magnitude of $\Delta(BCR)/\Delta(d/B)$ is minimum, and any further increase of the depth of reinforcement may not be considered economical.

4.3 Test Series I-C

Model tests in this series were conducted to determine the optimum nondimensional width ratio $(b/B)_{cr}$ for the geogrid layers for mobilization of the maximum bearing capacity ratio. The magnitudes of u/B and h/B were kept equal to 0.333. Also, the number of layers of reinforcement was equal to 4 since it was determined from Series I-B that $(d/B)_{cr}$ is equal

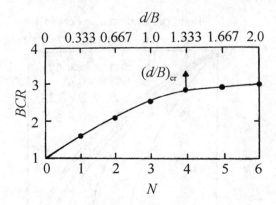

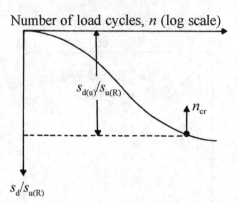

Fig. 7 Variation of BCR with N and d/B (Test Series I-B)

Fig. 9 Variation of foundation settlement due to dynamic load

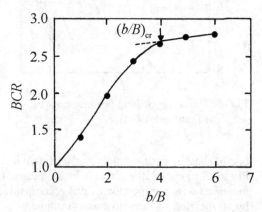

Fig. 8 Variation of BCR with b/B (Test Series I-C)

to about 1.32. The ratio b/B was varied as 1, 2, 3, 4, 5 and 6. The variation of BCR with b/B was obtained from the plots of q versus s in a manner similar to that discussed under Series I-B, and this is shown in Fig. 8. As can be seen from the figure, the magnitude of BCR increases with b/B to an approximate maximum at $b/B = (b/B)_{cr} \approx 4$ and remains practically constant thereafter.

5 MODEL TEST RESULTS OF PHASE II - DYNAMIC TESTS

The interpretation of the dynamic test results can be done in the following manner. Based on the present test results (also see Fig. 2), it

appears that, for a given value of $q_s/q_{u(R)}$ and $q_d/q_{u(R)}$, the nature of variation of $s_d/s_{u(R)}$ with n (where s_d = settlement due to dynamic load only; $s_{u(R)}$ = settlement at ultimate load $q_{u(R)}$ as obtained from static tests in Phase I for similar reinforcement conditions; and n = number of load cycle applications) will be as shown in Fig 9. The settlement increases with the increase of the number of load cycles (n) to almost a maximum value of $s_{d(u)}$ at $n = n_{cr}$. For $n > n_{cr}$, the slope $\Delta s_d/\Delta n$ is zero or very small.

As mentioned before, for all tests in this phase, u/B = h/B = 0.333; $b/B \approx (b/B)_{cr} = 4$; $d/B \approx (d/B)_{cr} = 1.32$. Figures 10, 11, and 12 show the plots of $s_d/s_{u(R)}$ versus n as obtained from Test Series II-A, II-B, and II-C, respectively. Also shown in these figures are the points which represent n_{cr} for each curve. Based on the experimental results shown in these figures, the following general conclusions can be drawn.

1. For a given value of $q_s/q_{u(R)}$ and number of load cycle applications, the magnitude of $s_d/s_{u(R)}$ increases with the increase of $q_d/q_{u(R)}$.

2. The magnitudes of $q_d/q_{u(R)}$ and n remaining constant, the value of $s_d/s_{u(R)}$ increases with the increase of $q_s/q_{u(R)}$.

3. For a given value of $q_d/q_{u(R)}$ and $q_s/q_{u(R)}$, the magnitude of n_{cr} is approximately the same.

4. For a given value of $q_s/q_{u(R)}$, the magnitude of n_{cr} decreases with an increase of $q_d/q_{u(R)}$. In the present study, $n_{cr} \approx 3 \times 10^5$ cycles when $q_d/q_{u(R)}$ is 4.3%. It decreases to about 1.75×10^5 cycles when $q_d/q_{u(R)}$ is 22.3%.

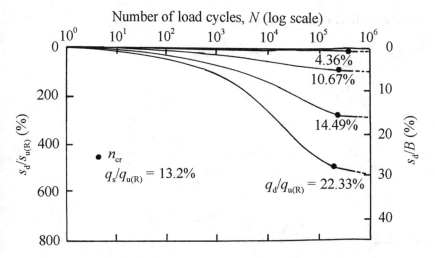

Fig. 10 Variation of $s_d/s_{u(R)}$ with n for $q_s/q_{u(R)}$ = 13.2% (Test Series II-A)

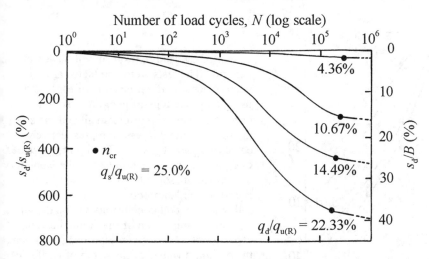

Fig. 11 Variation of $s_d/s_{u(R)}$ with n for $q_s/q_{u(R)}$ = 25.0% (Test Series II-B)

As shown in Figs. 2 and 9, the maximum settlement due to dynamic loading, $s_{d(u)}$, may be approximated to be equal to the settlement between n = 0 and n = n_{cr}. Using this definition, the variation of $s_{d(u)}/s_{u(R)}$ for various $q_s/q_{u(R)}$ and $q_d/q_{u(R)}$ combinations were calculated and are shown in Fig. 13. Based on this figure, it can be seen that with FS (= $q_{u(R)}/q_s$) varying between 7 and 3 and $q_d/q_{u(R)} \approx$ 10%, the magnitude of the maximum settlement due to dynamic load may be in the range of 5% to 20% of the width of the foundation. If $q_d/q_{u(R)}$

is increased to about 20%, the maximum settlement $s_{d(u)}$ for a similar range of FS increases to a range of 30% to 40% of the foundation width.

It appears reasonable to also expect that the magnitude of $s_{d(u)}/B$ will be a function of (a) the relative density of compaction of sand, (b) the stiffness of the geogrid used for reinforcement, and (c) the soil-geogrid interface friction angle. However, the present test limitations do not allow evaluation of the effect of the above parameters.

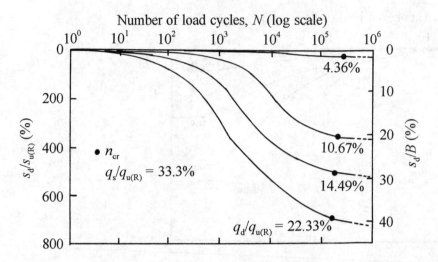

Fig. 12 Variation of $s_d/s_{u(R)}$ with n for $q_s/q_{u(R)}$ = 33.3% (Test Series II-C)

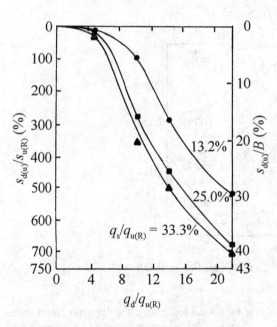

Fig. 13 Variation of $s_{d(u)}/s_{u(R)}$ with n for $q_d/q_{u(R)}$

6 CONCLUSIONS

A number of laboratory model tests were conducted to assess the cyclic load-induced settlement of a shallow square foundation supported by sand reinforced with layers of geogrid. The tests were conducted at one relative density of compaction of sand, and by using only one type of geogrid. For dynamic tests, the depth of reinforcement and the size of the reinforcement layers were, respectively, equal to d_{cr} and $b_{cr} \times b_{cr}$. The optimum values of d_{cr} and b_{cr} were determined in the laboratory by conducting static bearing capacity tests. Based on the laboratory model tests, the following general conclusions can be drawn.

1. For mobilization of maximum bearing capacity ratio for a given sand-geogrid system, the optimum values of the depth of reinforcement and the width of reinforcement are approximately equal to 1.32B and 4B, respectively.

2. For a given sustained load (q_s) and number of load cycles (n), the settlement due to dynamic loading increases with the increase in the magnitude of q_d.

3. For similar values of q_d and the number of load cycles, the dynamic load induced-settlement increases with the increase in the sustained load (q_s).

4. The maximum dynamic load-induced settlement of a foundations is a function of q_s, q_d, stiffness of the geogrid, and the degree of compaction of soil.

REFERENCES

Akinmusuru, J.O. & J.A. Akinbolande 1981.
Stability of loaded footings on reinforced soil.
*Journal of the Geotechnical Engineering
Division*, ASCE, 107:819-827.

Binquet, J. & K.L. Lee 1975. Bearing capacity
of reinforced earth slabs. *Journal of the
Geotechnical Engineering Division*, ASCE,
101:1257-1276.

Fragaszy, R.J. & E.C. Lawton 1984. Bearing
capacity of reinforced sand subgrades.
*Journal of the Geotechnical Engineering
Division*, ASCE, 101:1500-1507.

Guido, V.A., Biesiadecki, G.L. & M.L. Sullivan
1985. Bearing capacity of a geotextile-
reinforced foundation. *Proc. 11th ICSMFE*:
1777-1780. Rotterdam: Balkema.

Guido, V.A., Chang, D.K. & M.A. Sweeney
1986. Comparison of geogrid and geotextile
reinforced slabs. *Canadian Geotechnical
Journal*, 23:435-440.

Huang, C.C. & F. Tatusoka 1988. Prediction of
bearing capacity in level sandy ground rein-
forced with strip reinforcement. *Proceedings,
International Geotechnical Symposium on
Theory and Practice of Earth Reinforcement*,
Fukuoka, Japan:191-196.

Huang, C.C. & F. Tatusoka 1990. Bearing
capacity of reinforced horizontal sandy
ground. *Geotextiles and Geomembranes*,
9:51-82.

A study of sand penetration in clay slurry

S.L.Lee & T.S.Tan
National University of Singapore, Singapore

T.Inoue
Fukken Co. Ltd, Hiroshima, Japan

ABSTRACT: When sand is spread onto a clay slurry, the mechanics of sand penetration is a complicated problem and is studied in this paper via a model of a sphere settling in a slurry. The analytical results show that the initial impact velocity affects only the entry into the clay and the subsequent penetration of the particle is controlled by its buoyant weight versus the shear resistance from the slurry. This led to the development of a dimensionless parameter, which is the ratio of the shear resistance over the buoyant weight of the sand spread, to analyze the experimental data. This analysis indicates that the finer particles are trapped at the upper part while the larger particles penetrated more deeply. A weighted mean diameter was used in the analytical model to reflect the grain size distribution of the sand. The results from this was found to agree well with both experimental and field data.

1   INTRODUCTION

To use or treat a clay slurry deposit, sand often has to be added. This sand when spread will either be arrested near the surface or penetrate into the slurry. The behaviour of the resulting clay–sand mix is dependent on the way the sand has penetrated. Thus, to enable engineers to control the spreading and penetration of sand into a clay slurry, the mechanics needs to be understood and the important parameters identified.

The mechanics of the penetration of sand is complex as many variables are involved. In this paper, a preliminary experimental and analytical study of the mechanics of sand penetration is reported. The aim is to identify those parameters that are easily quantifiable in the field that control the penetration process. Two aspects are considered in the study, namely, whether the spread sand will penetrate and if so, how are the sand particles trapped.

2   EXPERIMENTAL STUDY

In this study, only one particular sand was used. The clay specimens used in the experiments were from the Changi South Bay Project (Karunaratne et al., 1990). The grain size distribution of the clay and sand used are shown in Fig. 1.

For the experiments, a clay slurry was poured into a perspex tube of 15 cm diameter, upto 50 cm in height and then filled up with water to just below the brim of the tube. Sand was then spread from the top of the tube. The penetration was monitored by measuring the density profiles using a γ-ray density gauge, which has been used previously for very soft clay (Tan et al., 1988). Measurements were also carried out 1 day later and after consolidation.   At the end of consolidation, the grain size distribution of the sand trapped at various heights were obtained by extracting a sample from each of these heights for a sieve analysis. To obtain the rheology and the shear strength of the clay slurry, the viscometric method and a penetration method proposed by Inoue (Inoue, 1990 and Inoue et al., 1990) were used.

The velocity of the sand just prior to entry into the clay is an important but difficult parameter to determine.   A series of tests were carried out to determine this velocity. For this, the same sand was spread into a cylinder containing water.   The entire settling process was then video-taped and then

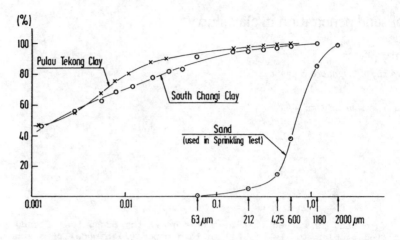

Fig. 1 Grain size distribution of clays and sand used in the experiments

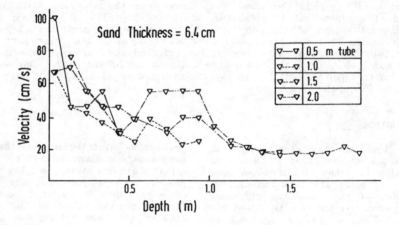

Fig. 2 Velocity of the front of settling sand mass

replayed frame by frame, where the time for each frame had been calibrated. The movement of the front of the settling sand particles with time was then tracked and the velocity estimated.

2.1 Settling Velocity

Right after spreading, the sand first gather in a mass but disperse subsequently as a result of the vortex and circulation. If the water depth is sufficient, the mass will reach the bottom fully dispersed. Due to this dispersion, the velocity of each particle is expected to be different and difficult to determine. Thus only the velocity of the wave front of the settling particles is determined as the motion of

these front particles is critical in determining the penetration of the sand. This is not illogical as once the front particles have penetrated the clay, the strength of the clay will be improved and the trailing particles will face higher resistance and thus are more likely to be arrested.

The accuracy of the velocity determined, especially right after spreading is not good because of the onset of turbulence. This problem is particularly acute when the tube is short. For longer tubes (height 1.5 m to 2.0 m), the velocity of the front is approaching some terminal velocity as shown in Fig. 2. Table 1 gives the velocities just prior to hitting the bottom under various spreading conditions. For the sand investigated in

56

this study, the terminal velocity is about 18 cm/s. Inoue (1990) has also shown that for a larger diameter tube, the same terminal velocity is obtained, the difference being that a slightly bigger depth is needed to reach it. In practice, considering the draught needed for most spreading equipment, it is expected that the spread sand will be able to reach this terminal velocity easily.

3   PROPOSED SAND PENETRATION MODEL

The motion of sand particles in a non-Newtonian fluid is a difficult problem. Even for uniform spheres, no rigorous solution is available. Even if the fluid is Newtonian, the motion of sand particles is still very complicated as the effect of turbulence is felt. Thus to obtain a rigorous solution of the motion

Table 1   Velocity of Front of Falling Sand

Thickness of Sand Spread (cm)	Velocity in cm/s				
	Depth of Column of Water (m)				
	0.5	0.75	1.0	1.5	2.0
2.2	25	20	18	18	18
3.2	32	22.5	18	18	18
6.4	33	26	23.5	18	18
9.5	33	26	24	21	18

of sand particles in a clay slurry is unlikely.   To understand the mechanics underlying the motion of such particles, and the interplay between the various parameters governing the motion, a simple model that can capture the essence of the mechanics is proposed.

As a first order approximation, the motion of a sphere in a non-Newtonian fluid is considered. The approach is essentially similar to that for the motion in a Newtonian fluid, the principal difference, following works by Ansley Smith (1967) and Hanks and Sen (1983), is to account for the non-Newtonian rheology of the fluid. The motion of a particle in the slurry is governed by:

$$\rho_s g\, V_s - \rho_f\, g\, V_s - F_D = (\rho_s V_s + \tfrac{1}{2} V_s \rho_f)\, \ddot{z}$$

$$(1)$$

where V_s is volume of the sand particle, ρ_s and ρ_f the density of solid and fluid respectively, F_D the drag force and $\ddot{z}$ the acceleration of the settling particle.

In Eq. 1, the left-hand side is the net force acting on the particle comprising the weight, the buoyancy and the drag force while the right-hand side is the inertia force and includes a virtual mass term.   According to Ansley and Smith (1967), a dynamic parameter proposed by them for a Bingham fluid has a strong correlation with the drag coefficient and the resulting drag force is given by:

$$F_d = \frac{7}{8}\,\pi^2 D^2 \tau_y + 3\pi D\eta \dot{z}$$

$$(2)$$

where D is the diameter of the particle, τ_y the yield stress and η the viscosity of the clay slurry.   Substituting Eq. 2 into Eq. 1 gives:

$$D^2 \left(\tfrac{1}{2}\,\rho_f + \rho_s \right) \ddot{z} + 18\eta \dot{z} = D^2 \left(\rho_s - \rho_f \right) g - \frac{21}{4}\pi D \tau_y$$

$$(3a)$$

The above equation accounts for a single particle.   When a group of particles together penetrate the slurry, it is expected that the drag force acting on them will be less than the sum of the drag on each individual particle.   This is similar to the idea of group efficiency in the analysis of piles group.   One way to account for this is to use a lower apparent shear stress, τ'_y.

$$D^2 \left(\tfrac{1}{2}\,\rho_f + \rho_s \right) \ddot{z} + 18\eta \dot{z} = D^2 \left(\rho_s - \rho_f \right) g - \frac{21}{4}\pi D \tau'_y$$

$$(3b)$$

This is a non-homogeneous linear ordinary differential equation with constant coefficients and a closed form solution can be easily obtained.   In Eq. 3, the right hand side reflects the balance between the buoyant weight and the opposing drag resulting from the yield stress.   If this term is positive, the gravitational load is larger than the shear resistance and the particle will not stop but penetrate to the bottom.   On the other hand, if it is negative, the particle will stop after some time.   Through the use of this simplified equation, we can investigate the mechanics of the penetration of sand into the clay slurry.   For convenience, Eq. 3 is normalized to give:

$$\frac{(\frac{1}{2}\rho_f + \rho_s)}{(\rho_s - \rho_f)} \frac{\ddot{z}}{g} + \frac{18\eta}{D^2(\rho_s - \rho_f)g} \dot{z} =$$

$$1 - \frac{21\pi}{4} \frac{\tau'_y}{D(\rho_s - \rho_f)g} \qquad (4)$$

4 PENETRATION OF A SAND PARTICLE

To calculate the motion of a particle in a clay slurry using Eqn. 4, the properties used are given in Table 2.

Table 2 Bingham properties for clay slurry used

w.c. (%)	τ'_y (N/m^2)	η (kg s/m^2)
200	1.425	0.00579
225	0.866	0.00259
250	0.714	0.00199
275	0.450	0.00060
300	0.273	0.00035

For τ'_y, a value of 10% is selected for the efficiency. This value was chosen after some back analysis. Fig. 3a shows the settling of a single particle in a clay slurry of 300%. The initial velocity used is 0.18 m/sec. In the figure, all the particles which have size .425 mm or greater penetrate to the bottom. The results for .212mm and .063mm which are not shown in the figure indicate that they are arrested at the surface. This agrees well with Fig. 3b which was obtained from the sand spreading experiments.

Fig. 4 shows the effect of the initial velocity. Since both curves become parallel after 0.2 - 0.4 second despite the large difference in the initial velocities, the effect is seen to be significant only at the initial stage. This suggests that the kinetic energy of the particles is dissipated by the clay slurry shortly after entry and subsequent penetration is affected more by the buoyant weight of the sand and the yield shear stress of the slurry.

To explore this further, a new parameter, K, equal to the right-hand side of Eq. 4 is now introduced:

$$K = 1 - \frac{21\pi}{4} \frac{\tau'_y}{D(\rho_s - \rho_f)g} \qquad (5)$$

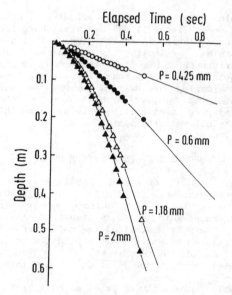

Fig. 3(a) Predicted penetration of a single particle in a clay slurry with a water content of 300%

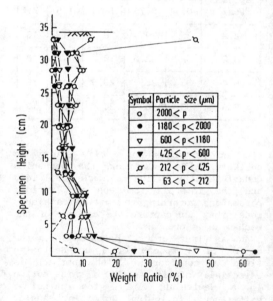

Symbol	Particle Size (μm)
o	2000 < p
●	1180 < p < 2000
▽	600 < p < 1180
▼	425 < p < 600
⌀	212 < p < 425
⌀	63 < p < 212

Fig. 3(b) Sand distribution in a clay slurry with a water content of 300%

If K is positive, the particle will keep on settling. If it is negative, the particle will be trapped at some depth depending on how negative is the value. Fig. 5 shows the trend of K calculated for water content in the range 200 – 300% and it can be seen that for the 2 mm particle, K is always positive which means that this particle will never stop in the clay slurry. On the other hand, in the case of 0.25 mm particle, since K is negative for all water content, it will stop and for the lower content, due to the large

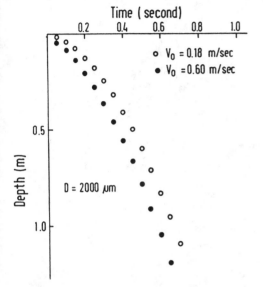

Fig. 4 Effects of initial velocity on sand penetration

negative value of K, it will stop near the surface. Thus the critical water content for a particular particle size can be estimated.

To examine this further, Fig. 5 is compared with the experimental results shown in Fig. 6. For all three cases, the same amount of sand, 6.4 cm thick, was spread. In the figure the weight ratio is the proportion of sand of a particular grain size range trapped at a certain height versus the total amount of sand of that size range. The results presented clearly follow the trend indicated in Fig. 5.

For example, Fig. 6a shows that for 250% water content, particles of size 600 μm or greater mostly penetrate to the bottom, whereas those smaller than 425 are trapped at the surface. Particles in the range 425μm < p < 600μm is clearly a transition range. Hence a portion of the particle in this range is trapped at the top while the rest penetrate to the bottom. Analytically, for the 250% slurry, Fig. 5 shows that the transition size is between 500 μm and 750μm which is very close to observations. For the 300% slurry, Fig. 6c shows that almost all the particles except the smallest size penetrate in agreement with the analytical results. Thus Eq. 5 can estimate the penetrability of sand particles and more importantly, the parameters in it are easily determined, even in the field.

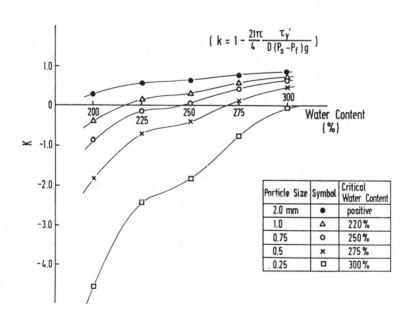

Fig. 5 k versus water content

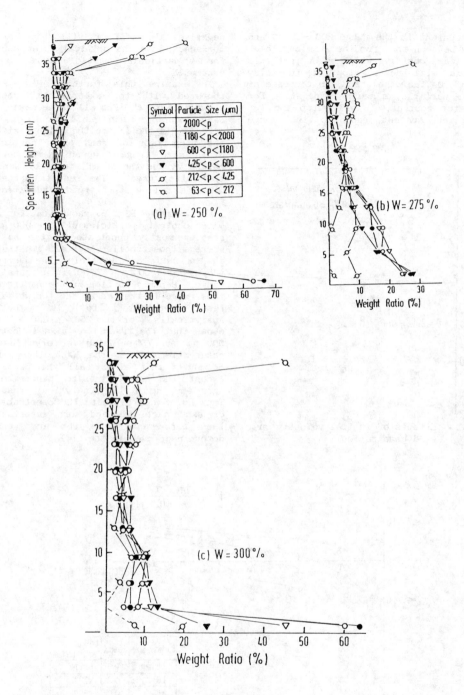

Fig. 6 Sand distribution in clay slurries with
varying water content

5   SAND PENETRATION RESULTS

Two points are clear from the model discussed above. First, the impact velocity of the falling sand affects only the initial penetration. Second, the subsequent penetration of the particles is controlled by Eq. 5 which is a balance between the drag resistance and the gravitational force of the particles. From these two considerations, the following parameter was chosen to analyse the result of the penetration:

$$P_1 = \frac{\tau'_y}{\gamma' T_s} \qquad (6)$$

Where γ' is the submerged density of sand, T_s, the thickness of sand and τ'_y, the shear strength of clay. This parameter which is suggested by the left hand side of Eq. 5 is physically meaningful and reflects the balance between the shear resistance and the buoyant weight of the sand spread.

To relate the distribution of sand in the clay, in particular, the amount trapped at the upper part to the parameter P_1, a practical parameter W_1/W_0 was chosen. W_1 is the weight of sand trapped at the upper portion and W_0 is the total weight of sand spread. To check the objectivity of this parameter, it is compared with the average penetration depth, $\bar{d}$. For a particular range of grain sizes in a test, if the distribution of

sand is divided into n segments and d_i is the depth for the ith segment and w_i is the weight of sand in it, the average penetration depth, $\bar{d}$ is estimated as follows:

$$\bar{d} = \frac{\Sigma(d_i \times w_i)}{\Sigma w_i} \qquad (7)$$

Fig. 7 shows the relation between W_1/W_0 and $\bar{d}/H_f$ where H_f is the final specimen height after consolidation for different ranges of grain size. Although there is still some scattering, W_1/W_0 is seen to be linearly correlated with $\bar{d}/H_f$ thus implying that the proposed parameter is sufficiently objective. W_1/W_0 is preferred since it is a more meaningful term for site control as it reflects the amount of sand that is trapped in the upper part of the clay. Fig. 7 also shows that sands of larger particle size penetrate nearly to the bottom, while the fine sands are mostly trapped in the upper part.

Fig. 8 shows the plot of P_1 versus W_1/W_0 and a reasonable linear relation can be seen. This suggests that the amount of sand trapped in the clay is indeed governed by the balance between the resistance of the clay and the buoyant weight of sand as suggested by the analytical model. This is not unexpected since once the particles enter the clay,

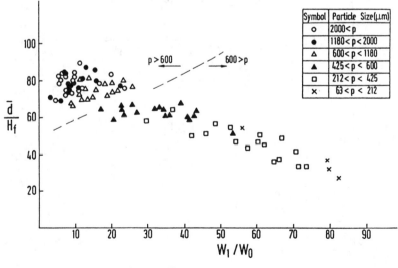

Symbol	Particle Size(μm)
o	2000 < p
●	1180 < p < 2000
Δ	600 < p < 1180
▲	425 < p < 600
□	212 < p < 425
×	63 < p < 212

Fig. 7 $\bar{d}/H_f$ versus W_1/W_0

61

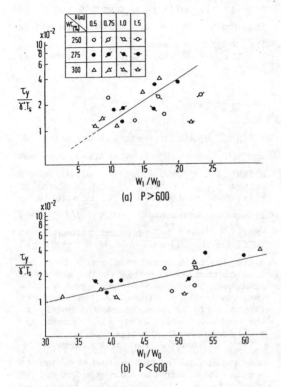

(a) P > 600

(b) P < 600

Fig. 8 $\tau_y / \gamma' T_s$ versus W_1 / W_0

the initial kinetic energy will be quickly dissipated, after which P_1 will govern the distribution as suggested by Eq. 4. Thus P_1 is an effective parameter to analyze sand penetration and more importantly, all the terms in it can be easily determined.

6   MODIFICATION OF THE SINGLE PARTICLE MODEL

An observation from Fig. 8 is that more of the finer particles are trapped at the upper part while the larger size particles penetrate more deeply and the overall penetration is an average of these values. Thus, a weighted mean parameter based on the grain size distribution curve shown in Fig. 1 is more appropriate for analysis.

Let Y be defined by Eq. 8 which takes into account the grain size distribution as shown in Fig. 1,

$$Y = \Sigma \ (K_i \ \beta_i) \qquad\qquad (8)$$

where K_i is the K value given by Eq. 5 calculated for a particular grain size and β_i is the fraction by weight of that grain size. Clearly, $\Sigma(\beta_i) = 1$. Fig. 9 shows the relation between Y and the water content of the clay slurry. The critical water content where Y becomes zero is about 250%.

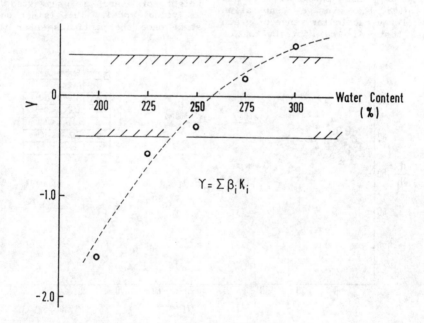

Fig. 9   K, the weighted average value of k versus water content

62

6.1 Penetration Results

For an overall impression of the influence of the various parameters, the results of the tests are divided into two broad categories, namely 'No Penetration' where most of the sand is seen to be trapped at the top and 'Penetration' where the sand is seen to penetrate into the clay column. This division is not precise as even in the case of 'No Penetration', it is expected that some sand particles will still penetrate. These results are summarized in Table 3.

Table 3: Summary of Sand Penetration Results

Water Content (%)	Sand Thickness (cm)	Sand Spreading Height (m)			
		0.5	0.75	1.0	1.5
200	3.2	0			
	6.4	0			
	9.5	0			
225	3.2	0			
250	2.2	0			
	3.2	X		0	
	6.4	X	O:X	X	0
	9.5	X			0
	12.0	X			0
275	2.2	0			
	3.2	X			
	6.4	X	X	X	0
	9.5	X			
285	3.2	X			
300	2.2	X			
	3.2	X			
	6.4	X	X	X	X

0 : No Penetration   X : Penetration

According to this table, sand is trapped at the surface if the water content is less than 250%. If this observation is compared to the analytical results in Fig. 9, it can be seen that Y is negative there which means that the shear resistance of the slurry is higher than the buoyant weight and thus the particles will stop.

The comparison clearly shows that by using a weighted mean diameter based on the grain size distribution curve, the analytical model is able to predict the water content at which the sand can be stopped. As the model is simple and the parameters involved can be easily determined, it provides engineers with a means to analyze and understand sand penetration.

7 CONCLUSIONS

When sand is spread onto a clay slurry, three parameters govern the penetration of sand, namely, the sand spreading height, the thickness of the sand spread and the shear strength of the clay slurry. To understand the mechanics of sand penetration, a simple model is developed as the complexity of the problem makes rigorous solution impossible. This model is based on the motion of a single particle in a non-Newtonian fluid. As a result of this analytical model, a dimensionless parameter was developed to analyze the experimental data. To reflect the amount of sand trapped, a weight ratio defined by the amount trapped at the upper part versus the amount of sand spread, (W_1/W_0), is introduced. The analysis shows that the overall penetration is an average of the penetration of the individual particles. This led to the use of a weighted mean diameter in the analytical model to reflect that the sand has a particular particle size distribution.

The motion of the sand in the water is also complicated as there is turbulence initially. But if the water column is sufficiently high, the sand will disperse and the wave front will reach a terminal velocity. In the field, the actual height to reach terminal velocity may be a bit higher, but from practical considerations, for example, the draught necessary for the barge to move freely, this height requirement should not be a problem.

A number of conclusions can be drawn from this study:

a. The analytical model is able to capture the mechanics of penetration of the sand. In particular, the model shows that the penetration is governed by the balance between the shear resistance and the buoyant weight of the particle while its impact velocity will only affect the initial entry.

b. A dimensionless parameter given by Eq. 6 which reflects the balance between the shear resistance and gravitational force is linearly correlated with (W_1/W_0), the ratio of

the amount of sand trapped at the

upper part versus the total amount spread.

c. It is found that the penetration is an average of the penetration of the individual particles. This led to the use of a weighted mean diameter in the analytical model to reflect the grain size distribution of the sand. The analytical results show that for the clay slurry examined, the critical water content to trap sand is around 250% which agrees well with experimental and field observations.

ACKNOWLEDGEMENTS

The research upon which this paper is based is partly funded by the Science Council of Singapore under the Research and Development Assistance Scheme Grant No. ST/86/05

REFERENCES

Ansley, R.W. and Smith, T.N. (1967), "Motion of spherical particles in a Bingham Plastic", Jour. of the Ameri. Inst. of the Chem. Engrs., Vol. 13, No. 6, pp. 1193-1196.

Hanks, R.W. and Sen, S. (1983), "The influence of yield stress and fluid rheology on particle drag coefficients", Proc. of the 8th Int. Conf. of Slurry Transport, San Francisco, pp. 71-80.

Inoue, T. (1990), "Behaviour and Treatment of Dredged Clay", Ph.D. Thesis, Dept. of Civil Engrg., National University of Singapore.

Inoue, T., Tan, T.S. and Lee, S.L. (1990), "An investigation of shear strength of slurry clay", Soils and Foundations. Vol. 30, No. 4, pp. 1 - 10.

Karunaratne, G.P., Yong, K.Y., Tan, T.S., Tan, S.A., Liang, K.M., Lee, S.L. and Vijiaratnam, A. (1990), "Layered clay-sand scheme reclamation at Changi South Bay", Proc. 10th South East Asian Geotechnical Conf., Vol. 1, pp. 71-76, Taipei, Rep. of China.

Tan, S.A., Tan, T.S., Ting, L.C., Yong, K.Y., Karunaratne, G.P. and Lee, S.L. (1988), "Determination of consolidation properties for very soft clay", ASTM Geotechnical Testing Journal, Vol. 11, No. 4, pp. 233-240.

2 Ground improvement techniques and reinforced earth

Prediction versus Performance in Geotechnical Engineering, Balasubramaniam et al. (eds)
© *1994 Balkema, Rotterdam, ISBN 90 5410 355 8*

Frictional behaviour between smooth and rough geomembranes, various soils and geotextiles

Thomas Kruse
Erdbaulaboratorium Ahlenberg, Herdecke, Germany

Thomas Voigt
Institut für Grundbau und Bodenmechanik, TUBS, Braunschweig, Germany

SUMMARY: To improve the environmental protection new waste disposal sites nowadays are covered at bottom and on top by waterproof sealings. Those sealings are made of a combination of mineralic soils and a layer of HDPE-material (High density Polyethylen). For the safety of the slopes stability and the durablility of the combined sealing, the friction behaviour between the different materials of the sealing is very important.

In many tests, the friction between geomembranes with rough and smooth surfaces, sand, silt, clay and geotextiles respectively has been investigated by using shear boxes of 10 cm by 10 cm and 30 cm by 30 cm length. Depending on the test results, recommendations are made for the shear-parameters in the investigation of the slope stability and other geotechnical calculations.

1.   Introduction

Combination seals composed of mineral sealing material, geomembranes as well as protective and drainage layers are used as basic or surface seals around waste dumps to protect the environment. In the construction of these combined seal systems on slopes, the shear behaviour between the individual system components plays an important role in the stability and durability of the seal overall.

In this paper, it is intended to present several basic aspects of the experimental theory and technique, together with results of experiments with regard to shear behaviour between different plastic sealing strips and cohesive mineral sealing materials, non-cohesive sand and geotextiles. Recommendations for the setting of shear parameters between geomembranes and different materials can be made on the basis of the results of these experiments.

2.   The Theory and Techniques of the Experiment

2.1   Basics

The following points should be included if the stability of combination seals on slopes is to be guaranteed:

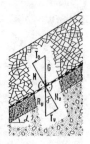

Acting forces at an element of a combinated sealing

- No tension in the geomembrane,

  if $T_o \leq T_u$

- No sliding, if

  slope angle $\alpha \leq$ angle of internal friction ϕ'

Figure 1.   Schematic representation of the forces acting in a combination seal

- The outer slope itself must be stable.
- The single components of the sealing system must be stable, both alone and together.
- It must also be proven that no lasting tensile force builds up (Figure 1). This requirement is met when the shear force T_o, which is produced in the sealing strip by dead weight and surcharges, is less than the backing and strengthening force T_u between sealing strip and mineral seal.

The transferable shear stress between sealing strip and the surrounding material can be obtained by the following equation:

$$\tau_{DB} = \sigma_N * \tan\delta' + a'$$

whereby:

τ_{DB} = shear strength between sealing strip and soil

σ_N = effective normal stress

δ' = friction angle between sealing strip and soil

a' = adhesion between sealing strip and soil.

Shear tests are normally carried out in shear test apparatus in order to ascertain the shear parameters δ' and a' between plastic sealing strips and other materials. The usual measurements of such apparatus are 10 cm x 10 cm and 30 cm x 30 cm. During the test the sealing strip is laid into the lower shear frame so that its closed surface, even under stress, lies exactly in the shear level created between the upper and the lower shear frames. (Figure 2.)

Each of the shear tests requires three partial tests to be carried out using different surcharges σ_N. The evaluation follows in shear stress -(τ)/normal stress diagrams (σ_N), whereby the shear parameters of the soil angle of internal friction ϕ' and cohesion c'

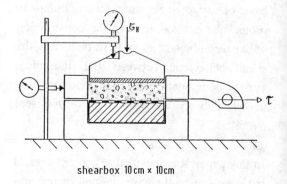

shearbox 10 cm x 10 cm

Figure 2.   Schematic representation of shear box with built-in sealing strip

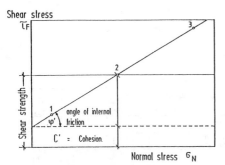

There is

for soil $\quad : \tau_F = \sigma_N \cdot \tan \varphi' + c'$

for geomembrane / soil :

$$\tau_{DB} = \sigma_N \cdot \tan \delta' + a'$$

i.e.

$\tau_F, \tau_{DB} =$	Shear strength
$\sigma_N =$	Normal stress
$\tan \varphi' / \tan \delta' =$	Coefficient of friction
$\varphi', \delta' =$	Angle of internal friction
$c', a' =$	Cohesion / Adhesion

Figure 3.   Shear stress / normal stress
diagram for a shear test

analogous to the shear parameter angle of
internal friction δ' and adhesion a' can be
seen.

In selecting the normal stress for the single
partial test, the later-expected loading should
be borne in mind:

For surface seals (maximum blanket 3
meters) and basic seals in construction e.g;
σ_N = 12.5, 25 and 50 kN/m².

For basic seals in the final state (filled site)
e.g; σ_N = 50, 100 and 200 kN/m².

3.   Tests and Test Results

The frictional behaviour between six HDPE

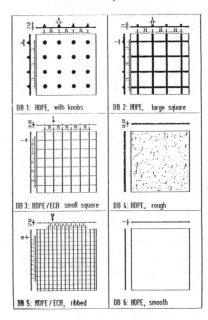

Figure 4.   Schematic   presentation   of
surface   structures   and
dimensions of the sealing strips
DB1 to DB6

(High - Density - Poly - Ethylen) plastic
sealing strips DB1 to DB6 with various
surface structures and different soils was
investigated. The surface structures and
measurements can be taken from Figure 4.

3.1   Tests with non-cohesive soils

For the tests with non-cohesive soils, uniform
medium, fine and coarse sand was used. The
grain size distribution lines for the sands
(labelled B1, B2 and B3) can be seen in
Figure 5.

The surcharge partial tests in the 10 x 10
shear box were carried out under σ_N = 50,

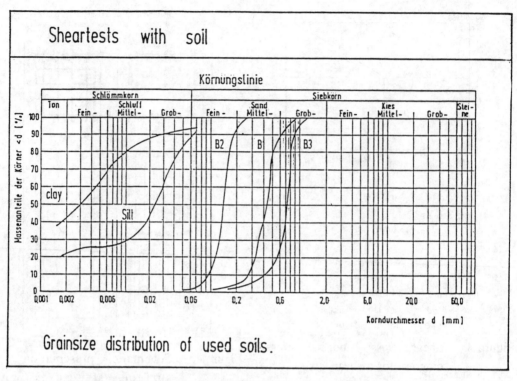

Figure 1.   Grain size distribution lines for the sand used (B1, B2 & B3) and the cohesive soils (clay and silt)

Sheartests with non-cohesiv soil					

Soil	B 1 medium sand		B 2 fine sand		B 3 coarse sand	
	δ', ρ'	$\eta_v \frac{\tan\delta'}{\tan\rho'}$	δ', ρ'	$\eta_v \frac{\tan\delta'}{\tan\rho'}$	δ', ρ'	$\eta_v \frac{\tan\delta'}{\tan\rho'}$
	[°]	[-]	[°]	[-]	[°]	[-]
Soil / Soil	44,65	1,00	37,95	1,00	45,57	1,00
Soil / DB 1	43,47	0,96	35,88	0,93	42,30	0,89
Soil / DB 2	45,85	1,04	33,56	0,85	42,89	0,91
Soil / DB 3	40,96	0,88	36,74	0,95	39,18	0,80
Soil / DB 4	40,96	0,68	34,22	0,87	42,30	0,89
Soil / DB 5	33,82	0,68	32,62	0,82	30,75	0,58
Soil / DB 6	32,13	0,64	29,90	0,74	28,37	0,53

Figure 6.   Results of the shear tests with various uniform sands and sealing strips

100 and 150 kN/m², at the test speed v = 1 mm/min. The angle of internal friction between soil and sealing strip was calculated in each case.

Linked coefficients of friction η_n = tan δ' / tan ρ' were given for better comparability of the test results. The test results are listed in Figure 6.

3.2   Tests with cohesive materials

The grain size distribution lines of the cohesive silt and clay soils are also shown in Figure 5. Firstly, a series of tests was carried out between the soils and the sealing strips

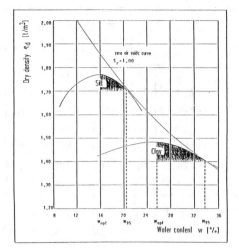

Figure 7. Proctor curves of the tested cohesive soils

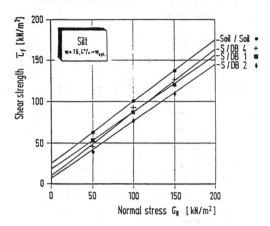

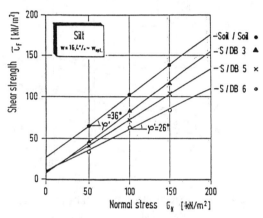

Figure 8. Shear strength and normal stress diagram for various sealing strips and silt

DB1 to DB6. The placement moisture of the soils tested represented almost the optimum water content and the accompanying degree of compaction of D_{pr} = 100 % (Figure 7).

The surcharges for the single part tests were σ_N = 50, 100 and 150 kN/m², the rate of shear v = 0.025 mm/min. All samples were consolidated approximately 24 hours before shearing off. It was proven that the consolidation phase was completed within this time by using time settlement curves. The shear stress / normal stress diagrams for the tests using silt and clay can be seen in Figures 8 and 9.

The test results are put together in Figures 10 and 11. What have been presented are the relations η_n of the transferable shear strength with various surcharges, since with increased surcharge the total share of the cohesion or adhesion in the overall shear strength will be reduced. By way of example the relations η_n

in surcharges of σ_N = 25 and 200 kN/m² has been used.

Thus:

$$\eta_v = \frac{\sigma_N * \tan\delta' + a'}{\sigma_N * \tan\rho' + c'}$$

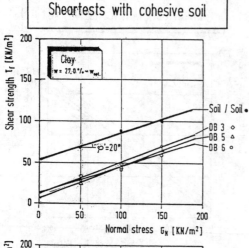

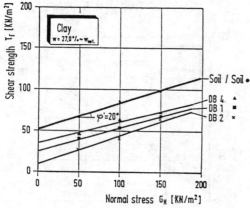

Figure 9.   Shear strength and normal stress diagram for various sealing strips and clay

Results of tests with silt				
	γ', δ' [°]	c', a' [KN/m²]	α_{f25} [-]	α_{f200} [-]
Soil / Soil	37,5	27,0	1,00	1,00
Soil / DB 1	33,1	19,5	0,77	0,83
Soil / DB 2	34,1	8,2	0,54	0,60
Soil / DB 3	35,7	8,9	0,58	0,87
Soil / DB 4	36,7	11,7	0,66	0,89
Soil / DB 5	31,6	9,7	0,55	0,74
Soil / DB 6	26,0	11,1	0,50	0,60

Figure 10.   Test results for silt
(w = 16.4 % = w_{opt})

Results of tests with clay				
	γ', δ' [°]	c', a' [KN/m²]	α_{f25} [-]	α_{f200} [-]
Soil / Soil	17,5	55,0	1,00	1,00
Soil / DB 1	15,5	25,0	0,51	0,68
Soil / DB 2	17,5	12,5	0,32	0,64
Soil / DB 3	18,8	12,5	0,33	0,68
Soil / DB 4	14,5	35,0	0,67	0,74
Soil / DB 5	18,5	12,5	0,33	0,67
Soil / DB 6	17,0	6,0	0,25	0,58

Figure 11.   Test results for clay
(w = 27 % = w_{opt})

3.2.1 Moisture content influence

In a further experiment, the influence of moisture content on shear behaviour between the smooth plastic sealing strip DB6 and silt and clay was investigated. The chosen moisture content for the silt lay at w = 16.1 %, 18.1. % and 20 %.

For the clay it was w = 26.1 %, 29.9 % and 32.9 %. The shear stress and normal stress diagrams for the surcharge range between 12.5 and 200 kN/m² are shown in Figures 12 and 13, respectively.

The changes in the shear parameters with rising moisture content are especially marked with the silt. The angle of internal friction between sealing strip and soil reached only approximately 2/3 of the angle of internal friction of the soil. Whilst the cohesion of the soils reduced markedly as the moisture content increased, the adhesion between soil and sealing strip was lost more or less completely.

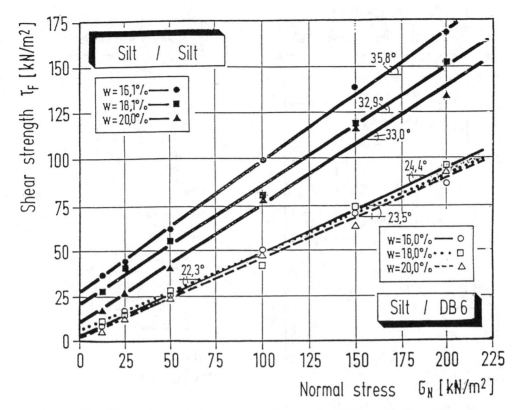

Figure 12. Shear strength between sealing strip DB6 and silt with various moisture contents

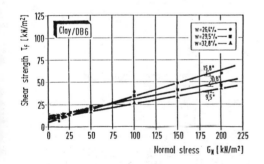

Figure 13. Shear strength between sealing strip DB6 and with clay various moisture contents

3.3 Tests with geotextiles

Two geotextiles were selected for the tests:

- Geotextile G1: strengthened felt material (Polyfelt TS 800)
- Geotextile G2 : woven material (Stabilenka 400)

Coarse and medium sands were taken as soil materials. The following variations were investigated:

Variation 1:   Soil / soil
                 B / B

Sheartest with cohesive soil

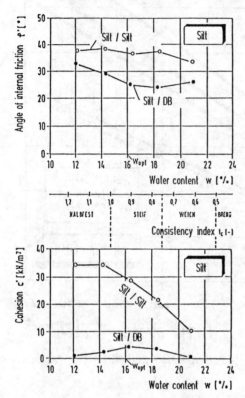

Internal friction angle and cohesion depending on
water content and changing consistency

Figure 14. Changes in the shear parameters
with silt dependent on moisture
content and consistency.

Variation 2:   Soil, geotextile / soil
                        B, G / B
Variation 3:   Soil, geotextile/geotextile,
                  soil
                        B, G /G, B
Variation 4:   Sealing strip / geotextile,
                  soil
                        DB / G, B

Sheartest with Geotextiles

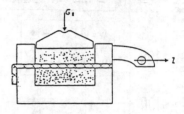

Type 2:

Soil / Geotextil , Soil.

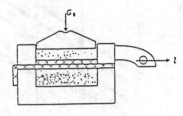

Type 3:

Soil , Geot. / Geot. , Soil

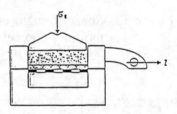

Type 4:

Geomembrane / Geot. / Soil

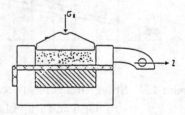

Type 5:

Geotextil / Soil

Figure 15. Schematic presentation of the
test variations

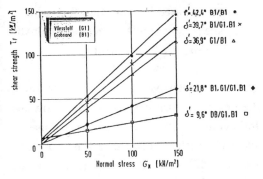

Results of different test variants

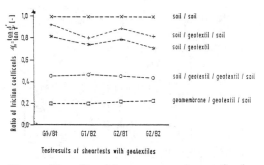

Testresults of shearlests with geotextiles

Figure 17. Graphic presentation of the coefficients of friction η_n

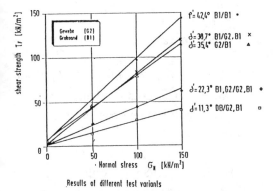

Results of different test variants

Figure 16. Shear strength / normal stress diagrams for the tests with coarse sand B1 and the geotextiles G1 and G2

The test results show that the slightest degree of friction can be transferred between geotextiles and the smooth sealing strip. In test variations 2 and 5, where the shear crack was between geotextile and soil, between 70 and 80 % of the shear strength of soil was reached. For better comparability of the results, the relevant coefficients of friction were calculated:

$$\eta_v = \tan\delta' / \tan\rho' \quad \text{(Figure 17)}.$$

Variation 5: Fixed ground, geotextile / soil

G / B

The position of the shear crack is marked by / in each case. Test variants 2 to 5 are shown in Figure 15. In order to facilitate presentation, the illustrations are not true to scale.

The shear strength / normal stress diagrams for the tests with coarse sand and the geotextiles G1 and G2 are shown in Figure 16.

Recommendations for internal angles of friction between non-cohesiv soil an geomembranes			
Geomembrane	Soil		
	Sand		Gravel
Profile height	Uniformity Coefficient U		
	U < 3	U > 3	
Smooth	$0,50\,\rho'$	$0,60\,\rho'$	$0,50\,\rho'$
0,5 – 1,0mm	$0,60\,\rho'$	$0,75\,\rho'$	$0,70\,\rho'$
1,0 – 2,0mm	$0,85\,\rho'$	$0,95\,\rho'$	$0,90\,\rho'$

Figure 18. Recommendation for angle of internal friction between cohesionless soils and sealing strips

Recommendation for angles of internal friction δ' and adhesion a' between geomembranes and cohesiv soil with different water content				
	Clay		Silt	
Watercontent	δ' [°]	a' [kN/m²]	δ' [°]	a' [kN/m²]
W_{opt}	0,9 f'	0,3 c'	0,65 f'	0,15 c'
W_{97}	0,8 f'	0,2 c'	0,65 f'	0
W_{95}	0,7 f'	0,2 c'	0,65 f'	0
with f' and c' at optimal Water content				

Figure 19. Recommendations for angles of internal friction δ' and adhesion a' between cohesive soils and sealing strips

Angle of internal friction between geotextiles and smooth HDPE - Geomembranes		
used Geotextil	δ' [°]	tan δ' [-]
Fibertex 600	8	0,14
Fibertex 300	10	0,18
Typar 3401	11	0,19
Polyfilter X	6	0,11
NW-Geotextile	9	0,16
Trevira 2125	10	0,18
Trevira 1135	12	0,21
Mirafi 500 S	10	0,18
Miradrain	6	0,11
Enkadrain	9	0,16
Geolon	9	0,16
Tensar DN 3 W	15	0,27
J-DRAIN 100	11	0,19

Figure 20. Angle of internal friction between smooth sealing strips and different geotextiles (Williams 1986 (2))

4. Recommendations for the use of shear parameters between plastic sealing strips and soil

The comprehensive tests, bearing in mind the degree of exactness possible, allow the following recommendations for the use of shear parameters between sealing strips and various different materials to be made (Figure 18 to 20).

5. References

(1) Balthaus / Meseck,(1986), Geomechanisches Verhalten von Kunststoffdichtungsbahnen, Bautechnik 64, S.58 - 63

(2) Williams, (1986), Bestimmung von Reibungswinkeln zwischen Geotextilien, Geomembranen und verwandten Produkten, III Int. Conf. Geotextiles Vienna, Austria, 1986

(3) Kruse, (1989), Standsicherheit von Kombinations- abdichtungen auf Deponieböschungen, Mitteilung des Inst.für Grundbau und Bodenmechanik, TU Braunschweig, Heft 29

Examples of rigid-plastic model application in predicting the mode of failure of reinforced-soil walls

D. Leśniewska
Institute of Hydroengineering IBW PAN, Gdańsk-Oliwa, Poland

B. Krieger & B. R. Thamm
Federal Highway Research Institute BASt, Bergisch-Gladbach, Germany

ABSTRACT: The paper presents comparison between experimental results taken from the tests of four different geotextile-reinforced full-scale walls constructed at the laboratory of the Federal Highway Research Institute in Germany and theoretical predictions of failure loads and slip lines' courses, obtained on the basis of rigid-plastic model of reinforced soil proposed by A. Sawicki.

1 INTRODUCTION

From the viewpoint of continuum mechanics reinforced soil is a two-component, homogeneous composite. The behaviour of this composite depends on the mechanical properties and interactive contribution of each component, Sawicki(1983a,b,c).

In practice the term "reinforced soil" indicates soil with some kind of uniform, regularly distributed inclusions (metal strips or rods, fibres, geogrids, geotextiles, geomembranes, etc).

Geotextiles have become recently the most popular reinforcing material used in reinforced-soil structures. To make the progress in understanding of such structures' performance, four different geotextile-reinforced full-scale walls were constructed and tested at the laboratory for static investigations of the Federal Highway Research Institute in Germany.

Each experimental wall was equipped with a large number of measuring instruments to monitor deformations at the front face of the structure, earth pressures and strains in the reinforcement.

One of the tests' aims was to observe the behaviour of the walls during extensive surface loading.

2 ASSUMPTIONS OF RIGID-PLASTIC MODEL OF REINFORCED SOIL

The basic assumptions of rigid-plastic model of reinforced soil are as follows:

1. Soil is rigid-plastic, obeying Coulomb-Mohr criterion. Associated flow rule is assumed to be valid.

2. Reinfocement is rigid-plastic, having strength only in tension, characterized by it's tensile strength and unit vector parallel to the reinforcement's geometrical orientation.

3. Both components of reinforced soil are perfectly mixed, which means that it is possible to select from the total volume of reinforced soil a representative elementary volume containing soil and reinforcement with volume much smaller than the total.

4. Mechanical behaviour of reinforced soil is described by two kinds of tensors: macrostress and plastic macrostrains tensors σ and ε and microstress and plastic micro-strains tensors σ^s, σ^r and ε^s, ε^r, which are respective averages over the representative elementary volume and its parts, filled by each constituent.

5. There is no sliding between soil and reinforcement:

$$\varepsilon = \varepsilon^r = \varepsilon^s \tag{1}$$

6. Following relation exists between macrostress and microstresses tensors:

$$\sigma = \eta_r \sigma^r + \eta_s \sigma^s \qquad (2)$$

where η_s and η_r are the volumetric ratios of soil and reinforcement.

3 PROBLEM OF BEARING CAPACITY OF REINFORCED SOIL SLOPE

Rigid-plastic model of reinforced soil, together with method of characteristics, gives the possibility to solve differential boundary value problem of bearing capacity of reinforced slope in plane strain state, Leśniewska(1988).

Solution of this problem is important from engineering point of view - it contains, amongst others, a value of limit load leading to the failure of the structure and a geometry of region, which becomes plastic under the limit load (determined by chracteristics net), Sawicki(1983), and characteristics' method, Sawicki, Leśniewska(1990).

Theoretically found limit load and failure mode (the shape of the plastic region limited by active slip line) may be easily verified experimentally in model or full scale tests. To determine both things (limit load and the slip line) one needs:

1. Geometry of a slope (or wall) - its' height h and inclination β.
2. Soil properties - cohesion c, angle of internal friction ϕ, unit weight γ.
3. Reinforcement's parameters - strength in tension R, volumetric ratio η_r, inclination to the horizontal θ.

These data are used by microcomputer program, based on characteristics' method.

4 SOIL PROPERTIES

The soil used for the fill was a gravely sand (SE according to DIN 18196, $C_u < 6$) with a Proctor density of 1.95 g/cm³ and an optimum moisture content of 8%. The angle of internal friction was determined from direct shear tests with normal stresses between 40 and 120 kN/m². These stresses correspond to an overburden soil pressure between 2 and 6m. From these test results an effective angle of internal friction of $\phi = 39°$ and apparent cohesion between 0 and 7.5 kPa were established. The average soil unit weight was 19.8 kN/m³.

5 EXPERIMENTAL STRUCTURES

5.1 *Websol wall*

The structure was 3.2m high and 4 meters wide (Fig.1, Thamm at al(1990)). The facing (Fig.1d) consisted of 0.18m thick prefabricated concrete panels of T-shape. The panels - 2.2m wide and 1.8m high - were placed in layers to form the facing. The facing connections were assured at four locations per panel, where the reinforcing strips were connected in four layers during construction with 8 fill lifts, each 0.40m high.

The wall was reinforced by PARAWEB strips, arranged as in Fig.1c. The strips were 80mm wide and 1.5mm thick and consisted of a polyester yarn core protected against mechanical damage by a matrix of polyethylen. The tensile failure force was measured by free tensile strip tests to give around 20kN at a failure strain of almost 11%. The length of the strips was chosen according to the German specifications for reinforced earth(1985) to be around 2.7m, that is 75% of the total height of the wall.

a)

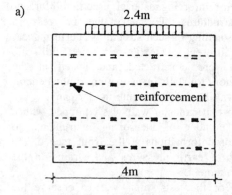

Fig.1 WEBSOL structure: a) cross-section

b)

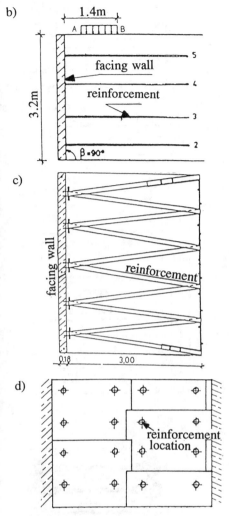

3.2m

1.4m

A [IIIIII] B

facing wall

reinforcement

5

4

3

2

β=90°

c)

facing wall

reinforcement

0.18   3.00

d)

reinforcement
location

Fig.1 WEBSOL structure:
b) profile, c) top view, d) facing wall.

5.2 Remutex wall

The structure was 3.6 m high and 3.1m wide
(Fig.2, Krieger at al(1992)). The facing (Fig.2d)
consisted of composite-shaped, three-
dimensional prefabricated concrete panels,
containing a kind of plant pots, which enable
covering the wall by climbing plants (Fig.2d).
The panels' dimensions are as follows: 1.1 x 1.0
x 0.4m. The weight of panel is about 450 kg.
The panels were placed in layers, forming the
wall sloping   at an average angle of 77.5°. It

was 9 layers, each 0.4m high.

The wall was reinforced by STABILENKA150
strips, laying in each layer as Fig.2c shows.
The strips were 0.3m wide and had tensile
strength, measured by free tensile strip tests,
equal to 148 kN/m. The length of the strips was
2.7m.

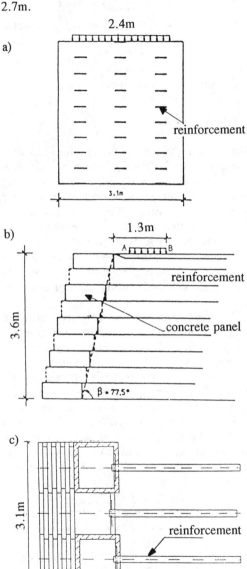

2.4m

a)

reinforcement

3.1m

b)

1.3m

A [IIIIII] B

reinforcement

concrete panel

3.6m

β=77.5°

c)

3.1m

reinforcement

Fig.2 REMUTEX structure: a) cross-section,
b) profile, c) top view

d)

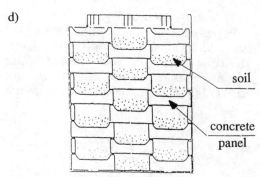

soil

concrete
panel

Fig.2 REMUTEX structure: d) facing wall.

b)

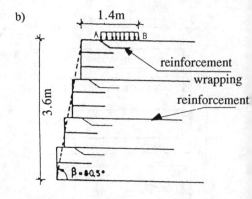

reinforcement

wrapping

reinforcement

5.3 *Tensar wall*

The structure belongs to the group of so called
"wrapped walls", (Fig.3, Thamm at al(1990)),
which means, that wall facing is obtained by
wrapping a layer of reinforcement. A wrapped
wall has been constructed up to a total height of
3.6m, consisting of 5 layers of reinforcement,
each between 0.8 and 1m in height. The wall
was constructed layer by layer by means of a
form work. Average inclination to the
horizontal, obtained during construction was
80.5°.

The reinforcement consisted of two different
geogrids. For the main reinforcement layers a
geogrid TENSAR SR2 (Fig.3d, strength in
tension 67 kN/m, ε_f = 11%) was used. The
additional reinforcement consisted of a geogrid
TENSAR SS2 (strength in tension 31.5 kN/m,
ε_f = 10.5%). These two different materials
were connected by a HDPE-rope. To prevent
the loss of soil from the front part   of the wall
through the grid, a TERRAM nonwoven was
placed along the face of the wall. The length of
reinforcement layer was 2.7m.

c)

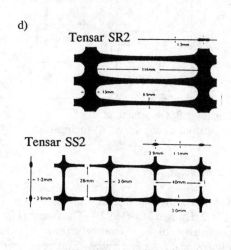

HDPE rope
Tensar SR2

d)

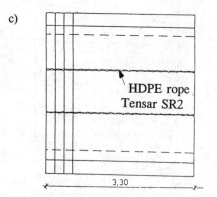

Tensar SR2

Tensar SS2

a)

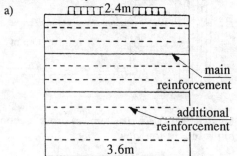

main
reinforcement

additional
reinforcement

3.6m

Fig.3 TENSAR structure: a) cross-section,
b) profile, c) top view, d) reinforcing grids

a)

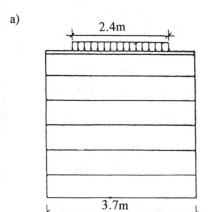

2.4m

3.7m

b)

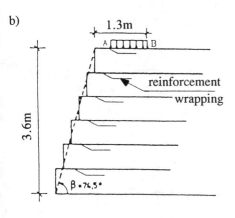

1.3m

A B

reinforcement

wrapping

3.6m

$\beta = 7\llcorner,5°$

c)

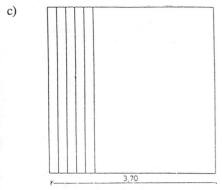

3.70

d)

STABILENKA 150

5.4 Stabilenka wall

The wrapped wall, 3.6m high, consisted of 6 layers, each 0.6m in height (Fig.4). The wall was contructed layer by layer by means of a form work. The average inclination from the horizontal obtained during construction was 77.5°.

The reinforcement consisted of STABILENKA 150 sheets - the tensile strength, measured by free tensile strip tests, was 148 kN/m. The length of the strips was 2.7m.

6 LOADING

After the end of construction the walls were loaded stepwise to the failure, one after another. At first a loading concrete slab was placed on the surface of each wall, at some distance from the wall facing, in the middle of the wall. A loading frame capable for a maximum load of 2000 kN was then placed on the concrete slab, using a steel profile IBP250 of 1.0m length to distribute the load. Value of load steps varied from 20 to 50 kN.

In order to monitor the behaviour of the walls during loading test, measurement equipment was placed within and outside each structure, allowing to measure between the others: horizontal earth pressures, forces within the strips, horizontal deformations of the wall facing, total vertical load placed on the surface of the wall and vertical settlements of loading slab.

Loading to failure took a lot of time: for WEBSOL and REMUTEX walls it was about 17 hours, for TENSAR 21 hours, for STABILENKA 22 hours without success.

It was found out that inside each structure, except STABILENKA, which didn't reach failure state at all, the slip created close to point B starting near the edge of the slab (Fig. 1-4). The shapes of these slip lines were precisely measured after destruction of the walls.

The destruction showed that failure of the three destroyed walls was caused by reinforcement breakage.

Fig.4 STABILENKA structure: a) cross-section, b) profile, c) top view, d) reinforcement

7 COMPARISON BETWEEN EXPERIMENT AND PREDICTIONS OF RIGID PLASTIC MODEL OF REINFORCED SOIL

At this moment, for theoretical analysis of WEBSOL, REMUTEX, TENSAR and STABILENKA walls, rigid-plastic model of reinforced soil, elaborated in IBW PAN, was used together with numerical program which gives solution of slope bearing capacity boundary value problem, Sawicki(1983), Sawicki, Leśniewska(1988,1991).

Solution of bearing capacity problem gives possibility to calculate for all four test walls the failure load and the slip line shape, which may be easily compared to experimental ones.

7.1 *Theoretical and experimental failure loads*

Table 1 shows the comparison between theoretical and actual failure loads for each experimental wall.

Table 1. Comparison between theoretical and actual failure loads

	WEBSOL	REMUTEX	TENSAR	STABILENKA
P(exp) [kN]	500	1500	1100	>>2000 *
P(theo) [kN]	609	1634	1421	4964

* STABILENKA under maximal load 2000kN didn't show any failure

Above comparison shows that the difference between theoretical and experimental failure loads is less then 20% (TENSAR). For REMUTEX it is 10% and for WEBSOL 2%. It is impossible to state the accuracy of predicting failure load for STABILENKA, because it was not destroyed, but one can be sure, looking at strains and stresses measurements, that actual failure load would be much greater than loading frame capacity (2000 kN).

Theoretical failure loads, obtained on the basis of rigid-plastic model of reinforced soil are slightly higher than actual - it may be caused by the assumption, made during calculations, that

a)

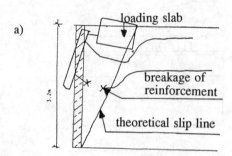

b)

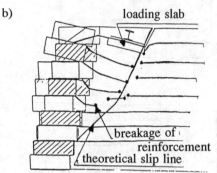

c)

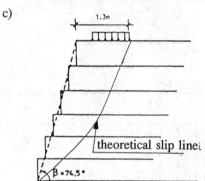

d)

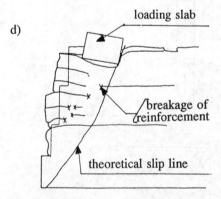

Fig.5 Comparison between theoretical and actual slip lines: a) WEBSOL wall, b) REMUTEX wall, c) STABILENKA wall, d) TENSAR wall.

all reinforcing elements inside experimental walls were broken. In fact, only 80-90% of them were destroyed. Taking this into account would improve the accuracy of predicting failure load, but even without this the accuracy is satisfactory. There may exist another reasons lowering the total accuracy of predicting failure loads, such as accuracy of determining soil and reinforcement parameters or boundary conditions, influencing behaviour of walls (for example rigid walls and rigid bottom of experimental basin).

7.2 Theoretical and actual slip lines

The theoretically obtained slip lines for investigated structures are compared in Fig. 5 with information gathered from measurements and construction. It can be seen that theoretical slip lines are in good agreement with experimental observations, especially in the upper portion. For geometrical conditions of four BASt walls, it seems reasonable to investigate the influence of experimental basin bottom on courses of lower parts of actual slip lines.

8 CONCLUSION

It seems to be proved that, on the basis of materials presented in this paper, rigid plastic model of reinforced soil predicts well the behaviour of reinforced soil structures at failure. It gives hope that, after further theoretical and experimental work, the model may be helpful in reinforced soil structure design.

REFERENCES

Sawicki, A. 1983. Plastic limit behaviour of reinforced earth, *Int.Geot.Engng., ASCE* 109, 7: 1000-1005.

Sawicki, A., Leśniewska, D. 1988. Limit analysis of reinforced slopes, *Geotextiles and Geomembranes* 7: 203-220.

Thamm, B.R., Krieger, B., Leśniewska, D. 1990. Full-scale test of a geotextile-reinforced soil wall, *Int.Reinforced Soil Conference*, Glasgow.

Thamm, B.R., Krieger, B., Krieger, J. 1990. Full-scale test on a geotextile reinforced retaining structure, *Geotextiles, Geomembranes and Related Products*, Den Hoedt(ed.), Balkema, Rotterdam.

Sawicki, A., Leśniewska D. 1991. Stability of fabric reinforced cohesive soil slopes, *Geotextiles and Geomembranes* 10: 125-146,

Krieger, B., Leśniewska, D., Thamm, B.R. 1992. Messungen an einer geotextilbewehrten Stutzwand mit Betonfertigteilelementen als Aussenhaut, *2. Internationaler Kongress Kunststoffe in der Geotechnik K-Geo*.

Performance evaluation of coal-ash treated subgrade in asphalt pavement

M. Nishi
Kobe University, Japan

N. Yoshida
Kyoto University, Japan

S. Kigoshi
Fujita Corporation, Osaka, Japan

H. Ohta
Hyogo Prefecture, Kobe, Japan

A. Maeda
Nippon Steel Corporation, Himeji, Japan

ABSTRACT: Coal-ash treated subgrade has been used in the test pavements constructed in three prefectural roads and its effectiveness is investigated. Performance of the test pavements is evaluated based on Falling Weight Deflectometer (FWD) test data. Effectiveness of coal-ash treated subgrade is discussed in terms of elastic modulus derived from a multi-layered elastic analysis. A non-linear elastic finite element analysis is also carried out and the results are compared with those of multi-layered elastic analysis and the field measurements. It is shown that coal-ash treated subgrade provides better pavement performance.

1 INTRODUCTION

Because of its fuel versatility and high combustion efficiency, a fluidized bed boiler has become popular in energy industry. The fluidized bed boiler employs limestone as a desulfurizing agent and produces waste of ashes which contain calcium sulfate (T.CaO) and calcium oxide (f.CaO) as shown in Table 1 (hereafter referred to as coal-ash). With increasing use of a fluidized bed boiler, more coal-ash waste is expected to be produced.

Recent environmental concern and viewpoints of waste disposal and recycling have dictated effective use of coal-ash. Greater efforts have been made to apply coal-ash to ground improvement, filling, road construction etc.

Applicability of coal-ash as a subgrade stabilizer for asphalt pavement has been investigated in this research. Coal-ash is used tentatively as coal-ash treated subgrade for asphalt pavement. The test pavement is constructed in three prefectural roads of Hyogo Prefecture, Japan. In order to evaluate performance of the asphalt pavement with a coal-ash treated subgrade, a comprehensive field investigation has been carried out together with laboratory tests.

This paper discusses the behavior of asphalt pavement with a coal-ash treated subgrade. Evaluation of the pavement performance is made based on field data obtained from Falling Weight Deflectometer (FWD) tests. Effectiveness of coal-ash treated subgrade is discussed in terms of elastic modulus derived from a multi-layered elastic analysis. Furthermore, a non-linear elastic finite element analysis is carried out and the results are compared with those of multi-layered elastic analysis and the field measurements.

2 COAL-ASH STABILIZER AND TEST PAVEMENTS

2.1 Coal-ash stabilizer

It is known that coal-ash exhibits "hydraulic" property because of its constituent substances and that it is generally angular, irregular shaped unlike fly ash, developing strong interlocking between particles. However, the hydraulic nature is often variable and insufficient for practical use. In order to overcome this shortcoming, a coal-ash stabilizer is developed by mixing calcium sulfate, calcium oxide and coal-ash as shown in Table 1 (often referred to as SASS) and is used in the test pavements described in the next section. Figure 1 shows the amount of SASS required for increasing CBR up to 20% for various subgrades of which water contents and CBR range from 19 to 59% and from 0.1 to 3.2%, respectively. For comparison, the amount of a conventional cement stabilizer required for the same conditions is also shown in the figure. It seems that both materials provide a similar stabilizing effect.

2.2 Test pavements

Test pavements were constructed in three prefectural roads, Wakasanankou Line, Misakayashiro Line and Nishiwakisanda Line, of Hyogo Prefecture for investigating the performance of asphalt pavement with a coal-ash treated subgrade. Figure 2 shows the plan view of the three test locations together with FWD test points. The section view of the test pavements are given in Figure 3. Each pavement section consists of an asphalt concrete surface layer, a crushed stone base-course, a coal-ash treated upper-subgrade and an untreated existing lower-

Table 1. Chemical composition of coal-ash and SASS

	Chemical composition (%)								
	SiO_2	Al_2O_3	T.CaO	MgO	Fe_2O_3	f.CaO	T.S	SO_3	Others
Coal-ash	25.4	17.8	15.5	0.6	3.5	2.9	2.7	6.2	25.4
SASS	15.2	10.7	30.7	0.7	2.1	25.0	3.5	8.8	3.3

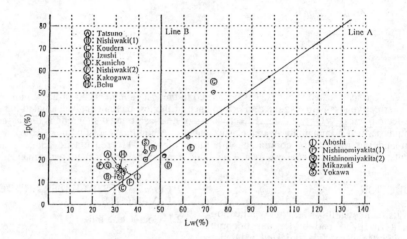

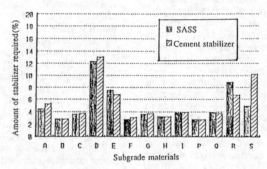

Fig. 1 Comparison between amount of SASS and cement stabilizer

subgrade, overlying hard ground. Conventional cement treated subgrade is also used in these test pavements for comparison.

3 FWD TEST AND PAVEMENT PERFORMANCE

3.1 FWD test

FWD test is a non-destructive one and becoming popular for evaluating the bearing capacity of asphalt pavement because of the simple operation, time-saving, etc. It consists of a circular loading plate, a falling weight and deflection sensors. Surface deflection, which is produced by impulse loading of a falling weight onto the loading plate, is measured at sampling points on the pavement surface using deflection sensors, and the shape of surface deflection is derived. In this study, KUAB-model 100 FWD system (JSCE, 1992) was employed and the falling load was 5 tf. The FWD test points and measuring direction are shown in Figure 2.

3.2 Test results and pavement performance

Figure 4 shows the maximum-minimum ranges of measured surface deflections for the three test pavements. The solid and dotted lines denote the

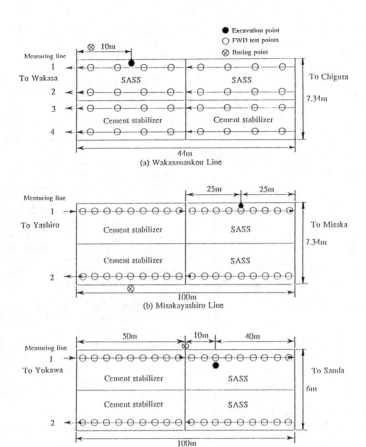

Fig. 2 Plan view of test locations

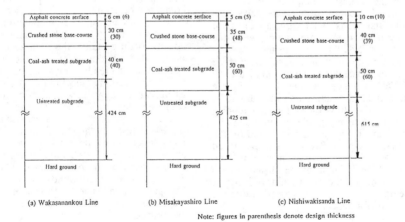

Fig. 3 Section view of test pavements

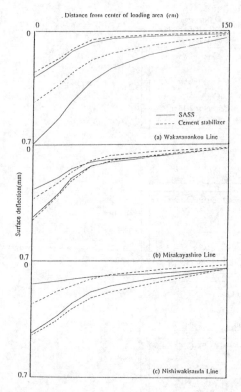

Fig. 4 Measured surface deflections on pavements
with coal-ash and cement treated subgrades

4 MULTI-LAYERED LINEAR ELASTIC ANALYSIS

Pavement performance can be evaluated based on the elastic modulus of each component layer determined from a multi-layered linear elastic analysis on FWD test data.

4.1 Multi-layered linear elastic analysis

A multi-layered linear elastic analysis is often carried out for evaluating behavior of asphalt pavement. Common assumptions made in a multi-layered elastic analysis are; that a pavement structure consists of finite number of horizontally semi-infinite component layers and that neighboring layers are fully contacted and no slip takes place at their boundary and so on. In this study, additional assumptions are made; that a pavement structure is axisymmetric linear elastic, that the number of component layers is smaller than five and that external load is applied vertically onto the pavement surface as uniform circular load.

The computation is carried out using the program code developed by Himeno (1989). The vertical load is 5 tf and the Poisson's ratios are 0.4 for surface layer and 0.35 for base-course and subgrade, referring to the field measurements and previous case studies. The elastic modulus of the hard ground is assumed as 150000 kgf/cm^2.

4.2 Determination of elastic moduli from FWD test data

The elastic moduli of the surface layer, base-course, coal-ash treated subgrade and untreated subgrade of the test pavements are determined through deflection matching. The deflection matching is conducted using a trial and error technique which iteratively compares the computed surface deflections with the field measurements. The computation begins with read-in of input data which contains external load and its loading radius, the number of component layers, their thickness and Poisson's ratios, etc. Then, multi-layered linear elastic analyses of the test pavement sections (see Figure 3) are carried out assuming initial values of elastic moduli for the component layers. The computed surface deflection is then compared with the deflection derived from the FWD test. If the the difference between the computed and measured deflections is greater than an allowable value, 0.01 mm, another computation is carried out with new values of elastic moduli. The computation continues until the difference becomes smaller than the allowable value. In this way, elastic modulus for each component layer is determined.

For coal-ash treated subgrade, Equivalent CBR is also calculated using the following empirical equation and compared with the design CBR:

$$CBR = E / 100 \qquad (1)$$

surface deflection of the test pavements with coal-ash treated subgrade and cement treated subgrade, respectively. It is seen from Figure 4 that the shape of surface deflection is similar between the pavements with coal-ash treated and cement treated subgrades. In the Wakasanankou Line, however, relatively large deflections seem to take place at the SASS pavements. This can be due to the poor condition, resulting from water seepage, of the initial ground in the SASS pavement sites.

The maximum values of measured surface deflections under the center of loading plate are 0.7 mm for the Wakasanankou Line and 0.5 mm for the Misakayashiro and Nishiwakisanda Lines, as can be seen in Figure 4. The allowable surface deflections, which would not cause any cracks or fatigue failure in asphalt pavement, are calculated, based on the Design Manual of Asphalt Pavements (Japan Road Association, 1988), as 1.3 mm for the Wakasanankou Line, 0.9 mm for the Misakayashiro Line and 0.6 mm for the Nishiwakisanda Line. Thus, it seems that coal-ash treated subgrade provides better performance.

where E is elastic modulus in kgf/cm^2.

4.3 Results

The computed elastic moduli for the component layers are summarized in Table 2. Note that the analysis is carried out for three different thicknesses of untreated subgrade as shown in Table 2. It is seen that the elastic moduli computed for coal-ash treated subgrade are larger than those for untreated subgrade at the Misakayashiro and Nishiwakisanda Lines and that very small values are estimated at the Wakasanankou Line. Large values are obtained for the base-course at the Nishiwakisanda and

Misakayashiro Lines. The thickness of untreated subgrade seems not to affect the estimate of elastic moduli of the other layers.

Regarding CBR, the values of Design CBR are computed as 14.6%, 12.9% and 15.7% for the Wakasanankou, Misakayashiro and Nishiwakisanda Lines, respectively, following the Design Manual of Asphalt Pavements. It is seen from Table 2 that the values of Equivalent CBR are larger than the Design values for the Misakayashiro and Nishiwakisanda Lines. In the Wakasanankou Line the Equivalent CBR values do not meet the Design values. This may result from the poor condition of the initial ground, as can be speculated from the deflection measurements given in Figure 4. Smaller values of

Table 2. Back-determined elastic moduli for component layers

(a) Wakasanankou Line

Thickness of untreated subgrade	Surface layer	Base-course	Coal-ash treated subgrade	Untreated subgrade	Hard ground
Infinite	160000~110000	2600~2250	350~325 (3.5~3.3)	2300~2000	
Measured thickness (Figure 3)	200000~112000	2200~2100	650~350 (6.5~3.5)	1350~1100	150000
Half of measured thickness	170000~115000	2350~2200	650~300 (6.5~3.0)	1200~ 800	150000

(b) Misakayashiro Line

Thickness of untreated subgrade	Surface layer	Base-course	Coal-ash treated subgrade	Untreated subgrade	Hard ground
Infinite	160000~155000	7000~2600	2300~1900 (23~19)	2300~2000	
Measured thickness (Figure 3)	160000~155000	8000~2800	2800~1500 (28~15)	1700~1300	150000
Half of measured thickness	160000~153000	7200~2700	3000~1500 (30~15)	1400~ 950	150000

(c) Nishiwakisanda Line

Thickness of untreated subgrade	Surface layer	Base-course	Coal-ash treated subgrade	Untreated subgrade	Hard ground
Infinite	190000~ 40000	19000~2800	4500~1000 (45~10)	1700	
Measured thickness (Figure 3)	220000~ 60000	23000~2750	2500~1150 (25~12)	1250	150000
Half of measured thickness	220000~ 58000	23000~2450	2400~1800 (24~18)	900~ 800	150000

Note: figures in parenthesis denote equivalent CBR

Equivalent CBR are estimated for some locations in the Nishiwakisanda Line, suggesting that a mere application of linear elastic analysis to asphalt pavement with a stiff layer in its lower part may be misleading.

5 NON-LINEAR ELASTIC ANALYSIS USING FINITE ELEMENT METHOD

A non-linear elastic finite element method is employed for a more realistic analysis of asphalt pavement which takes into account stress-dependency of the resilient characteristics of base-course and subgrade materials. Material characterization is made based on laboratory test on samples taken from the test sites.

5.1 Material characterization

Resilient characteristics, deformation modulus and Poisson's ratio, of base-course, treated and untreated subgrade materials were determined from repeated loading triaxial compression tests on samples taken from the sites (Nishi and Tanimoto, 1978). The test specimens were 10 cm diameter and 20 cm high for base-course and treated subgrade materials and 5 cm diameter and 11 cm high for untreated subgrade material. The loading condition used in the test is summarized in Table 3.

Previous experimental and theoretical studies (Zhou et al., 1990; Allaart et al., 1990; Nishi et al., 1992) have indicated that base-course and subgrade materials exhibit non-linear response, and expressions for describing their resilient characteristics have been proposed. In the following, those expressions are used.

5.1.1 Asphalt concrete surface layer

The stiffness modulus of surface layer material is temperature-dependent as shown in Figure 5 which is derived from laboratory tests such as repeated loading flexure test (Nishi, 1982). The stiffness modulus is determined from the temperature measurements during the FWD test and the stiffness modulus-temperature relation given in Figure 5.

The Poisson's ratio is assumed as 0.4 for all analyses.

5.1.2 Crushed stone base-course

The test results are presented in Figure 6 in a form of resilient deformation modulus versus mean principal stress. Based on the previous studies, the following expressions are used for representing the resilient deformation modulus and Poisson's ratio:

$$M_r = K \, P^M \tag{2}$$

$$v_r = A_0 + A_1 \eta + A_2 \eta^2 \tag{3}$$

where p and q are mean principal stress and deviator stress in kgf/cm^2, respectively, and η is stress ratio (q/p). K, A_0, A_1 and A_2 are experimental constants whose values are given in Table 4. Straight lines in the figure are drawn using Equation (2).

The elastic moduli determined from multi-layered elastic analyses are also shown in the figure. It can be seen that the elastic moduli from multi-layered analysis are larger than the experimental values.

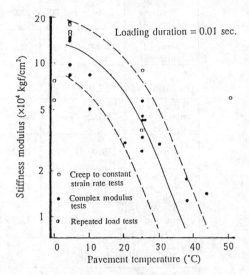

Fig. 5 Stiffness moduli obtained from various tests

Table 3. Loading conditions

	Number of loading	Loading duration (sec.)	Repeated deviator stress (kgf/cm^2)	Mean principal stress (kgf/cm^2)
Base-course material	>3000	0.3	0.3~3.6	0.8, 1.2, 2.1, 3.4
Treated subgrade material	100~300	0.1	0.3~0.9	0.1, 0.2, 0.3, 0.4
Untreated subgrade material	> 100	0.1	0.12~0.36	0.1, 0.2, 0.3, 0.4

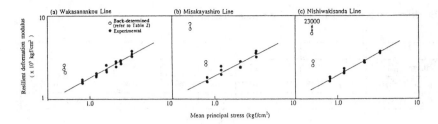

Fig. 6 Resilient deformation moduli of base-course materials

Table 4. Experimental constants for base-course materials

	K	M	A_0	A_1	A_2
Wakasanankou	1810.24	0.529	0.0530	0.3983	-0.1330
Misakayashiro	1925.74	0.482	0.0504	0.4108	-0.1373
Nishiwakisanda	1932.54	0.526	0.0662	0.4328	-0.1808

Reason of such large discrepancy is not clear yet but adoption of a longer duration time of loading in the laboratory test and of constant values of Poisson's ratio in the multi-layered elastic analysis certainly contribute to this discrepancy.

5.1.3 Coal-ash treated subgrade

Figure 7 shows the relation between resilient deformation modulus and mean principal stress. The resilient deformation modulus and Poisson's ratio are expressed as

$$M_r = \frac{1}{A + B q} + C p + D \tag{4}$$

$$v_r = A_0 + A_1 \eta + A_2 \eta^2 \tag{5}$$

where A, B, C, D, A_0, A_1 and A_2 are experimental constants whose values are given in Table 5. Curves in the figure are drawn using Equation (4). Also shown in the figure are the values determined from multi-layered elastic analyses. It is seen that elastic moduli from multi-layered analyses are smaller than the experimental values.

5.1.4 Untreated subgrade

The test results are shown in Figure 8. As in the case of coal-ash treated subgrade, Equations (4) and (5) are used for this layer. The experimental constants are given in Table 6. It is seen from Figure 8 that the elastic moduli from the multi-layered analysis are larger than the experimental values.

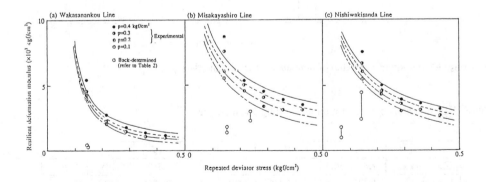

Fig. 7 Resilient deformation moduli of treated subgrade materials

91

Table 5. Experimental constants for treated subgrade materials

	A	B	C	D	A_0	A_1	A_2
Wakasanankou	-1.914×10^{-4}	1.458×10^{-3}	2355	-447	0.0328	0.2811	-0.0604
Misakayashiro	5.928×10^{-5}	3.894×10^{-4}	5798	-869	0.0046	0.2688	-0.0486
Nishiwakisanda	7.055×10^{-5}	3.338×10^{-4}	4003	-973	0.1067	0.2085	-0.0397

5.2 Non-linear elastic finite element analysis

A non-linear elastic finite element analysis is carried out taking into account non-linearity of the resilient characteristics of base-course and subgrade materials described in the previous section.

An axisymmetric condition is assumed; the radius is 140 cm and the depth is variable depending on the thickness of untreated subgrade in the test pavements. Each pavement section is discretized by eight-node rectangular isoparametric elements of seventy to a hundred and ninety depending on the thickness of untreated subgrade. The lateral boundary is fixed in the horizontal direction and the bottom boundary in both the horizontal and vertical directions. The radius of circular loading area and the magnitude of loading pressure are 15 cm and 7.08 kgf/cm², respectively.

5.3 Results

The computed surface deflection of the test pavements for three Lines are summarized in Figure 9 together with those obtained from the FWD test. It can be seen that significant discrepancy exists between the computed and measured deflections at the Misakayashiro and Nishiwakisanda Lines.

In the following, the effects of the thickness and resilient deformation modulus of untreated subgrade and of the stiffness of surface layer and the resilient modulus of base-course are investigated in order to understand the cause of this discrepancy.

5.3.1 Effect of the thickness and resilient deformation modulus of untreated subgrade

The analysis was conducted with three different thicknesses of untreated subgrade and with a

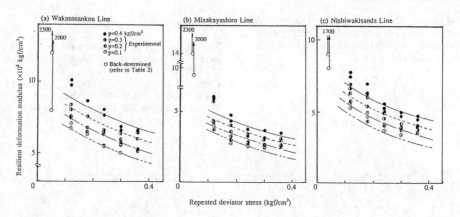

Fig. 8 Resilient deformation moduli of untreated subgrade materials

Table 6. Experimental constants for untreated subgrade materials

	A	B	C	D	A_0	A_1	A_2
Wakasanankou	7.479×10^{-4}	7.313×10^{-4}	724	-600	0.149	0.195	-0.047
Misakayashiro	1.175×10^{-3}	1.557×10^{-3}	481	-436	0.095	0.278	-0.065
Nishiwakisanda	1.177×10^{-3}	2.976×10^{-3}	644	-236	0.099	0.267	-0.064

constant thickness of 100 cm but three different magnitudes of the resilient deformation modulus. The results are given in Figure 10. It is seen that decreasing the thickness of untreated subgrade and increasing the resilient modulus result in a smaller discrepancy between measured and computed deflections and that these changes do not affect the shape of surface deflection. Comparing Figures 9 and 10, however, a significant decrease in the thickness of untreated subgrade increases the curvature of surface deflection near the loading area.

5.3.2 Effect of the stiffness of surface layer and the resilient modulus of base-course

The computation was conducted with the stiffness of surface layer based on the temperature measured 5 cm below the pavement surface and with the resilient modulus of base-course material increased by 50%. Figure 11 shows the results. It is seen that these changes in the temperature of surface layer and the resilient modulus of base-course material affect surface deflection near the loading area and that the curvature of surface deflection decreases with increasing the stiffness of surface layer and the resilient modulus of base-course.

From the above consideration and referring to

Fig. 9 Comparison between computation results and FWD data

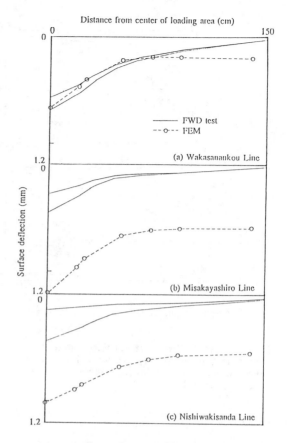

Fig. 10. Effects of thickness and resilient deformation modulus of untreated subgrade

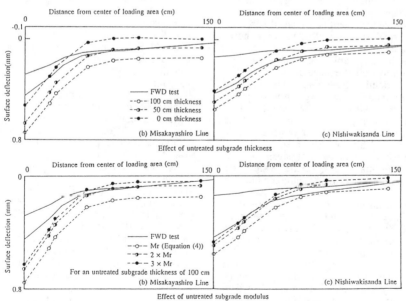

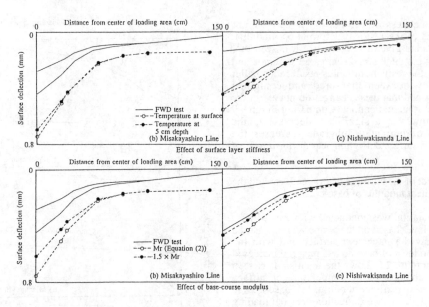

Fig. 11. Effects of surface layer stiffness and base-course modulus

Figures 6 and 8, it seems reasonable to think that the resilient moduli of base-course and untreated subgrade materials derived from laboratory tests underestimate the field moduli.

6 CONCLUSIONS

Coal-ash treated subgrade has been used tentatively in the test pavements of three prefectural roads. Performance of the pavement is evaluated based on the FWD test. Analysis of the FWD test data shows that surface deflections measured in the test are smaller than the allowable values and that the pavement performs well as expected.

The elastic moduli of base-course and subgrade are back-determined by applying a multi-layered linear elastic analysis to the FWD test data. It is shown, from a comparative study of these computed elastic moduli, that larger values of elastic moduli are obtained for coal-ash treated subgrade than for untreated one at the two test pavements. At the remaining test pavement, smaller values are obtained not only for coal-treated subgrade but also for base-course.

Non-linear relations of resilient deformation modulus of base-course and untreated subgrade materials derived from laboratory tests underestimate the elastic moduli determined from a multi-layered linear elastic analysis.

Non-linear elastic analysis tends to underestimate surface deflections observed in the test pavements. This is probably due to underestimated resilient moduli for base-course and untreated subgrade materials mentioned above. Nevertheless, the results suggest that a coal-ash stabilizer serves as a better pavement material.

REFERENCES

Allaart, A., Galjaard, P.J. and Vos, E., 1990. Computation of road structures with granular base courses. Proc., 3rd Int. Conf. on Bearing Capacity of Roads and Airfields, Vol. 2, pp. 791-803.

Himeno, K., 1989. Multi-linear elastic analysis of pavement structure using personal computer. Asphalt, Vol. 32, No. 161, pp. 65-72. (in Japanese).

Japan Road Association, 1988. Design Manual of Asphalt Pavements. pp. 197. (in Japanese).

Japan Society of Civil Engineers, 1992. Evaluation methods of pavement functions. pp. 40. (in Japanese).

Nishi, M. and Tanimoto, K., 1978. A consideration in determining elastic properties of granular materials under laboratory dynamic loading. Proc., 5th Japan Earthquake Engineering Symp., pp. 737-744.

Nishi, M., 1982. Structural Analysis and Design of Flexible Pavement. Dr. Engng. Thesis, Kyoto University.

Nishi, M., Kawabata, K. and Iida, Y., 1991. Behavior of asphalt pavements in circular road tests and its analyses. Proc. of JSCE, No. 426/V-14, pp. 101-110. (in Japanese).

Zhou, H., Hicks, R.G. and Bell, C.A., 1990. Development of a backcalculation program and its verification. Proc., 3rd Int. Conf. on Bearing Capacity of Roads and Airfields, Vol. 1, pp. 391-400.

Prediction versus Performance in Geotechnical Engineering, Balasubramaniam et al. (eds)
© 1994 Balkema, Rotterdam, ISBN 90 5410 355 8

Research and application of the retaining structure of deep cement-mixed piles

Borong Ye, Mingsheng Shi & Meikun Xu
Institute of 3rd Navigation Engineering Bureau, Shanghai, People's Republic of China

ABSTRACT: As compared with the others, the retaining structure of cement-mixed piles has advantage of low cost, high efficiency, non-vibration, noiseless, non-pollution, short work period and simple construction equipment, etc., so it has been successfully used in Shanghai in more than ten projects of foundation pits. This paper, through the practical example in engineering, introduces the design and construction technique of the retaining structure of cement-mixed piles and some technical problems which we must be aware of in the construction, hence advances this method further.

1 INTRODUCTION

The cement-mixed method began in America and was brought into Japan in 1950s. After the cement-mixed cure method(CMC) was successfully exploited and developed by Japan Port and Harbour Research Institute, etc. in 1974, this method has become most commonly used in the stabilization of soft soil foundation. It was formally used in engineering by China in 1980 and the satisfying stabilization effect and good social and economic benefits have been obtained. This method is applicable to the stabilization of mud, mucky soil and some other soils such as clayey soil and silty clay whose water content is high and bearing capacity is not more than 120kPa. In order to develop the technique further and broaden the application, this method was used to construct the retaining structure of foundation pit, and has been successfully used in more than ten projects of foundation pits in Shanghai in the middle of 1980s. The maximum area and the maximum longitudinal span of excavation of the foundation pit of Sun plaza building and the maximum depth of Gaoqiao Water Pump Station have reached 8300m², 130m and 9m, respectively. The maximum length of the cement-mixed piles used in the project reached 14.5m.

The basic principle of cement-soil stabilization is that a series of chemical reactions between cement and soil take place The cement-soil is gradually strengthened, and the process is shown in Fig. 1 --

Chemical Reactions between Cement and Soil.

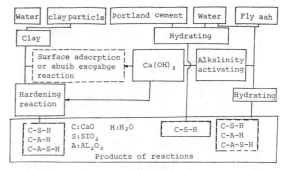

Fig. 1 Chemical Reactions between Cement and Soil

The properties of cement soil are related to the properties of soil, cement-soil ratio and admixture used, etc. The 425# ordinary Portland cement and Cinder cement are usually used in engineering and the cement-soil ratio is 7%-15%. The unconfined compression strength q_u of cement-soil increases with the increase of cement-soil ratio and its age(as shown in Fig. 2). So it is suitable to use the strength of three months age as the standard strength of cement-soil. In order to improve the performance of cement-soil and to raise its strength, some admixtures such as calcium lignosulphonate, gypsum, sodium chloride and sodium sulfate, etc. can be used.

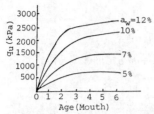

Fig. 2 Relationship Between Strength
Age and Cement-Soil Ratio

The water content of soil influences the unconfined compression strength greatly. The higher the water content, the lower the value of qu. The organic substances in soil keep the soil in relatively high plastisity, expansibility and water-solubility, and relatively low permeability, and make the soil acidic, which hinder the hydrous and hydrolytic reactions of cement and decrease the strength of cement-soil.

In coastal area, the seawater has some influence on the strength and deformation of cement-soil. The results of tests indicate that the higher the concentration of seawater is, the higher the early strength of cement-soil immersed in seawater and the earlier the cement-soil is damaged by its own expansion. The physical and mechanical properties of cement-soil are following: the unit weight of cement is higher than that of ordinary soft clay, the specific gravity of cement-soil is about 4% higher than that of soft soil. The water content of cement-soil is lower than that of soft soil, the unconfined compression strength q_u is 700-2000 KPa, the tensile strength is 0.15 - 0.25 q_u, the cohesion c is 0.2 - 0.3 q_u, the internal friction angle is $20°$ - $30C$? the coefficient of permeability K is 10^{-7} - 10^{-8} cm/s and the modulus of deformation is 120-150 q_u. It can be seen from above that the unit weight of cement-soil is similar to that of soft soil and, however, the cement-soil has higher strength, lower compressibility and excellent water-proof property. Therefore, the cement-soil structure is a good retaining structure for foundation pit.

2 DESIGN OF THE RETAINING STRUCTURE OF CEMENT-MIXED PILES

As described above, the C. M. pile is a brittle material with certain rigidity, its tensile strength is much lower than its compressive strength. Therefore, in engineering, its advantage of relatively

high compressive strength ought to be sufficiently utilized and its disadvantage of tensile strength ought to be avoided. In order to ensure the strength of the retaining structure, the lap width of 20cm is used(as shown in Fig. 3). The good linkage between the costa and the vertical walls must be ensured.

The steps of the design of the retaining structure of cement-mixed piles are as follows: First, the length of piles, the width of retaining structure and the size of latticed store-house are assumed, according to the properties of soil and the depth of excavation of foundation pit, then the assumed values are modified through a series of checking calculations. It is necessary that every factor of safety ought to satisfy the demands of design. Suming up the existing projects, the length of pile can be designed from the following:

$$L = 1.6 - 2.0H$$

in which H=depth of excavation of foundation pit.

The width of retaining structure can be designed from the following formula:

$$B = 0.7 - 0.95H$$

2.1 Calculation of granary pressure

Within the latticed storehouse, the earth pressure acting on side wall can be calculated as the granary pressure. In the place of lap, the maximum tensile stress and the maximum shear stress produced by above earth pressure ought to be lower than the tensile strength and the shear strength, respectively.

$$Px= \frac{\gamma F(1- \frac{C'u}{\gamma F})}{f'u} \left\{ 1-\exp(\frac{-f'\lambda auZ}{F}) \right\} \quad (1)$$

in which Px = horizontal earth pressure acting on side wall
γ = unit weight of woil
F = area of horizontal section of storehouse
u = circumference of inside wall of storehouse
c' = cohesion between soil and wall of storehouse
f' = coefficient of friction between soil and wall of storehouse
λ = friction angle between soil and wall of storehouse

λa = coefficient of active earth pressure

z = depth of calculation, z=b ctg(45 -∅/2)

b = width of storehouse

∅ = equivalent friction angle of soil

in which γ_i = unit weight of each soil in the range of length of pile

h_i = thickness of each soil in the range of length of pile

ϕ_i = internal friction angle of each soil in the range of length of pile

C_i = cohesion of each soil in the range of length of pile

2.2 Calculation of earth pressure acting on retaining structure

1) Active earth pressure --- The active earth pressure above the level of underground water is

$$P_a=(q_0+\gamma h)\,tg^2(45°-\emptyset_q/2) \qquad (2)$$

The active earth pressure from the level of underground water to the excavation surface is

$$P_a=(q_0+h_1+\gamma'h')\,tg^2(45°-\emptyset_q/2)+h'\gamma_w \qquad (3)$$

in which q_0 = external load acting on earth surface

γ = weighted average of unit weight of soils in the range of length of pile

γ' = weighted average of buoyant unit weight of soils in the range of length of pile

h = depth of calculation under earth surface

h_1 = distance between earth surface and level of underground water

h' = depth of calculation under level of underground water

γ_w = unit weight of water

$\emptyset_q$ = weighted equivalent internal friction angle of soils in range of length of pile

$\emptyset_q = \Sigma\phi qih_i/\Sigma h_i$

ϕ_{qi} = equivalent internal friction angle of each soil

h_i = thickness of each soil

Based on many years of experience, the lateral earth pressure may be calculated according to the principle in which the earth pressure of each soil will be equal, the equivalent internal friction angle is used instead of the indexes of shear strength and the cohesion C_i need not be considered once again. The equivalent formula is

$$\Sigma\gamma_ih_itg^2(45°\pm\emptyset_{qi}/2)=\Sigma\gamma_ih_itg^2(45°\pm\phi_i/2)$$
$$\pm2C_itg(45°\pm\phi_i/2) \qquad (4)$$

The distribution of earth pressure above the excavation surface is trapezoidal and is rectangular under the excavation surface, but the earth pressure in the bottom can not be lower than the hydrostatic pressure produced by underground water.

2) Passive earth pressure --- The passive earth pressure can be calculated according to the following formula:

$$P_p=\gamma h_2 tg^2(45°+\emptyset_q/2) \qquad (5)$$

in which h_2 = depth from surface of foundation pit.

2.3 Checking calculation of stability of retaining structure --- In order to ensure the stability of retaining structure, following checking calculations are needed. The safety factors ought to satisfy the demands of design.

1) Checking calculation of sliding stability --- safety factor kg=1.3

2) Checking calculation of overturn stability --- safety factor K_0=1.4

3) Checking calcualtion of overall stability --- safety factor K_s=1.3

4) Checking calculation of anti-seepage stability --- safety factor K_p=1.3

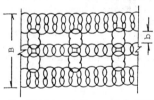

Fig. 3 Latticed Compound Structure

3 CONSTRUCTION TECHNOLOGY

The flow chart of construction technology of cement-mixed pile is shown in Fig. 4. To ensure the quality, the interval of construction of two piles which are mutrally lapped can not exceed one week and the error of location can not exceed 3 cm. The speed of prestirred sinking is controlled by current monitor, the working current not larger than 70A. To the

depth of design, the stirrer begin to blo-
wout cement while swirling, lifting up at
the designing speed. In order to ensure
that the cement paddle in cement-mixed
pile is quantitatively sufficient and

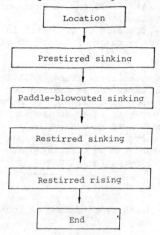

Fig. 4 Flow chaft of tech.

uniform, the error of speed can not be
more than ±10cm/min. The perpendicularity
of cement-mixed piles must be ensured,
the error cannot be more than 5%.   In
the process of construction, the inspection of quality ought to be reinforced and
the original data ought to be carefully
recorded, such as position of pile, working current, voltage, speed of sinking,
time of sinking, speed of rising, time of
rising, time of conveying paddle, quantity of cement used, quantity of admixture
used and the water-cement ratio of cement
paddle. The abnormal phenomenon appearing
in the process of construction ought to
be recorded in detail. The record error
of depth can not be more than 5cm and the
record error of time can not be more than
5s.

4 IN-SITU OBSERVATION AND CENTRIFUGE
MODEL TESTS

Sun Plaza Building is in Shanghai Hong-
qiao Developing Area. The soil data are as
follows: the top soil is miscellaneous
fill of 1.7m, the rest successively are
brown yellow loam of 2.3m, grey plastic
flow mucky loam of 4.1m, grey plastic
flow mucky clay of 9.7m and grey soft plastic loam of 20.9m, etc. The depth of
excavation of foundation pit is 4.6 - 6.7
m; the area of excavation is 8300m²; the
width of the retaining structure of cement-

mixed piles is 3.9 - 6.2m; the length of
piles is 8.3 - 13m and the total axial
length of the retaining structure is
413m. The northern wall was chosen as
the observation section where the depth
of excavation is 6.7m, the width of the
retaining structure is 6.2m and the length of piles is 13m.

The results of observation are as follows:

4.1 Earth pressure

The results measured are shown in Fig. 5.
The earth pressure, before the excavation,
is close to linear distribution, i.e.,
the earth pressure at rest, the coefficient of lateral earth pressure k_o is
0.662. With the excavation of the foundation pit, the retaining structure shifted to the foundation pit, and the pressure acting on active area decreased to
active earth pressure and the passive
earth pressure also appeared in passive
area. By comparing, the earth pressure
measured in active area is closed to the
values calculated and their distribution
patterns are basically identical, which
indicate that the hypothesis and parameters used in the design are reasonable.
In upper part, the passive earth pressure
measured is close to the values calculated, but, in the lower part, is much
lesser than the values calculated. The
reason is that the displacement of the
retaining structure is relatively low.
For practical purposes, the formula is
revised here;

$$\lambda = K[\zeta + (\lambda_p - \zeta)X]$$
$$X = [2(S/S_{pr}) - S/S_{pr})^2]^{1/3}$$

in which λ = coefficient of passive earth pressure corresponding
to S/S_{pr}

ζ = coefficient of lateral earth pressure at rest

λ_p = coefficient of passive earth pressure

S = actual displacement of wall

S_{pr} = displacement of wall corresponding to limit equilib-
rium

X = coefficient of reduction

K = coefficient relating to
properties of soil and rigidity of wall

4.2 Pore water pressure

The pore water pressure meters were installed at the back of the retaining structure, the values measured are close to the
hydrostatic pressure. It reasons that
the excavation of the foundation pit does
not cause the underground water behind
the wall to runoff enormously. The retaining structure of deep cement-mixed
piles has a good performance of seepage-
proof and water partition.

4.3 Horizontal displacement of the top
    of retaining structure

The horizontal displacement increases with
the excavation. During excavation and
shortly right after, the displacement near
the excavation point sharply increased and
tended to be stable for a week. A maximum
displacement of 18.2cm was noticed in the
central part of the side of foundation pit;
the wall shifted   and at the same
time inclined towards the foundation pit at
an inclination rate of about 3%.

4.4 Settlement of the top

The settlement increased slowly with
excavation, the most was at the central
point which was 6.4cm.

4.5 Centrifuge model tests

In order to check for reasonable design,
the centrifuge model tests were carried
out based on the design scheme.
The value of N=100(which expresses that the
model is equivalent in size to the actual
structure simulated when the acceleration
is 100g) was used according to the size
of model box. The foundation soil which
was undistrubed soil was saturated and
precompressed, then, consolidated in the
centrifuge to make its physical indexes
close to that of undisturbed soil. The
results of tests indicate that the
displacement of the wall is slight, from
4cm on top to 15 cm at the bottom. The
height of upheaval of foundation pit is 2
cm and the settlement on the ground of
active area was 7 - 8cm which is in
conformity with the results of finite
element method. The retaining structure was
therefore considered stable.

5 DEVELOPMENT IN FUTURE

5.1 Strengthening and improving
     antibend ability

In Japan,   section steel and   steel
sheet pile are used at present to make a
composite retaining structure of cement-
soil, i.e., the SMW Method, it is mainly
considered as a bent structural member in
design.

In China, bamboo, instead of steel, is
inserted into cement-soil to make the
composite structure, because the area of
contact   surface is increased and, at
the same time, the elastic modulus of
bamboo is much lesser than that of steel,
thus, the bamboo can work more harmoniou-
sly with cement-soil than steel. On the
other hand, bamboo is less expensive, thus
reduces the cost. The results of a comparative test in a project indicated that
without bamboo, the horizontal cracking of
the piles occurs when the horizontal
pushing force of 3t acted on it while with
bamboo, there was no cracking even at 10t
force.

5.2 Adjusting earth pressure acting on
     the C.M. retaining structure

The bored piles can be set up behind the
retaining structure, which penetrate through
active sliding surface into stable layer in
the lower part. With its barrier effect, the
active earth pressure acting on the
retaining structure decreases. The degree of

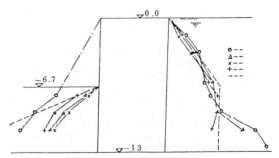

Exca. of 1st block
finished at Jul. 8th
   -- obser. Jun. 14th
   -- obser. Jul. 4th
   -- obser. Jul. 16th
 - -- calculation

Fig. 5 Earth pressure of A-1 block

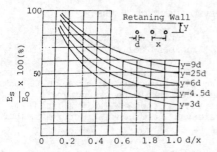

E_S -- Earth Pressure with barries effect

E_O -- Earth Pressure without barrier effect

Fig. 6 Barrier Effect of Bored Piles

barrier effect is inversely proportional to the distance between the retaining structure and bored piles and the distance between bored piles. The results of tests (Wataribe Yasaku) are shown in Fig. 6.
In front of the retaining structure, some anti-slide piles can be inserted or soil can be stabilized to increase the passive earth pressure.

5.3 Changing the shape of the cement-mixed retaining structure

The retaining structure with successive arch shape can be used. In the linking place of arch, the filling piles which can bear the force transferred by the cement-mixed piles may be used. With its effect, the cement-mixed piles will mainly bear pressure.

6 CONCLUSIONS

With the practical application of the retaining structure of cement-mixed piles, the following conclusions are reached:
1. This method has following advantages:
   a. low cost, compared with sheet pile wall, the cost is 30% to 40% lower.
   b. high efficiency. It ensures the safety of underground buildings and lines and creates good construction environment.
   c. during construction, there is no vibration, noise and pollution etc.
   d. simple equipment and short construction period.
2. The results of in situ observation and centrifuge model tests indicate that the design according to gravity retaining structure is in conformity with the pra-

cticalities, the parameters used in the design are reasonable and the construction method of lap width of 20cm is successful.

REFERENCES

Guogun Zhou, 1988. Deep Mixing Method. Foundation Treatment Handbook.
Shiede Zhang, 1991. Stability of the Retaining Structure of Cement-Mixed Piles --- Geotechnical Centrifuge Model Tests. Proceedings of the 2nd National Conference of Geotechnical Centrifuge Simulating Technique, held at Shanghai, June 14-17, 1991.
Weiming Cai, 1990. New Development of Retaining Technique.
Changming Hu, 1991. Experimental Studies and Application of Deep Mixing Retaining Wall.
Tuqiau Zhang, 1991. Experimental Studies of Seawater's Eroding Cement Soil.
Wataribe Yasaku. Proceedings of 17th Conference of Academic Lectures of Japanses Civil Engineering Society
Borong Ye, 1987. Research of Earth Pressure on the Steel Sheet Pile Retaining Wall of a Wet Dock of Xingang. Tianjing Soft Foundation.

Computerized grouting design and control

M. Guillaud & J. P. Hamelin
Solétanche Entreprise Nanterre, France

ABSTRACT : Anyone who has been involved in a grouting project will retain the sad memory of overflowing shelves of files containing countless shift reports, grouting records, etc., and days or weeks on the drawing board, preparing the summary graphics needed to estimate the ground response and workmanship. Without mentioning the problem of preparing the design in the first place, and abstracting grout quantities, or last-minute amendments to deal with surprise discoveries.
This was the challenge which prompted the leading specialist companies to equip themselves with electronic aids which quickly became standard tools in dealing with the requirements of major grouting jobs. They are present at every level of operations, from the initial design (for setting out borehole patterns and calculating grout quantities) to on-site control of the grout pumps and data collection, to verification of the quality of the finished job and work rates.
The paper describes the components of the computer system for the typical grouting project.

1. INTRODUCTION

Every soil improvement project based on grouting requires prior knowledge of the following points (as in figure 1, illustrating a typical job) :

(a) Physical properties to be produced in the treated soil (chiefly strength and imperviousness).

(b) Geometry of treated zone (cross section, volume, boundaries).

(c) Access (working areas at ground level, in adits, shafts, etc.).

(d) Positions and routes of buried obstacles in the working zone.

(e) Original in situ properties of soil.

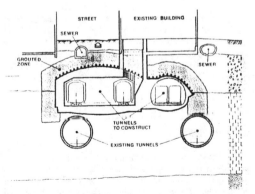

Fig.1. Typical grouting design for a tunnelling project.

Points (a) and (b) are design criteria, (c) and (d) are geometric data relating to the site which are readily measurable but may be difficult to translate into meaningful drawings. Point (e) on the other hand represents a body of geotechnical parameters usually acquired by the site investigations and in situ and laboratory tests.

From this data combined with a good dose of experience acquired on previous similar projects, the engineer :

(f) Specifies the grouts to be used for the job.

(g) Designs appropriate grout hole patterns (in which the grout rheology also has an impact on hole spacing.

(h) Determines the main grouting criteria (grout quantities, grouting pressure and flow rate)

Grouting is not an exact science - far from it, since the host medium is never the same but always totally anisotropic, and grout flow does not obey any simple law. It is an empirical science, and it is only by analysing results as grouting proceeds that the engineer can judge the efficacy of the treatment and correct or amend certain parameters under points (f), (g) or (h) if necessary.

An average-size grouting job involves several hundred boreholes, each grouted in 10 to 20 stages with each stage grouted two or three times. The total amount of data to be analysed with respect to the three main critera - grout volume, flow rate and pressure - may be up to several tens of thousand items.

Only five or six years ago, human processing of such amounts of data occupied two or three engineers and as many draughtsmen every day and the job was still only partially done.

But now, computers are offering a priceless aid to those who were wise enough to develop specific programs tailored to site engineers' needs, and more importantly, they permit real-time analysis of ongoing grouting work, an assurance of reliability and quality in soil grouting treatment.

2 PRIME AREAS FOR COMPUTER AID

It is in respect to the three families of parameters (e) (soil properties), (g) (grout hole pattern) and chiefly (h) (grouting criteria) that computers offer the greatest assistance in the design stage. The way in which they aid real-time works monitoring and quality control will be discussed later.

2.1. Soil Properties

All underground engineering projects are of course based on prior site investigation by means of a few conventional core borings and in situ and laboratory tests. But the soil grouting project must have accurate information on :

a) composition, position and thickness of the different strata, and

b) the permeability (or grain size) of each layer for each grout hole. It was for this reason that continuous drilling parameters recording started to appear in the eighties, the data acquisition apparatus being first analogue type, then digital (as in the ENPASOL system). Today's third generation ENPASOL3 is nothing less than a complete workstation mounted on the drilling rig, capable of displaying eight plain or composite drilling parameters in real time, and yielding a log of the soils strata encountered by the bit, whose accuracy is measured in centimeters (Fig. 2.).

ELISE (Edition de Log Intelligent par Système Expert) is an

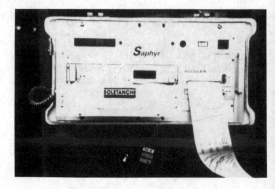

Fig. 4. SAPHYR system

expert system which uses the ENPASOL3 data as input to further refine it and present it in even more easily assimilable form. For example, a rule base can be set up for a given site enabling it to recognise the strata and log the hole concurrently with the drilling (Fig. 3.).

It has also become possible to measure soil permeability by means of a real-time hydraulic parameter acquisition system (SAPHYR) comprising pressure and flow sensors, a control and data acquisition unit and an interpretation program (Fig. 4.). A conventional permeability test (say, Lefranc test in alluvium) will thus immediately yield a curve interpreted in terms of permeabitily coefficients.

2.2. Grout Hole Pattern

As already mentioned, designing grout hole patterns is a time-consuming, meticulous task involving a host of geometric constraints (zone to be treated, working areas, location of buried obstacles, machine type), geotechnical parameters (soil types, radius of grout penetration) and physical factors (grout rheology). In addition, design changes frequently become necessary in the course of the work through, for example, changes in working areas or project route, different soil layer geometry, etc.).

The CASTAUR program (Conception Assistée des Auréoles d'Injection) has been developed for the most difficult situations in built-up areas. It enables the designer to work on a complex model describing all the components of the job in three-dimensions. It has recently been re-written in object-oriented C language and comprises three functional units :
1. 3DME modeller
2. Automatic borehole fan generator
3. Borehole CAD and graphic grouting results monitor.

2.2.1. 3DME Modeller

The first step is for the geometry of the complete project site to be described by a series of elementary volumes generated by projecting a profile which may be variable - along a director line (fig. 5.). The volumes thus generated are then assembled according to type to represent the project environment. The resulting model may be complex since the various obstacles and buildings present may be intimately intermingled. Inputting the data may take time but it is amply justified by the subsequent results.

Fig. 2. ENPASOL system attached on a drilling rig.

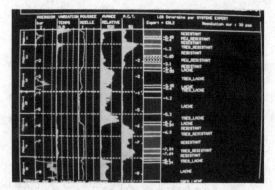

Fig. 3. Print-out ELISE expert-system

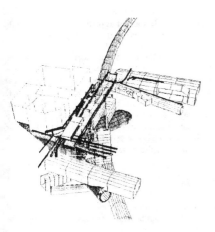

Fig. 5. 3-D representation with CASTAUR modeller

2.2.2. Automatic Borehole Fan Positioner

This unit automatically draws cross sections through the basic volumes. After the first section, which is usually horizontal or vertical, the subsequent ones are positioned on the basis of spacing criteria referring to the volumes and soil properties (fig. 6). The scene specific to each fan is then calculated.

2.2.3. Automatic Borehole Fan Design

This unit is a tool for quickly producing the borehole patterns on the fans previously specified. The first hole is positioned by the designer, and then the program assists him in placing the remainder. Various strategies are possible : the holes may radiate about a common axis, or be aimed to hit specified points on a cut-off, etc . Then all the holes are assigned grouting sleeve valves to suit the surrounding soil, the arrangement being optimised with reference to neighbouring holes and fans (fig. 6).
Summaries can be printed out for quick examination of multiple scenarios before deciding on the technical and economic feasibility of any particular variant.

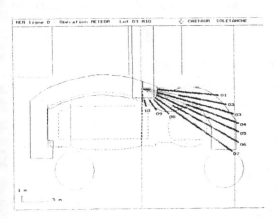

Fig. 6. CASTAUR, Borehole fan positioning.

2.3. GROUTING CRITERIA

All parameters relevant to grouting operations are fed into a program named CHAIRLOC (Computerised Help for Analysing the Injection Results and Localizing Optimal Corrections), which is a relational data base. CASTAUR feeds CHAIRLOC with the description of each grouting (soil type and volume, percentage of voids to be treated) while the project engineer specifies the design grouting criteria, grout type, flow pressure and volume, bursting pressure of cement sheath surrounding a grout valve, maximum amount of grout to cause bursting, etc. CHAIRLOC then prepares the programs which will control the grout pumps. The grouting instructions generated by these two programs are loaded at the start of each grouting stage, and the grout pump operator can interactively amend these control variables.

2.4. GROUT PUMP CONTROL (fig. 7, 8, 9)

The SINNUS 3 system (Système d'Injection Numerique Solétanche, 3rd Generation) was developed for automated grout pump control. The components are
- pressure and flow sensors on each grout pump.
- ARCSINNUS (Acquisition, Regulation, Control) units for rapid data acquisition and electronic delivery and pressure control of two pumps. Each unit has an Assembler-programmed controller. The various data acquisition and control tasks and communications with the central unit are organised around a real -time core.

Fig. 7. Control Desk

Fig. 8. Grout pumps and ARCSINNUS system

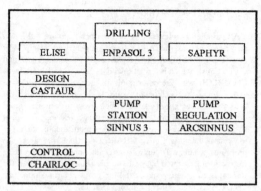

	DRILLING	
ELISE	ENPASOL 3	SAPHYR
DESIGN		
CASTAUR		
	PUMP STATION SINNUS 3	PUMP REGULATION ARCSINNUS
CONTROL		
CHAIRLOC		

Fig. 9. Computerised control and logging system.

- SINNUS 3 central unit comprising an industrial PC computer, graphic printer and stabilised power supply. The supervision program is written in Turbo Pascal plus a multi-task utility.

SINNUS 3 takes charge of data acquisition, corrections for head loss, real-time display on large graphics screen, pump grout delivery and pressure control ; it is programmed to stop the pumps according to a variety of criteria such as grout quantity, refusal pressure, excess flow, etc. A detailed report is prepared for each grout pass and stored on the hard disk for subsequent transfer to the CHAIRLOC data base.

The SINNUS 3 sytem represents very serious advantages in jobsite operations. The pumps are continually monitored and the data in memory is reliable and standardised. Electronic control produces very smooth operation of the reciprocating pumps conventionally used, even at very small deliveries. Further advantages are that a single man can attend up to 12 grout lines, and time lost in proceeding from one grout stage to the next is reduced, with the result that overall jobsite productivitiy is distinctly improved.

2.5. QUALITY CONTROL

The data recorded by SINNUS 3 is checked and fed into the CHAIRLOC data base at each working area. This makes it possible to print out a full set of reports necessary for daily monitoring of progress in terms of both quality and cost.

Quality control is enhanced by the many possibilities of sorting the data according to a variety of criteria such as duration of grouting of each pass, mean grout flow, final pressure, maximum pressure, total grout take, etc. One important criterion which can be estimated is the Lugeon Equivalent evaluated from the flow/pressure ratio and used to compare grouting passes injected at different flows or pressures. This data can also be transferred to CASTAUR and plotted on the grout hole fan sections (fig 10.).

3. CONCLUSION

Microelectronics have now become a vital part of all grouting work. The resulting improvements chiefly cover the following areas :

- improved cost, through increased grout pump efficiency and site staff productivity.
- improved grouting quality in that quality control is exercised at all levels - at the pump and in the office. Quality monitoring can be done frequently and deals in calibrated, objective data. And it can be done quickly, so that the grouting reflects actual soil response.
- improved liaison, in that reports are clear and coherent, and issued regularly.

These techniques are rejuvenating the image of grouting jobs, which are now entering the "industrial" era. Grouting is still a difficult art demanding much insight and experience, but by relieving the site staff of burdensome paperwork, computers enable them to give the whole of their attention to achieving the prime goal of delivering a good job.

REFERENCES

- Hass G., Hetuin E., La Fonta J-G., Hamelin J-P., Drilling parameters recording : developing new trends, International Conference Geotechnics and Computers Paris, Sept. 29, 1992.

- Buton Ph., Perru J-L., La Fonta J-G., Hamelin J-P., Computerised monitoring on grouting sites. International Conference Geotechnics and Computers Paris, Sept. 29, 1992.

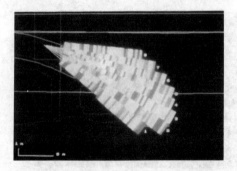

Fig. 10. Visualisation of grout flow, pressure and volume for checking a treated zone.

Prediction versus Performance in Geotechnical Engineering, Balasubramaniam et al. (eds)
© *1994 Balkema, Rotterdam, ISBN 90 5410 355 8*

Prediction versus performance on the behavior of soft clay ground improved by vertical drains

K.H.Xie, Q.Y.Pan, G.X.Zeng & X.R.Zhu
Geotechnical Engineering Institute, Zhejiang University, Hangzhou, People's Republic of China

ABSTRACT: To assess the adequacy of the technique of vertical drains combined with surcharge precompression for improving the soft clay stratum of more than 30m thick beneath airport runway, a field test was conducted at the Ningfeng plain, southwestern Ningbo city, where Ningbo airport, the first airport on such weak ground in China was to be constructed. In this paper, the performance of the test and the prediction of the consolidation behaviour of the ground by finite element method are described. A nonlinear stress - strain relationship taking account of dilatancy of clay is proposed based on laboratory tests and used in the prediction. Back analyses are also performed for determining some relevant soil parameters. The efficiencies of the ground improvement technique and the prediction by FEM are illustrated and emphasized.

1 INTRODUCTION

In order to meet the need of economic development of China, many airports have to be constructed on the ground of very weak and highly compressible soil near coastal cities. One of the main difficulties of building airport on such soft clay ground is to satisfy the strict demand on deformation of an airport runway. The technique of vertical drains combined with surcharge precompression may be used to improve such weak ground since it can eliminate the settlement of subsoil to a desired degree before the construction. The successful engineering practices of the constructions of more than two airports in China have shown that it is a very economical and effective technique (Pan et al.(1991)).

However, when Ningbo airport was to be constructed, no such engineering experience was available because it was the first airport in China which would be built on a soft clay stratum of more than 30m thick. A field test was therefore carried out to assess the adequacy of the technique of vertical drains combined with surcharge precompression for improving the soft clay ground of Ningbo airport runway which was 45m wide and 2500m long. The performance of the test and the prediction of the behaviour of the ground are described in the paper.

2 SITE AND SOIL CONDITIONS

The site of the field surcharge precompression test was located at Ningfeng plain, the southwest of Ningbo city, where the Ningbo airport was to be constructed. In this area, the geological conditions are very complicated and marine deposit extend to great depth including soft clay layers of more than 30m thick.

Conventional laboratory tests were carried out to investigate the engineering properties of the soils. It was shown that the top soil layer of the site was weathered while all other layers were normally consolidated. The distributions of soil layers in the test site and their main properties are shown in Table 1.

3 FIELD SURCHARGE PRECOMPRESSION TEST

Fig.1 shows the cross-section of the loading fill and the arrangement of the vertical drains in the field surcharge precompression test.

The vertical drains are sand wicks of 7 cm diameter and installed by the displacement method in a triangular pattern at 1.4m spacing. The drains are 20m in length, without penetrating the main soft clay stratum (layer 4, see Table 1).

On the surface of the ground, a sand blanket of 0.35m thick and then two layers

Table 1. The main engineering properties of soils in the test site.

Layer No.	Depth (m)	w (%)	e_o	γ (kN/m^3)	w_l (%)	I_p (%)	k_v	k_h
							($\times 10^{-8}$cm/s)	
1	0.0	33.3	0.92	19.0	43.2	17.4	2.08	3.03
2	1.5	52.8	1.45	17.2	45.0	17.5	7.52	10.30
3	4.0	49.0	1.34	17.5	39.5	15.5	8.61	5.09
4	7.5	48.9	1.35	17.4	40.9	16.9	9.88	7.18
5	30.0	35.8	1.02	18.9	33.6	11.5	-	-
	32.5							

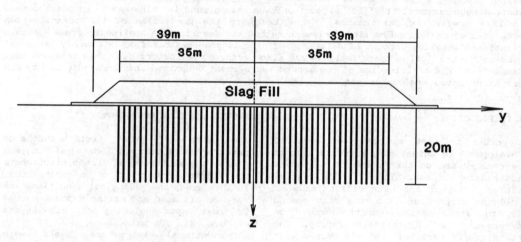

Fig.1   The cross section of the fill and the arrangement of the vertical drains in the surcharge precompression test

of slag fill of a total height of 3.65m are placed as loading.   The first layer of slag fill is compacted with an average unit weight of 17.7 kN/m^3 and a thickness of 1.3m.   The average unit weight of the second layer of the fill is 15.7 kN/m^3. The loading schedule is shown in Fig.2.

The widths of the top and the bottom of the fill are, respectively, 70m and 78m which were assumed to be wide enough to ensure that there were no undesirable differential settlement along y-direction of the runway of 45m wide.

The preloading test was well instrumented with settlement gauges and piezometers. The pore water pressures as well as settlements were observed and recorded at different points during entire period of

testing.   Some of the observed results are presented together with the ones of finite element analysis in section 6.

4 RELATIONSHIP OF STRESS AND STRAIN

To achieve a realistic prediction, the main deformation properties of soft clay such as dilatancy should be taken into consideration. Relevant laboratory tests were therefore conducted on the undisturbed soils from the test site to study the relationships of stress and strain.

The dilatancy of soft clay is usually distinguished by the fact that its volume will decrease with the increase of shear stress if it is normally consolidated and may be studied by the CIP test carried out

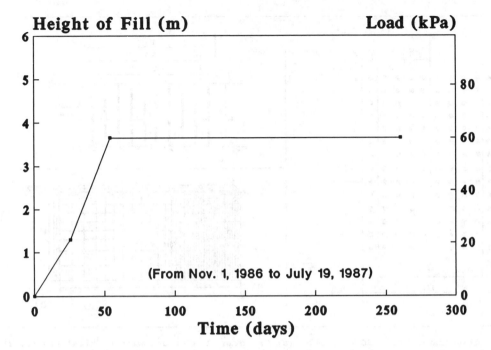

Fig.2 Loading schedule in the filed test

on a triaxial apparatus under drainage
condition. During the compression process
of CIP test, the mean effective stress P is
kept constant by means of decreasing the
cell pressure, and the volume change of
sample thus measured is therefore entirely
caused by shear stress.

Following stress and strain relationship
has been found based on the results of the
laboratory tests:

$$\varepsilon_v = C_a \ln \frac{P}{P_o} + b(M-M_o) \qquad (1a)$$

$$\varepsilon_s = C_b M \ln \frac{P}{P_o} + \frac{c(M-M_o)}{1-d(M-M_o)} \qquad (1b)$$

in which, ε_v and ε_s are volume strain and
shear strain, respectively; C_a =
$0.434C_c/(1+e_o)$; C_c, the compressibility
index; e_o, the initial void ratio; b, the
dilatancy index; c, the coefficient of
shear modulus; d, the coefficient of the
ratio of ultimate stress; $C_b = 2C_a/M_o/3$;

$M = q/P$; q and p, the shear stress and mean
effective stress, respectively; $M_o = 3(1-k_o)/(1+2k_o)$, the initial value of M; $k_o = 1-$

$\sin\phi$, the coefficient of earth pressure at
rest; ϕ, the effective angle of internal
friction; P_o, the initial value of P.

It can be seen from Eq.(1) that the
volume strain is not only   a function of
mean effective stress but also   a function
of shear stress, and the shear strain is
not only related to shear stress but also
affected by the mean effective stress. The
nonlinear and the dilatancy properties of
soil are thus   included in the stress
strain relationship.

5 PROCEDURE OF PREDICTION

5.1 Consolidation analysis by FEM

The consolidation behaviour of a soft clay
ground improved by vertical drains is
usually predicted by single well
consolidation theories (Barron (1948),
Hansbo (1981), Xie (1987), Zeng et
al.(1989)). However, the prediction cannot
take into consideration the true
distributions of load and vertical drains
as well as the complicated soil properties.
Finite element method has therefore been
employed to predict the consolidation
behaviour of the tested ground.

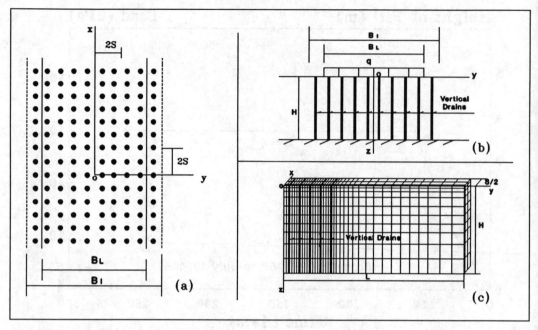

Fig.3 Illustration of PDSS approach. (a) The plan of a ground with vertical drains. (b) The cross section. (c) The F. E. idealization

Fig.3 illustrates the schedule of analyzing a ground with vertical drains by PDSS approach (Xie (1987), Zeng et al. (1987)) in which the circular section of a drain is replaced by a square one of the same area and the deformation of the ground along the direction of x-axis is assumed to be zero while the seepage remains in three dimensional (i.e., Plane Deformation, Spatial Seepage). The advantage of such approach has been discussed recently in detail by Cheung et al. (1991).

For the problem of vertical drains concerned in this paper, the PDSS approach was applied. The triangular arrangement of vertical drains was converted into the equivalent square pattern by changing the drain spacing from $S\Delta$ (i.e. 1.4m) into $S\Delta/1.074$ (i.e. 1.3m) and increasing the number of vertical drains so that the area improved by vertical drains was almost the same as the original. Both the rationality and the advantage of reducing computation work of such simplification have been shown by Xie (1987). The total thickness of soil layer, H (see Fig.3), was taken as 35m and the base assumed to be impervious. A total of 1440 elements and 2457 nodes were then used in finite element idealization similar to Fig.3(c).

The stress-strain relationship introduced above was adopted in the analysis for each soil layer except for the top one which was assumed to be elastic material with an elastic modulus of 4000kPa and a Poison's ratio of 0.3 since it was weathered. All relevant parameters used in the analysis are listed in Table 2.

5.2 Back analysis

It is well known that the coefficient of permeability of soil measured from laboratory test is smaller than the one obtained from in situ tests. This is also confirmed by the finite element analysis performed by the authors since the pore pressure calculated is greater than the ones observed if the values of coefficients of permeability of soils listed in Table 1 are used.

Back analysis is therefore carried out based on finite element analysis to determine the coefficients of permeability of soils. The results are shown in Table 2. It can be seen from Table 1 and 2 that the back calculated values are about 2-10 times of the previous ones.

As one of the advantages of the PDSS approach, the non ideal behaviour such as well resistance of vertical drains can be simply taken into consideration, even as the drains are partially penetrated. Because the coefficient of permeability of the sand used for the wicks was in the order of $10^{-2} - 10^{-3}$ cm/s and the drains

Table 2. The parameters of soils used in the prediction.

Layer No.	C_a	b	c	d	ϕ (degree)	k_v ($\times 10^{-7}$cm/s)	k_h
1	–	–	–	–	30.0	3.04	2.08
2	0.068	0.095	0.075	1.53	28.5	8.24	6.12
3	0.070	0.098	0.080	1.45	29.5	1.72	1.02
4	0.065	0.091	0.041	1.48	29.0	9.88	7.18
5	0.055	0.085	0.036	1.40	31.0	1.00	1.00

were so slender, the great influence of well resistance on the consolidation of the ground may be expected, and therefore, well resistance is taken into consideration in the analysis.

The coefficient of permeability of sand which is required for such analysis is also determined by back analysis and the value thus obtained is 0.005 cm/s.

6 COMPARISONS AND DISCUSSIONS OF THE OBSERVED AND PREDICTED RESULTS

6.1 Settlements

Figs.4 and 5 show the development of the settlement with time at the centre of

loading fill (y=0) and at the edge of runway (y=22.5m) on the ground surface (z=0), respectively. It can be seen that the values observed in the field test and predicted by FEM are in good agreements. The fact that the settlement of more than 80cm can be eliminated within about 8 months demonstrates that the technique of vertical drains combined with surcharge precompression is economical and effective in improving the soft clay ground of Ningbo Airport runway.

Figures 6 and 7 are the calculated distributions of settlements at different depths when t = 54 days, i.e. the loading fill reaches its final height (3.65m), and t = 180 days, respectively. It can be seen that the settlements at same depth are almost same as y ≤ 25m, indicating that

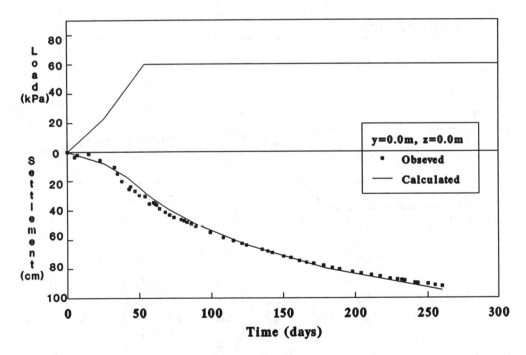

Fig.4   Development of settlement with time at the centre of fill

109

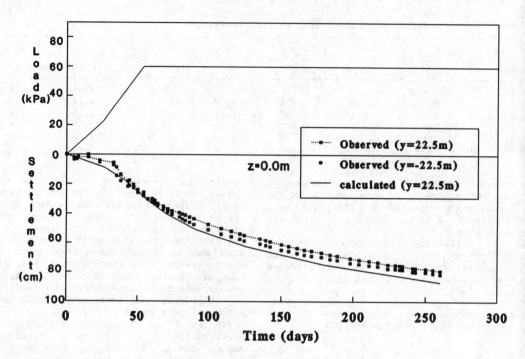

Fig.5 Development of settlement with time at the edge of runway

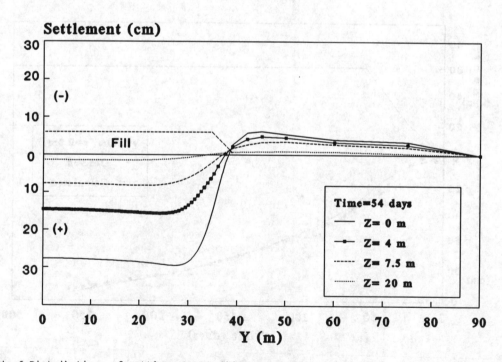

Fig.6 Distributions of settlements at different depths (t=54days)

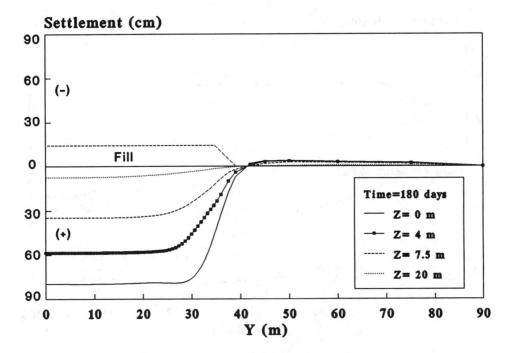

Fig.7 Distributions of settlements at different depths (t=180days)

there seems to be no differential settlement within the range of the width of runway and that the width of loading fill is properly designed to satisfy the demand on deformations of the runway. Besides, it is interesting to see that the settlement of the soft clay stratum under vertical drains (i.e. at z=20m) is very small although the underlying stratum is 15m thick.

6.2 Pore Water Pressure

The observed and calculated pore water pressures at z = 2.1m and z = 16.7m on the centre line of loading fill (y=0) are shown in Figure 8 from which good agreements of the two results and the efficiency of vertical drains in accelerating the rate of consolidation of soft clay ground, can be observed.

Figs.9 and 10 show the calculated distributions of pore water pressure in the ground. It can be seen that the pore pressures within the zone of vertical drains are distributed in a wave shape because of the presence of vertical drains. If vertical drains are of infinite permeability, i.e. drains are ideal without well resistance, the pore pressures in

drains should be zero. However, the two figures show that the difference between the pore pressure in vertical drain and the one at the centre of spacing is small, within the range of 0 (as z=20m) to 30% (as 2.1m ≤ z ≤ 7.5m). This indicates that the effect of well resistance in the case is very significant.

7 CONCLUSIONS

The following conclusions may be summarized from the paper:
1. Vertical drains combined with surcharge precompression is an economical and effective technique for improving soft clay ground.
2. The PDSS approach can be applied to analyze the consolidation behaviour of a soft clay ground improved by vertical drains to achieve more realistic prediction.
3. The non-ideal behaviour of vertical drain, such as well resistance, and the main deformation properties of soft clay such as dilatancy, etc, should be taken into consideration in prediction. The relevant parameters which are difficult to obtain from available tests may be determined by back analysis.

111

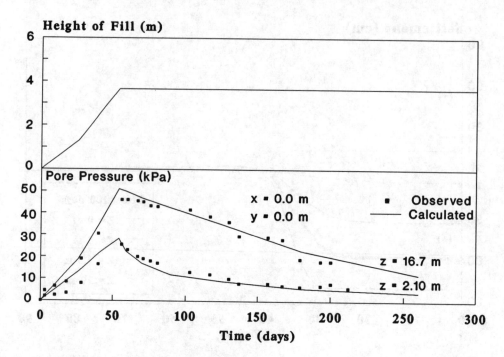

Fig.8 Development of pore pressures with time

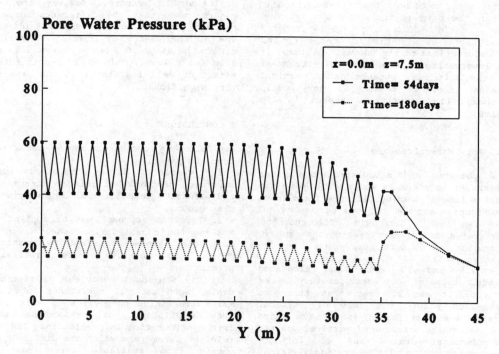

Fig.9 Distributions of pore pressures at different times (z=7.5m)

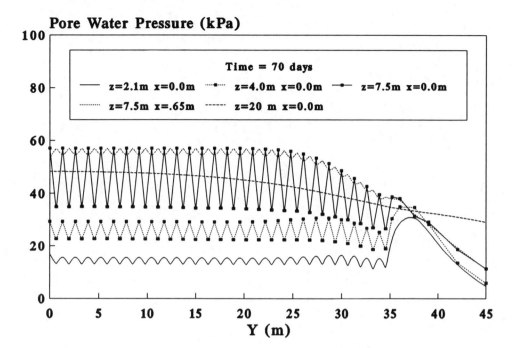

Fig.10 Distributions of pore pressure in the zone of vertical drains (t=70days)

ACKNOWLEDGEMENT

The part of the subject is financially supported by the China National Natural Science Foundation No.59009506.   This is gratefully acknowledged.

REFERENCES

Barron, R.A. 1948. Consolidation of fine grained soils by drain wells. Trans. ASCE, Vol.113: 718-742.
Cheung, Y.K., Lee, P.K.K. and Xie, K.H. 1991.   Some remarks on two and three dimensional consolidation analysis of sand-drained ground. Computers and Geotechnics 12: 73-87.
Hansbo, S. 1981. Consolidation of fine-grained soils by prefabricated drains. Proc. 10th ICSMFE, Vol.3: 677-682.
Pan, Q.Y., Zhu, X.R. and Xie, K.H. 1991. Some aspects of surcharge precompression on ground with sand drains. Chinese Journal of Geotechnical Engineering, Vol.13, No.2: 1-12.
Xie, K.H. 1987. Sand drained ground: analytical and numerical solutions of consolidation and optimal design. Ph.D. dissertation, Zhejiang University.
Zeng, G.X., Xie, K.H. 1989.   New Development of the vertical drain theories.
Proc. 12th ICSMFE, Vol.2: 1435-1438.
Zeng, G.X., Xie, K.H. and Shi, Z.Y. 1987. Consolidation analysis of sand-drained ground by FEM. Proc. 8th ARCSMFE, Vol.1: 139-142.

Prediction versus Performance in Geotechnical Engineering, Balasubramaniam et al. (eds)
© *1994 Balkema, Rotterdam, ISBN 90 5410 355 8*

Case studies of reinforced ground with micropiling and other improvement technique

Masashi Kamon
Disaster Prevention Research Institute, Kyoto University, Uji, Japan

ABSTRACT: Soil reinforcement method is a relatively new ground improvement technique which strengthens weak area of ground using inclusions whose material properties are known. In this paper two different types of soil reinforcement techniques are introduced. They are micropiling reinforcement and deep mixing method or DMM. In the first type, a case study involving micropile reinforcement, which is one of the representative methods of the natural slope reinforcement, is examined. In the case study described here, the performance during construction is evaluated numerically. A case study which utilizes the DMM was also discussed. According to the large increase in the level of the strength improved by DMM, the soil properties improved should be considered as similar with soil reinforcement. Also numerically discussed are the performance of the composite ground involving DMM.

1 INTRODUCTION

The need to improve unfavorable ground conditions to make it suitable for purposes of construction poses a new challenge to construction engineers. The ground improvement technique, which is one of the most important fields in foundation engineering, is in correspondence to various types of ground. That is to search for the safe and prompt operational techniques in order to accomplish all required purposes for construction. Consequently, it is considered possible that overall cost of construction can be reduced due to the inducement of the justifiable ground improvement technique.

When a ground improvement method is selected, the concepts of ground improvement have to be considered in conjunction with such factors as site conditions, soil properties and economic conditions in order to investigate the proper method. Main considerations in process selection can be categorized as follows: (Kamon and Bergado, 1991):

1. Reinforcement method
Strengthen weak areas of soil using other materials---(reinforcing materials)

2. Soil properties improving method
Improve properties of soil or ground by cement grouting and/or admixture--(soli-

dification)
High intensity densification, soil compaction---(densification)
Removal of water from the soil---(dewatering)

3. Replacement method
Replace the weak soil with good quality soil or light weight materials

Reinforcement as one of the ground improvement techniques has advanced intensively in its applications in the past several years. This technique can be applied to various types of soils ranging from cohesive to granular soils. For example, improving soft clays and alluvial deposits by geotextile, geogrid and other inclusions are considered effective while improving natural soils that cannot be excavated, micropiling or nailing are used.

In the reinforcement technique, many types of inclusions are introduced into the ground. Their functions are mainly tensile and/or compressive reinforcement by the material properties of inclusions and the interactional force induced between soil and inclusion. In the soil properties improvement method, the weak soil properties themselves are greatly improved. In particular, the improved strength obtained by solidification sometimes becomes so high that solidified parts are like piles. Therefore, the structural stability of these improved parts should be evaluated and this method can be considered as one of the reinforcement techniques.

2 EVALUATION OF GROUND IMPROVEMENT EFFECT

Evaluation of the improvement effect should be always made. It can be done using several exploring methods and they depend upon their improving mechanism. Firstly, we have to estimate the change of soil properties during and after execution of ground improvement works. Secondly, the field monitoring data should be accumulated and then, their compatibility with the ground improvement principles and the design procedure will be analyzed. After these data would be compared with predicted values, we obtain finally the improved safety factor and confirm the accurate knowledge for the improvement technique. Instrumentation techniques for soft clay ground are summarized in Table 1.

Fig.1 shows a configuration of many kinds of instruments in the monitoring case of embankment work. We can divide the monitoring strategy of the soft ground into three areas. Area I is for

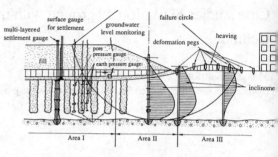

Fig. 1   Example of Instrumentation

the settlement control, Area II is for the stability problem and Area III is for the effect on the surrounding structures. The horizontal deformation measured by inclinometers and deformation pegs is very important for the slope stability.

In order to establish the true prediction, it is necessary that we accumulate the well documented analysis based on the observed data. Because inaccuracy of prediction causes mainly from complicated ground condition, instrumentation and monitoring data analyzed are the most important. The following factors are emphasized:

1. Rational estimation technique for soil properties after improvement
2. Susceptibility of model test in order to evaluate the improvement effect
3. Estimation technique for design procedure
4. Geotechnical instrumentation, in particular, field monitoring and mechanism of settlement

3 CASE STUDY - 1

As a first case study, an example of the countermeasure for the slope failure, which took place during cutting a slope for the backyard of the express highway bridge pier, was introduced. A micropiling reinforced soil method was adopted. By comparing the in situ measured data and the numerical analyzed results, the ground confining effect by the micropiling reinforced soil method was examined.

3.1 Object site and geotechnical conditions

Japan Highway Public Corporation has constructed 27 km express highway at south of Kyoto Prefecture. In the mountainous area of this highway route, an elevated bridge was designed and the slopes were cut. The geological condi-

Table 1   Instrumentation for Evaluation of Improvement Effect

Subjects of instrumentation	Evaluation objects	Technique and equipment	Soil improvement method objected
Estimation of soil properties change	Strength increased by consolidation Water content decreased	Labo test by TWS In situ test by vane & CPT	Preloading, Vertical drain, Compacted sand piles, DMM, etc.
Monitoring of Settlement	Comparing with design values Prediction of final settlement Check of consolidation rate	Surface gauge for settlement Multi-layered settlement gauge	
Monitoring of lateral deformation	Check of shearing deformation Prediction of failure Prevention effect against deformation	Surface peg Surface extensometer Inclinometer	
Monitoring of pore water pressure	Check of drain effect Check of effective stress increased	Pore water pressure gauge Ground water measurement	
Monitoring of earth pressure	Check of applied vertical and horizontal loads	Earth pressure gauge Strain gauge	Reinforcement method, DMM, etc.
Monitoring of acceleration	Measurement of acceleration at earthquake	Seismic gauge	DMM, Anti-liquefaction method
Miscellaneous	Stress & deformation of inclusion Behavior of adjacent structure & ground Level & quality of groundwater	Stress & strain gauge Settlement gauge Inclinometer Water level gauge Quality check	Reinforcement method Almost all methods Grouting

Note: TWS means thin-wall sampling method
CPT means cone penetration test

116

tion of this site consists of the sedimentary rock at the upper strata (N-value varies from 20 to 60) and slate at the lower strata (N-value around 60). Since they are affected by tectonic forces, many cracks in the fracture zone and the part of argillization by the weathering were observed. The bearing capacity of the ground is enough for the bridge foundation, however, some risk for the slope failure existed. This bridge pier site had very limited area (Fig.2) as the abutment foundation was planned closely on the excavated slope for the pier.

Immediately after slope excavation for the pier of bridge was carried out, a slope failure took place. Because the reduction of slope angle was not possible due to the limited area condition, the part of the excavated volume was buried and micropile reinforcement as the remedial work was adopted to stabilize the slide. The layout plan and the cross sectional diagram of the slope stability work, which was designed by the conventional method, are shown in Fig.2. After the completion of the upper half of the reinforced slope, slope excavation were resumed. The micropiles were installed immediately after each step of the excavation. The construction sequence of the reinforced slope are shown in Fig.3.

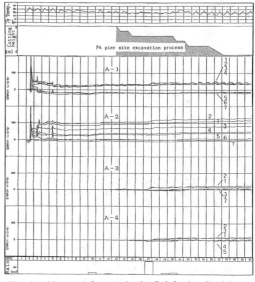

Fig. 3   Excavating work sequence

3.2 Monitoring results

In monitoring the slope stability, the following activities are carried out.
1. Induced stress in the inclusions
2. Ground movement

Induced stress in the inclusions was measured by strain gauges fixed on both sides, along the axial direction of the inclusions for the determination of axial stresses. The displacement of the slope was determined by a borehole inclinometer and the reading were taken at intervals of two to six hours. The layout of instrumentation is also shown in Fig.2. The value of stress monitored at the permanent slope A-1 to A-4 are shown in Fig.4. The stress change during three months are shown in Fig.5. The maximum stress in the individual steel bar moves gradually from the end near the slope to the other end. This shows either the potential rupture surface moves inwards, or slippage of the bars takes place as

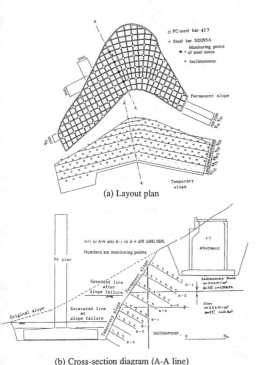

(a) Layout plan

(b) Cross-section diagram (A-A line)

Fig. 2   General Feature of Slope Stability Work

Fig. 4   Measured Stresses in the Reinforcing Steel Bars

117

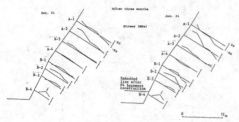

Fig. 5   Measured Axial Stress in the Reinforcing Bars

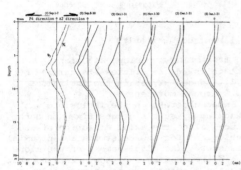

Fig. 6   Deformation Data in the Slope

time elapsed. The monitoring results also show an increase in stresses in the steel bar following the rainfall, also shown in Fig.4. The stresses in the capping beams are not shown here, but it has to be realized that the results would be affected by and at the same time supply information for the structural design of the capping beams. The lateral movement of the slope are shown in Fig.6. The maximum displacement took place at the boundary of the stratum.

3.3 Numerical analysis

The slopes under the following three different conditions are analyzed as equilibrium field problems:
  1. Unreinforced natural slope (Slope A);
  2. Reinforced slope, with the actual construction sequence closely simulated (Slope B); and
  3. Reinforced slope, with a different construction sequence as an ideal pattern (Slope C) (The excavations and the installations of micropiles are to be carried out continuously.)

3.3.1 FE method of reinforced slope

The reinforced slope is modeled as composed of continuum media made up of finite number of triangular elements. The reinforcing steel bars as inclusions are treated as one-dimensional bar

element, and the capping beams as beam elements. The finite element mesh is shown in Fig.7.

Material constants of the ground were decided by the SPT results only. Excavation of the slope involves incremental changes in stresses, and the behavior of

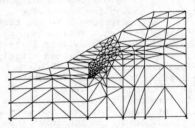

Fig. 7   Finite Element Mesh of the Slope

the soil and rock materials is both non-linear and stress dependent. Due to the lack of information about the material, the analyses here are carried out by assuming an elasto-plastic behavior of the material which obeys von Mises yield function. Neither the elastic constants nor the Poisson's ratios are assigned new values after each steps in the analysis. The bar and beam elements are assumed to follow the linear stress-strain relation-ship. The face between the micropiles and soils are characterized as perfectly rough with no possibility of slippage was considered. Since the mortar around the steel bar cannot sustain large tensile forces, its role as an element is neglected in this particular analysis. That is, the tensile force acting on the micropiles is assumed to be fully sustained by the reinforcing steel bar. The material properties of the elements are summarized as Table 2.

The initial stress of slope is established by the gravity loading applied to all elements. The stress analysis due to the successive steps of excavation

Table 2   Material Constants for Case 1

(1) Soil Element

	Sedimentary Rock	Slate
bulk density (kN/m³)	20	25
angle of internal friction (degree)	70-80	35
cohesive strength (kPa)	20	60, 80
elastic modulus (MPa)	615	2500
Poisson's ratio	0.3	0.2
unconfined compressive strength (kPa)	40	2000
elastic modulus after yielding (kPa)	0.0	1000

(2) Inclusions

Reinforcing bar	PC-steel bar φ23	Steel bar SD295A
cross-sectional area (cm²)	4.15	7.9
elastic modulus (MPa)	210 000	210 000

Capping beam	upper slope	lower slope
cross-sectional area (cm²)	0.15	0.05
elastic modulus (MPa)	10 000	10 000
moment of inertia (m²)	0.00028	0.00001

followed as the construction sequences as shown in Fig.3. The stresses of Slope A and Slope B are carried out by simulating the actual steps as closely as possible to the site. Slope C is the case of the earlier inclusion of micropiles at the upper part of the slope, as to investigate the rapid confining effect.

3.3.2 Numerical results

The principle stress axes of the slope are reoriented during the excavation and the greater stress concentration is observed at the slope toe as shown in Fig.8. It is found that the confining effect is achieved by reinforcing the slope. That is, the minimum principal stresses near the reinforced slope surface are increased while those of the maximum principal axes are reduced when compared to the unreinforced slope. Also, the orientation of the principal axes by the excavation is less affected in the case of the reinforced slope. By comparing the analyzed results of Slope B and Slope C, it can be concluded that the later shows better confining effect.

The axial stresses in the reinforcing steel bars are illustrated in Fig.9. The tensile stresses in the bars are larger than those at the upper slope in Slope B. This delineates the concentration shear stress at the toe of the slope which is sustained partly by the reinforcing bars. By an earlier inclusion of the bars, only those in the upper slope are subjected to large tensile while in the lower slope are subjected to only negligible amount of forces.

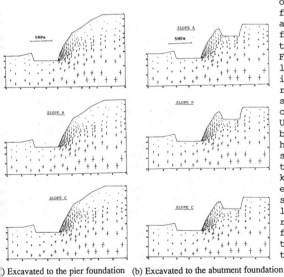

(a) Excavated to the pier foundation (b) Excavated to the abutment foundation

Fig. 8   Principle Stress Distributions in the Slopes

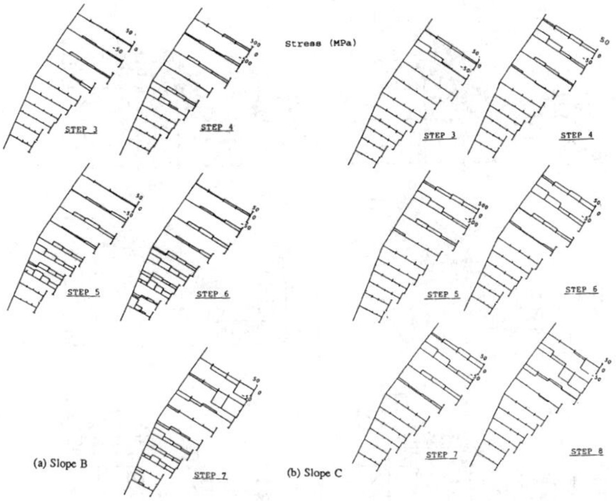

Fig. 9   Calculated Axial Stresses in the Reinforcing Bars

The analyzed values of displacement of the slope are shown in Fig.10. The slope has a movement with its lower part being displaced inward in the initial step of analysis. As the excavation is proceeded, however, the slope gradually moves outward. The reinforced slope with micropiles is efficient in constraining the slope movement at the last step of the excavation. The reinforced slope is believed to have increasing effect in constraining large movement. Thus, Slope C shows such an effect more than Slope B.

The stress around the slope may be large enough to cause local failure to the soil even if a satisfactory value of safety factor against the catastrophic failure is obtainable. Therefore, it is also necessary to investigate the local factor of safety of the element around the reinforced and unreinforced slopes. Fig.11 gives the shaded regions where local factor of safety is less than unity in the slope. Such a region needs to be reinforced. The region in the reinforced slope is drastically reduced when compared with the unreinforced slope. Upon the completion of the construction, both Slope B and Slope C are found to have their regions with local factor of safety greater than unity. Usually, with the principal stresses distribution known, a potential rupture surface can be estimated and the overall factor of safety determined. According to the limiting equilibrium analysis on this reinforced slope, the overall safety factor is greater than unity. Therefore, the reinforced slope is safe from both the local and catastrophic failure.

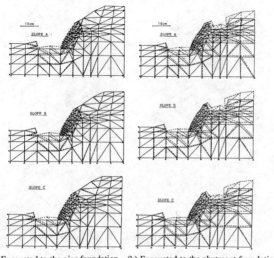

(a) Excavated to the pier foundation   (b) Excavated to the abutment foundation

Fig. 10   Deformation of Mesh in the Slopes

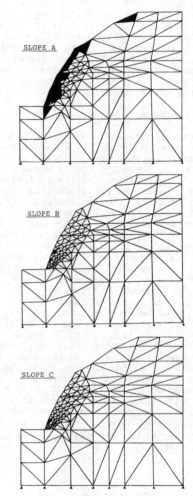

Fig. 11   Local Factor of Safety in the Slopes
(shaded regions; Fs is less than unity.)

3.4 Discussions

The measured stress induced in the reinforcing bars varies slightly from the analyzed values. The discrepancy could be attributed to the release of stress prior to burying the volume of slope immediately after failure at the site, or the modeling of the micro-pile may need to be reconsidered. The effects of full loading are considered throughout the analysis which may not be at the site. Furthermore, the excavations and installations of micropiles at the site were carried out continuously without waiting for the step to achieve its equilibrium, and such a fact is almost impossible to simulate in the analysis. However, from the results at the mid-points of most reinforcing bars, when the final step in Slope B is compared with the final diagram in Fig. 5 an agreement is shown. Certain amount of equilibrium is believed to have been achieved at this step of monitoring.

This simple numerical analysis was conducted under a very limited condition. Nevertheless, calculated results coincided with the measured values.

The applicability of the micropiling reinforced soil method to the natural slope stability is considered as follows:

1. High confining effect of micro-pile has brought the stable steep slope.

2. Light weight machines are suitable to access to the slope.

3. Even rough and bumpy surface of in situ slope is acceptable.

4 CASE STUDY - 2

As a second case study, an application of DMM to the reinforcement of the soft clay ground for the embankment work was investigated. DMM is the in situ mixing soft soils with admixture to form piles, walls, grids or blocks in the ground. It has been developed and applied extensively since 1970's. Quick lime was initially used as admixtures, but now portland cement is more popularly used in both slurry state and dry powdered state in Japan. Since the soft clay ground is improved vertically due to the mechanical mixing effort, a group of the solidified piles is obtained in the ground. The formation of the improving type mainly depends on area improvement ratio, a_s. When a_s is less than 79%, the formation is like a pile or wall group and a block formation is obtained at more than 80%. The strength and bearing capacity of the improved ground can be increased and the compressibility can be extremely reduced by increasing the admixture content and area improvement ratio. It is considered that the ground improved by DMM is a composite ground. The deformation characteristics of solidified piles quite differ from the original clay soil. The solidified piles become too brittle to evaluate as the soil material. Therefore, the performance as the pile foundation should be justified, and so the role of DMM is considered as the reinforcement of the soft clay ground.

4.1 Performance of solidified piles for reinforcement in numerical model analysis

In a numerical analysis three dimensional (3D) method is the most reliable for the behavior of these composite ground. However, the 3D method requires a large capacity computer and the real site conditions are not always clear. Two dimensional (2D) method is often used in order to simplify the complex analytical problems. In the case of composite ground, the deformation pattern analyzed by 2D models differs from the in situ condition because the model is postulated under a plane strain condition for the wall type improved ground ignoring the remaining unimproved soil among piles. Therefore, the compatibility of the 2D-FEM models is examined here by comparing with 3D-model behavior (Kamon et al, 1987 and 1988a).

Three types of modified 2D-model with different pile properties are analyzed, namely:
1. 2D-model(1) means that center position of each pile is fixed and diameter of the pile is reduced in proportion to a_s.
2. 2D-model(2) means that diameter of pile is same with real one, but the stiffness of pile is reduced in proportion to a_s.
3. 2D-model(3) means that the uniform ground is assumed whose stiffness is increased in proportion to a_s from the original ground.

The loading conditions are illustrated in Fig.12. The depth of improved model ground is 10m and both vertical and horizontal loads are applied. Four kinds of area improvement ratio, a_s, namely 17, 40, 63 and 79% are selected. Area ratio 79% is that when solidified piles are tangent to each other.

The plan of the 3D-finite element mesh is shown in Fig.13. The patterns of the solidified pile group are calculated in

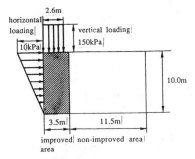

Fig. 12   Vertical Section of the Model Ground
Improved by DMM

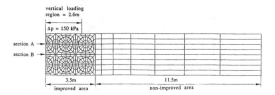

Fig. 13   Plan of 3D Model Mesh
(Area improvement ratio is 40%)

Table 3   Material Constants for Model Analysis

	Solidified pile by DMM	Original ground
3D-model 2D-model(1)	Ec=400MPa, v=0.10, γ=7kN/m³	Es=10MPa v=0.40 γ=7kN/m³
2D-model(2) 2D-model(3)	Ec is increased in proportion to Area improvement ratio(a_s) v and γ are constant	Es=10MPa v=0.4 γ=7kN/m³

triangular and square spacing. This figure is an example of square pattern of piles and 40% area improvement ratio.

The analyses are carried out by assuming a linear elastic behavior and material constants used are shown in Table 3.

Some examples of calculated results are shown in Figs.14 and 15. Surface settlement is illustrated along the sections A and B of 3D-model in Fig.13. The average settlement pattern is obviously similar with the 2D-model(1) result. Stress concentration ratio has the maximum value from 13 to 18 at few meters below the top surface of the soft ground and gradually decreases with the depth.

By comparing the results, we found that the 2D-model can be used to simulate the in situ performance of the composite ground improved by DMM.

4.2 Object site and monitoring

Ministry of Construction engages in the

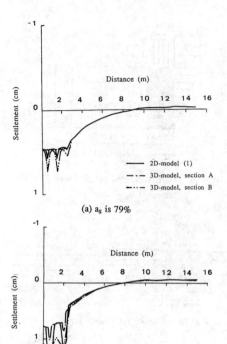

(a) a_S is 79%

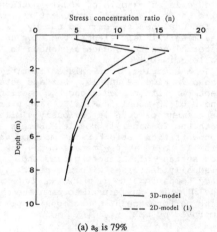

(b) a_S is 40%

Fig. 14   Surface Settlement of the Model Ground
(Square pile spacing)

Stress concentration ratio (n)

(a) a_S is 79%

Fig. 15   Stress Concentration Ratio with Depth
(Square pile spacing)

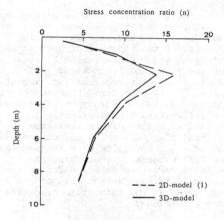

(b) a_S is 40%

Fig. 15   Stress Concentration Ratio with Depth
(Square pile spacing)

control of river maintenance in Japan. River dyke rebuilding is one of the effective means. In order to support the embankment load on the soft clay ground, DMM was adopted at Tega River, which is a branch of Tone River in Chiba Prefecture. The geological condition of this site consists of the alluvial deposit. The upper strata is very soft clay layer (qu-value around 20kPa) and continues homogeneously to 17m depth. The lower strata is silt layer (qu-value gradually increases to 70kPa) 4m thick. Under the alluvial deposit the hard diluvial sand deposit continues. According to the field exploration, the depth to be improved by DMM was decided at 17m due to economic reasons. Although the shear strength of the alluvial silt soil increases with depth and it is not enough to support the full embankment load, the foundation is considered as a floating type.

When monitoring the embankment stability, the measurement of the settlement and lateral movement are carried out. Induced stress on the solidified pile and unimproved ground was measured by pressure gauges laid on both top surface. The settlement of the embankment was determined by settlement plates and the lateral movement was measured by borehole inclinometers. The layout of instrumentation is shown in Fig.16.

4.3 FE method of reinforced soft clay ground

The soft clay ground reinforced by solidified piles is modeled as composing of continuum media made up of finite number of rectangular elements. The

122

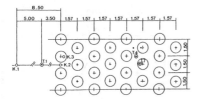

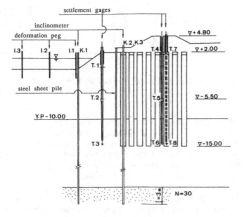

(a) Layout plan of DMM and Instrumentation

(b) Cross-section of Embankment

Fig. 16   General Feature of Embankment Reinforced by DMM

characteristics of solidified pile in the modified 2D-models are same as mentioned above. The elastic modulus of solidified piles as inclusions is 200 times the original ground. The finite element mesh is shown in Fig.17 and the area improvement ratio is 33% in this case.

Material constants of the original and improved ground are decided by laboratory tests. The stiffness of solidified pile in 2D-model is modified due to each analytical pattern, but Poisson's ratio fixed to a constant value. The modified elastic modulus, E' is defined as follows:

$$E'=E_i\frac{a_sD}{\phi}+E_c(1-\frac{a_sD}{\phi})$$

where E'; modified elastic modulus of improved ground, E_i; elastic modulus of solidified pile, D; spacing of solidified piles, and ϕ; diameter of solidified pile E_c; elastic modulus of original clay.

The material properties of the elements are summarized in Table 4.

4.4 Numerical results and discussions

The initial stress of the ground is also established by the gravity loading applied to all the finite elements.

The analysis of stress induced by the embankment work directly affects vertical settlement. According to the numerical results, the maximum principle stress distribution clearly shows the concentration of stresses in the solidified piles as illustrated in Fig. 18.

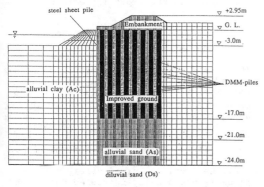

Fig. 17   Numerical Model Mesh of the Soft Ground

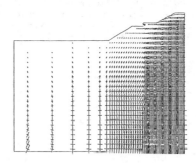

Fig. 18   Vectors of Maximum Principle Stress (2D model(1))

Table 4   Material Constants for Case 2

parameters \ grounds	Solidified pile by DMM			Original ground		
	2D-model(1)	2D-model(2)	2D-model(3)	allu.clay	allu.sand	dil.sand
unit weight γ (kN/m³)	14.1	14.1	14.1	14.1	18.0	18.0
elastic modulus E(MPa)	200	105	67	1.0	5.0	10
Poisson's ratio ν	0.01	0.01	0.01	0.40	0.30	0.30

123

Fig.19 shows the deformation vector of the ground. Settlement values (8 to 13 cm) coincided well with the observed in situ results. Stress concentration ratio, n is 15.6 in the in situ data instead of 19.6 in 2D-model(1) and 13.1 in 2D-model(2). These large values of n may mean the pile groups should be considered as the foundation structures rather than the composite ground.

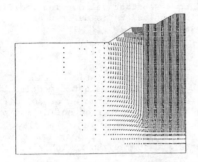

Fig. 19   Vectors of Deformation (2D model(1))

The patterns of lateral movement at the toe of the embankment (K2 and K3 points) and at the river base (K1 point) are shown in Figs.20 and 21. The shape of movement is quite similar between calculated values and observed value. Because the area improvement ratio is 33% in this site, the calculated lateral movement of the case of 2D-model(1), which is modified as that the smaller diameter of pile is installed, will be expected to deform largely in lateral direction. Nevertheless, the values of lateral movement observed are less than the other models.

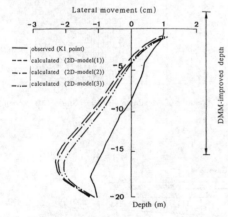

Fig. 20   Lateral Movement at K1 Point

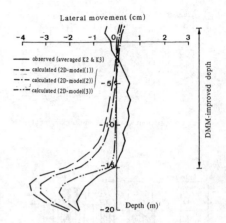

Fig. 21   Averaged Lateral Movement at K2 and K3 Points

Although there are some discrepancies shown in these figures and simple 2D-models are used in the numerical analyses, calculated performance of reinforced soft clay ground by DMM coincides with the measured in situ values.

Though it is really important to predict the performance analyzed by the ground improvement effect, the harmony of the accuracy between the analyzing method and the properties of ground improved is requested. Therefore, it should be aimed for getting the promising performance by an appropriate model analysis because it is still very difficult to get the perfect evaluation of the improving effect of the site.

5 CONCLUSIONS

The case studies of the micropiling reinforced slope and deep mixing foundation soil were introduced in this paper. The major findings from this current study are as follows;

1. By comparing the results of reinforced slopes with an unreinforced slope, the confining effect of reinforced slope is clearly shown.

2. The local factor of safety near the slope is drastically increased due to the introduction of the reinforcement.

3. Two dimensional approximation in the numerical analysis of composite ground is acceptable and revealed by the monitoring results.

4. The construction sequence has a great influence in the stress distribution on both the soil and reinforcing bars.

5. Upon careful designing of the construction sequence, better confining effect including the sharing of stresses and movement constraining may be obtained,

and thus may admit economic design by
reducing the total number of the reinforcing materials installed.

REFERENCES

JSSMFE (1986), Reinforced Soil Method,
   Library of Soil and Foundation
   Engineering - 29, 430p. (in Japanese).
Kamon, M. et al. (1987), Vertical
   stresses analysis of composite improved
   ground by DMM, Proc. 42nd Annual
   Meeting of JSCE, pp.806-807 (in
   Japanese).
Kamon, M. et al. (1988a), 3-Dimension
   numerical deformation analysis for
   horizontal load in pile-shaped improved
   soil, Proc. 23rd Annual Meeting on
   SMFE, JSSMFE, pp.2271-2272 (in
   Japanese).
Kamon, M. et al. (1988b), Ground
   confining effect by reinforced soil
   method, Jour., Material Science, Vol.
   37, No. 422, pp.1289-1293 (in
   Japanese).
Kamon, M. and Itoh, Y.(1991), Case study
   of slope stability by micropiling
   reinforced soil method, Proc. Internt.,
   Symp. on Natural Disaster Reduction and
   Civil Engineering, JSCE, pp.367-376.
Kamon, M. and Bergado, D.T. (1991),
   Ground improvement techniques, Theme
   Lecture of 8th Asian Reg. Conf. on
   SMFE, Bangkok.

3 Embankments, excavations and buried structures

Permissible dimensions of unsheeted wall sections of excavations and trenches

P. Bilz
Dresden University of Technology, Institute of Geotechnics, Germany

ABSTRACT: Unsheeted sections of structures for securing walls of excavations and trenches must not exceed specific dimensions for reasons of safety. A method of approximation is presented for the calculation of all types of sections which are unsheeted due to the technology and/or the design.

Introduction

The boundary of excavations and service trenches must get their required slope or appropriate sheeting according to DIN 4124. The different kinds of sheeting have more or less large unsheeted sections due to technology and/or design, i. e., sections in which the excavation wall remains without a support by a specific sheeting element for shorter or longer periods of time. The larger such a section is and the more unfavourable the strength properties of the in-situ building ground are, the greater is the potential danger with respect to a partial or total collapse of the excavation wall. The danger to man and material resulting from that must be kept within justifiable limits or excluded under certain condition.
Among the simple, conventional types of sheeting (designated as standard sheeting) (cf. VOTH, 1984, inter alia) the horizontal sheeting has an unsheeted section of up to two batten widths above the excavation or trench bottom advancing with digging (figure 1a). Analogous conditions apply to soldier piles with horizontal sheeting (Berlin sheeting) (cf. VOTH, 1984, inter alia). Especially for the securing

of service trenches there have been increasingly developed devices and techniques reducing the high (in many cases manual) expenditure of work as well as the wear of material and raising the level of mechanization of these partial processes.
The variety of the specific **developments** by companies (cf. SCHLICK, 1986, UFFMANN, 1985; VOTH, 1984, inter alia) can be grouped as follows:

. Preliminary assembly of the horizontal sheeting and insertion of so-called shetting cages into the completely excavated trench section which is stable without a support for a short period of time, the trench wall being directly supported by battens or plates.
. Sinking of sheeting plates in sliding rails with the advance of excavating
. Horizontal advance of sheeting systems or sliding supporting walls in the course of continuous partial processes of excavating, sheeting, pipelining and partial backfilling.

With the mechanized types of sheeting mentioned, due to the technology in all cases unsheeted sections of different size occur (up to the whole trench

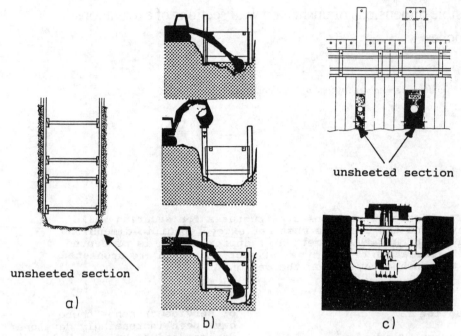

unsheeted section

unsheeted section

a)

b)

c)

Figure 1   Examples of technologically demanded unsheeted
            sections

            a horizontal sheeting
            b steel plate sheeting with sliding rails
            c crossing lines or conduits

wall for a short period of time). Moreover - as also with the vertical standard sheeting - there occur possibilities of leaving partial sections unsheeted, e. g., in order to pass crossing lines or conduits through.   Figures 1a to 1c schematically show selected examples of the position of unsheeted sections like that. The unsheeted sections at excavations and service trenches in soil types having a high inherent stability (especially in the form of cohesion) may assume extreme dimensions, if there is used only a support of the wall in vertical direction by soldier piles, bored piles (figure 2) or different elements placed in front of the wall which are horizontally braced (WANG and SALMASSIAN, 1978) or if sheeting cages without battens or plates are used. Since there has not been available a uniform calculation method for proving the stability of such unsupported parts of walls, the author was induced to develop a more practicable method of

Figure 2 Example of unsheeted sections intended by design (boredpile wall with a distance between the piles)

130

approximation (BILZ, 1987; BILZ and BÖHME, 1988; BILZ, 1989) and to make it accessible to the user by preparing it appropriately.

Fracture Pattern and Calculation Model

The formation of the figures of fracture for unsheeted parts was observed by model experiments on a scale of 1:10. Besides the variation of the geometrical parameters of the unsheeted parts (width b and height h), the density and the water content of the medium sand was varied, in order to determine the shear strength, inclusive of a cohesion share **resulting** from capillary forces (BILZ, 1983; BILZ and VIEWEG, 1992). The contours of the figures of fracture determined in this way are schematically represented in figure 3.

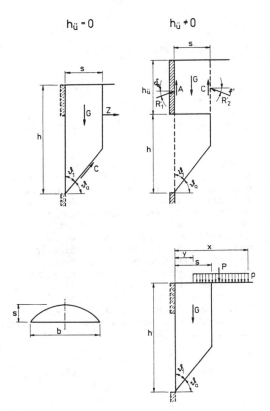

Figure 4   Calculation model

Taking into account the calculation models offered in the literature for proving the stability of diaphragm wall sections as well as the information about the formation of horizontal and vertical arching effects (TOTH, 1978) behind the sheeting elements, the evaluation of the model experiments led to a calculation model according to figure 4, in which the quantities required for comparing the applied and resisting forces are stated.
For unsheeted sections extending to the ground surface ($h_ü=0$), stability is proved by means of **equation** (1), see figures 3 and 4:

$$F=\frac{G\sin\vartheta_1\tan\varphi'+c'A+\sigma_z b(h-s\tan\frac{\vartheta_a}{2})\sin\vartheta_1}{G\sin\vartheta_1}$$

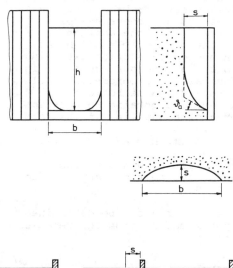

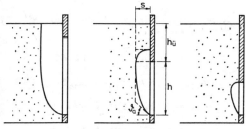

Figure 3   Fracture pattern

with

$$G=\frac{2}{3}\gamma bs(h-\frac{2}{5}s\tan\vartheta_a)$$

$$A=\frac{2}{3}bs\frac{1}{\cos\vartheta_a}$$

$$\vartheta_a=45°+\frac{\varphi'}{2}$$

$$\sigma_z=2c'\tan\vartheta_1$$

$$\vartheta_1=45°-\frac{\varphi'}{2}$$

The decisive value min F is obtained by variation of the camber s within the range $0 < s < b/\tan\Phi'$. For unsheeted sections not extending to the ground surface ($h_{ü} > 0$), according to equation (1) an additional force has to be added to the weight G, which is essentially dependent on the overlying height $h_{ü}$, on cohesion c', on adhesion a as well as on the angle of wall fricition δ_a. This additional force G* may be approximately determinded according to equation (2), see figure 4:

$$G^*=\frac{2}{3}\gamma bsh_{ü}-abh_{ü}-2c'h_{ü}\sqrt{\frac{b^2}{4}+s^2}-\gamma K_{ah}h_{ü}^2\cdot$$

$$\cdot\sqrt{\frac{b^2}{4}+s^2}(\tan\varphi'+\tan\vartheta_a)\qquad(2)$$

with a < c'/2 (adhesion)
δ_a (angle of wall fricition)
K_{ah} (active-earth-pressure coefficient)

If G* > 0 there is the danger of break-through of the overlying masses up to the ground surface. Comparative investigations have shown that for real shear strength parameters φ' and c' the aformentioned case becomes decisive only at b > 2 m. Surcharges may be included in accordance with figure 4. The additional force P is determined by means of equation (3) and added to G according to equation (1)

$$P = p (s - y) b \qquad (3)$$

The camber s **has** to be varied within the range 0 < s < x.
This has to be examinded, however, whether we will get s > h/tan ϑ_a . In this case, one should calculate with

$$\overline{\vartheta}_a=\arctan(\frac{h}{s})$$

and

$$\overline{\vartheta}_1=90°-\overline{\vartheta}_a$$

Application to the Design Practice and the Construction

The permissible free height h of the unsheeted sections may be plotted in **appropriate** diagrams in dependence on their width b, the soil parameters (γ, φ', c') an, if required, different safety factors F. For that, equation (1) is transformed with respect to perm. h.
On closer examination of these graphics, examples of which ha- ving been chosen for figure 5, one notices that from a width b* on there remains perm. h = const. Hence follows the possibility to construct other graphics for the dependence b* = f(tan φ',c',γ,F) in the form b*=k*.c'. Figure 6 presents analogous examples. Finally, for the estimation of the permissible height of an unsheeted section extending to the ground surface (e.g. in the case of a bored-pile wall with a distance between the piles) one obtains an equation of a straight line in the form

$$\text{perm. h = k . b*} \qquad (4)$$

the existing width of the unsheeted section **has** to be b > b*.

The factor k can be determined approximately from the relation (5):

$$k = a - b . \tan \varphi' + c . \tan^2 \varphi' \quad (5)$$

$$a_{1.1} = 7.24, \quad b_{1.1} = 11.60$$

$$a_{1.3} = 6.97, \quad b_{1.3} = 11.50$$

$$c_{1.1} = 8.125 \text{ for } F = 1.1$$

$$c_{1.3} = 7.875 \text{ for } F = 1.3$$

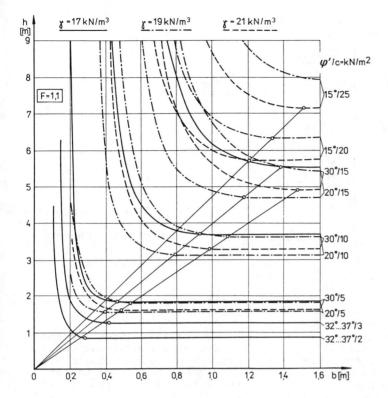

Figure 5 Dependence of the permissible height on the width of the unsheeted section at $h_\ddot{u} = 0$

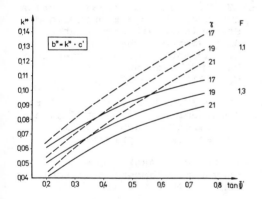

Figure 6   Presentation of the function b* = k*.c'

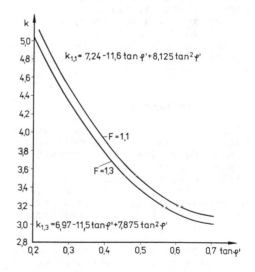

Figure 7   Presentation of the function k = f(F, tan φ')

In figure 7 the function k = f(tan φ', F) is represented. For unsheeted sections not extending to the ground surface (e.g. in cases of standard sheeting as well as crossing pipes or conduits)

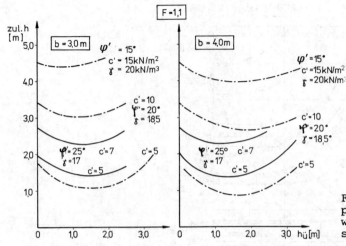

Figure 8 Dependence of the permissible height on the width of the unsheeted section at $h_ü > 0$

graphics may be constructed, too; they are more extensive, however. Selected examples are shown in figure 8. A noticeable influence is exerted by an overlying $h_ü$ only at friction angles $\varphi' < 20°$. Decisive are the value of the cohesion c' as well as the width of the unsheeted section b. In cases of unsheeted sections with an upper limit set by sheeting elements, for $h_ü > 0$ the reduction of perm. h compared with $h_ü = 0$ is greater; this applies especially to the range $1 < h_ü < 4$ m.

Concluding Remarks

with the aid of the approximation method presented it is possible to determine the permissible dimensions of unsheeted sections required by technology or intended by design when walls of excavations or trenches are supported. For soil parameters relevant to practice ($15°<\varphi'<35°$; $2< c'< 25$ kN/m²; $17<\gamma < 21$ kN/m³) it is recommended that user-friendly graphics be used, examples of which being given.

REFERENCES

BILZ, P. (1983): Abschätzung der Kohäsion nichtbindiger Locker-gesteine. Bauplanung - Bautechnik 37, 7, 316-317/320.

BILZ, P. (1989): Effektive Konstruktionen zum Verbau von Baugruben- und Grabenwänden - Standsicherheit unverbauter Bereiche - Wiss. Ztschrf. TU Dresden 38, H. 4, 177 - 181.

BILZ, P./BÖHME, K. (1988): Berechnungsverfahren zum Nachweis der Standsicherheit unverbauter Baugrubenbereiche. Beitrag zur XI. Grundbautagung Dresden 1988. Bauforschung-Baupraxis, Heft 218, 31 - 33.

BILZ, P./VIEWEG, J. (1992): Zur Größe der Kohäsion von Sanden und deren Zeitabhängigkeit. Geotechnik (in print).

SCHLICK, H. (1986): Moderne Grabenverbaumethoden. Baumaschinentechnik, 10, 431 - 446.

TOTH, L. (1978): Gewölbebewirkung bei einer aufgelösten Bohrpfahlwand und Ermittlung der lichten Abstände der Bohrpfähle. Bautechnik 55, 6, 181 - 188.

UFFMANN, P. (1985): Bauausführung im Kanalbau. Tiefbau, Ingenieurbau, Straßenbau 12, 712 - 715.

Prediction versus performance of a granular pavement tested with the Accelerated Loading Facility (ALF)

Binh Vuong
Australian Road Research Board Limited, Nunawading, Vic., Australia

ABSTRACT : In this paper a computing model for pavement analysis (NONCIRL) is used to predict the performance of a granular pavement tested with the Accelerated Loading Facility (ALF) complemented by data obtained from repeated load triaxial testing. The program has been incorporated into the proposed revision of the design of overlays in the AUSTROADS Design Guide and also into various rut depth models. The effects of material non-linearity, loading configuration, transverse distribution of load, loading history, environment and surface conditions were also considered. It is shown that the AUSTROADS subgrade strain criterion overestimated the pavement life because the subgrade deformation was not critical in this case. The Rut Depth model, which can take into account deformations in all pavement layers and in the subgrade, could predict closely the pavement deformation measured throughout the ALF loading test.

1. INTRODUCTION

At the Australian Road Research Board (ARRB), research has been directed towards applying the theory of material behaviour to computer modelling and analyses of pavement structures with the aim of identifying parameters which have crucial influence on the performance and management of Australian road pavements. This would enable the development of optimum pavement designs and more effective management pavement models as set out in the AUSTROADS Strategy for Pavement Research and Development (1992).

The research program has been supported with the Accelerated Loading Facility (ALF). This loading facility is a relocatable road testing device, which can apply controlled full-scale wheel loads to sections of real pavement. The program has also provided reliable data for laboratory material properties and field performance, which can be used for verification/selection of pavement performance models.

In this paper, emphasis is placed on models for predicting rut depth, which is the major factor influencing performance of granular pavements. For this purpose, an AUSTROADS performance model and a rut depth model, which are incorporated in the computing program NONCIRL, are used to predict the performance of a granular pavement tested with ALF. Comparisons of the predicted and measured performance allow the selection of the most appropriate performance models for the structural evaluation and design of granular pavements.

2. DETAILS OF ALF LOADING TEST

The trial at Benalla, Victoria, was the second accelerated

full-scale loading test in the ongoing ALF program. Details of the ALF program to date are contained in Sharp (1991). The trial was conducted on a newly-constructed section of the Hume Freeway, the major link between Melbourne and Sydney, before it was opened to traffic.

2.1 Pavement Configuration

The pavement was an unbound crushed rock structure with a surface seal built to high standards under strict quality control. The pavement structure consisted of the following layers:

(i)   surfacing: double application sprayed chip seal
(ii) base: 200 mm of crushed rock (Class 2A)
(iii) upper sub-base: 200 mm of crushed rock (Class 3)
(iv) lower sub-base: 170 mm of ripped sandstone/siltstone (design soaked CBR = 8)
(v)   imported subgrade: 300 mm of siltstone (design soaked CBR = 4)
(vi) subgrade: siltstone

The design traffic loading was about $2.1*10^7$ ESAs (1 ESA = 40 kN dual wheel load).

2.2 Loading Conditions

The loading applied to the pavement is given in Table I, together with the transverse loading distribution applied for each load, whilst the loading history is given in Table II.

2.3 Environmental and Surfacing Conditions

The test was carried out over a period of 8 months, thus

TABLE I: CHACTERISTICS OF THE APPLIED LOADS

Load (kN)	Radius of Load (mm)	Tyre Pressure (kPa)	Distance Between Wheel Centres (mm)	Transverse Loading Distribution	
				Total Width (mm)	Standard Deviation (mm)
40	107.4	552	375.0	1400	110
60	117.6	690	375.0	1400	110
80	135.8	690	375.0	1000	50

TABLE II: LOADING HISTORY

Loading Stage	Duration (week)	Load (kN)	Applied Cycles	Accumulated Loading Cycles
1*	1	40	13 429	13 429
2*	1	60	16 690	30 119
3	2-4	80	169 053	199 172
4*	5	40	2 236	201 408
5-8	5-34	80	1 200 574	1 401 982

* Stages 1,2 and 4 essentially shake down loadings

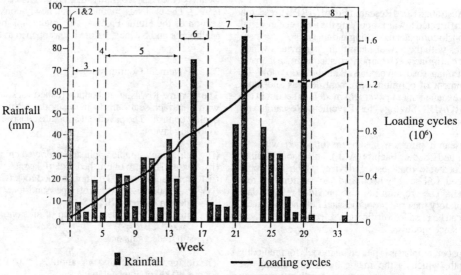

Figure 1 - Rainfall recorded in the Benalla ALF trial

enabling some investigation of the pavement sensitivity to environmental conditions, i.e. to changes in moisture content within the pavement. However, the moisture contents in the pavement were not fully recorded throughout the loading test and therefore they had to be interpolated from parameters influencing the moisture condition in the pavement such as rainfall, surface condition and temperature. High rainfall and bad surface condition (with cracks and pot-holes) would allow more water to penetrate into the pavement, whereas high temperatures would increase the evaporation rate of excess water on the surface, thus reducing the penetration rate of water into the pavement. These parameters were fully recorded throughout the ALF loading test and have been reported in Kadar (1986).

Figure 1 shows the variation in rainfall throughout the loading period. It can conveniently be divided into 3

TABLE III: LOADING STAGES CONSIDERED IN THE ANALYSES

Loading Stage	Temp. Rating	Surf. Cond. Rating	Rain. Rating	Mois. Cond. Rating	Load (kN)	Applied Cycles	Accumulated Loading Cycles
1*	Low	Good	High	Ave.	40	13 429	13 429
2*	Low	Good	High	Ave.	60	16 690	30 119
3	Low	Ave.	Ave.	Ave.	80	169 053	199 172
4*	Low	Ave.	Ave.	Ave.	40	2 308	201 480
5	Low	Bad	High	High	80	530 520	732 000
6	High	Bad	Low	Low	80	160 000	892 000
7	High	Bad	High	High	80	289 500	1 181 500
8	High	Bad	Low	Low	80	220 482	1 401 982

* Stages 1,2 and 4 essentially shake down loadings

conditions: "low" (for periods without rain or occasional showers with weekly rainfall of less than 5 mm), "average" (for periods of prolonged rain with weekly rainfall between 5 and 15 mm) and "high" (for periods of prolonged rain with weekly rainfall greater than 15 mm). Table III shows the rainfall ratings for all the loading periods.

The temperatures for all periods can also be conveniently divided into two conditions: "low" (with a weekly mean minimum temperature of -2° and weekly mean maximum temperature of 15°C) and "high" (with a weekly mean minimum temperature of 5°and weekly mean maximum temperature of 35°C). The temperature ratings for all loading periods are given in Table III.

Deterioration of the seal surface was observed at the end of the third week, when transverse cracks appeared. The surface condition was fairly bad after the 7th week (of trafficking 350,000 cycles) and the surface seal was frequently repaired after this time. It is also convenient to divide the surface condition into three conditions: "good" (no cracking), "average" (with small cracks), "bad" (with large cracks and pot-holes). The surface condition ratings for all loading periods are given in Table III.

2.4   Moisture Conditions in the Pavement

The moisture condition at a particular loading period can be determined based on the ratings of rainfall, surface condition and temperature recorded in the period concerned, as one of three principal conditions: "high" (for the combination of "high" rainfall/"bad" surface), "average" (for the combination of "average" rainfall/ "bad" surface or "high" rainfall/"average" surface/"high" temperature) and "low" (for the combination of "low" rainfall/"bad to good" surface or "average" rainfall/"good" surface/"high" temperature). For simplicity, marginal moisture conditions produced by other combinations were individually checked and rated to the closest principal condition. Using this method, the moisture condition ratings for all loading periods were assigned as tabulated in Table III.

The moisture contents in the Benalla pavement were

recorded over two periods, including a period of 4 weeks between the 17th and 20th week and a period of 4 weeks between the 29th and 32nd week (Kadar 1986). The first period had "average" rainfall (prolonged rains with weekly rainfalls of about 8 to 10 mm), "bad" surface condition and "high" temperature; therefore, the moisture condition of this period can be regarded as the "average" moisture condition. The second period had "low" rainfall (occasional showers with weekly rainfalls less than 3 mm) and "bad" surface condition; therefore, the moisture condition of this period can be regarded as the "low" moisture condition.

Moisture content (%)

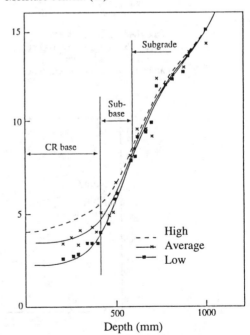

Figure 2 - Variation of moisture content in the pavement

137

The representative moisture contents recorded in the two periods are plotted against depth in Figure 2. The results showed that the moisture contents in the granular basecourse for the two conditions were different (about 1 per cent), but that there were no significant changes in moisture content in the lower sub-base and the subgrade as these materials were virtually impermeable and below zone of seasonal moisture variation.

Since there were no data for a "high" moisture condition, it was assumed that the moisture contents followed the trend shown as the dotted line in Figure 2.

Based on the loading sequences applied in the ALF trial and the three selected moisture conditions, the loading history can be divided into 8 combinations of load and environmental condition, as shown in Table III.

3. MATERIAL PROPERTIES

Vuong (1987) carried out a study on the mechanical response properties of the materials obtained from the Benalla pavement test section. The resilient moduli and Poisson's ratios and the plastic properties of the materials tested are described briefly below.

3.1 Base and Upper Sub-base Crushed Rock

The resilient modulus, E_r (MPa), for the base and sub-base crushed rock can be expressed in the form:

$$E_r = E_1 \ \{\sigma_m / \sigma_{ref}\}^n \quad (1)$$

where σ_m = $1/3 \ (\sigma_1 + \sigma_2 + \sigma_3)$
is the mean stress (in kPa)
σ_{ref} is the reference stress of 100 kPa
n is the material constant
E_1 can be dependent on stress ratios

E_1 was found to increase slightly with increasing stress ratios and to reach a maximum value at a stress ratio of about 7 to 10.

Figure 3 shows the variation of E_1 (maximum) and n with moisture content and sample densification under increasing number of loading cycles.

Poisson's ratios for the crushed rock were found to be high (i.e. in the range between 0.45 to 0.65) and were influenced by deviator stress near failure condition, particularly at high moisture content. However, values of Poisson's ratios must not be greater than 0.50 if elastic layer theory is to be applied to the data. Therefore the Poisson's ratio was taken to be 0.50 in this analysis.

Permanent deformation of the crushed rock can be expressed in the form of (from Kenis et al. 1981 and Monismith, 1976):

$$\varepsilon_p = \varepsilon_1 . (\mu / \alpha) . N^\alpha = \text{plastic strain} \quad (2)$$

where ε_1 is the resilient vertical strain
N is the number of loading cycles
μ and α are the material parameters

Figure 4 shows the variation of m and a values for different moisture contents.

3.2 Sub-base and Subgrade Materials

Resilient modulus, E_r (MPa), for the lower sub-base and the subgrade can be related to the deviator stress, q (kPa), and to the number of loading cycles. It can be expressed in the forms of:

$$E_r = E_{max} (1 - q/300)^P \quad (3\text{-}a)$$

and $E_r \geq E_{min} \quad (3\text{-}b)$

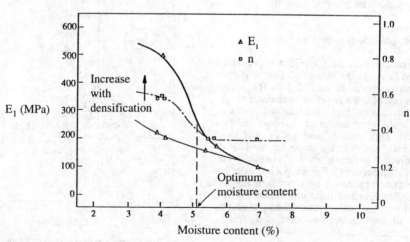

Figure 3 - Variation of the parameters E1 and n with moisture contents for the crushed rock base

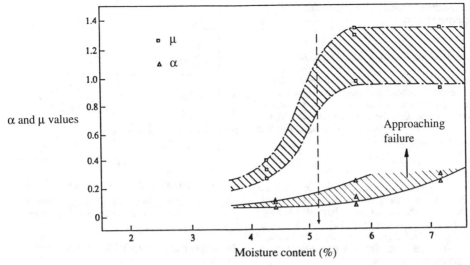

α and μ values

Figure 4 - Variation of the parameters m and a with moisture contents for the crushed rock base

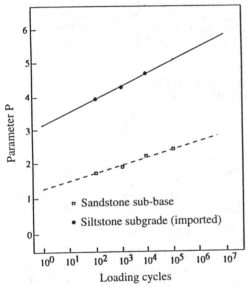

Parameter P

Figure 5 - Variation of the parameter p with loading cycles for the sandstone sub-base and siltstone subgrade

where E_{max} and E_{min} are material constants for a given moisture content and density and p is dependent on the number of loading cycles.

The values of E_{max} and E_{min} were determined as 450 MPa and 120 MPa for the lower sub-base at an average moisture content of 7%, and as 220 MPa and 60 MPa for the imported subgrade at a moisture content of 8.4%. Values of p for both materials were found to vary with loading cycles as shown in Figure 5.

The Poisson's ratio for the sub-base sandstone/siltstone was found from laboratory tests at field moisture condition to be constant at 0.40, whereas for the imported subgrade was found to vary between 0.25 and 0.35 and to depend on the confining stresses. However within the experimental errors, the Poisson's ratio of the imported subgrade can be assumed to be a constant of 0.30. Plastic deformation of the sub-base and the subgrade can also be expressed using Eqn (2). The average values for m and a were determined as 0.09 and 0.07 for the sub-base, and as 0.06 and 0.05 for the imported subgrade.

The natural siltstone subgrade was not tested. The field moisture contents within the natural subgrade were found to vary from 11 to 19%. Therefore, it may be expected that the modulus of this layer would be smaller than that of the imported subgrade. However, in this analysis it was assumed that the natural siltstone subgrade and the imported subgrade had the same moduli.

4. PERFORMANCE ANALYSIS USING NONCIRL

In order to carry out structural analyses of flexible pavements, the current AUSTROADS Pavement Design Guide (1992) uses a computer program CIRCLY (Wardle 1974), which is based on linear elastic theory. To make this program more versatile, it has been modified to NONCIRL (Vuong 1991) for analysing pavements having non-linear material characteristics. NONCIRL uses CIRCLY in the iterative process to calculate layer moduli of pavement layers and subgrades for given modulus-stress relationships. It also considers correction for tensile stresses in unbound layers and for overburden static pressures.

This study considered two performance models incorporated in NONCIRL, namely the AUSTROADS subgrade strain criterion and the rut depth model. They are briefly described below.

4.1 AUSTROADS Subgrade Strain Criterion

In the current AUSTROADS Pavement Design Guide (1992), the relevant distress mode for granular pavements with a thin bituminous surfacing is the permanent deformation developed in the subgrade. It also adopts a subgrade strain criterion to calculate the design life of the pavement, which is expressed as:

$$N = (8511/\varepsilon)^{7.14} \qquad (4)$$

where N is the number of cycles of the design loads and ε is the maximum subgrade elastic strain (in units of micro-strain) at the top of the subgrade calculated with CIRCLY.

However, this method does not predict the amount of permanent deformation anticipated after a given number of load applications.

4.2 Rut Depth Model

A discussion on the selection of some rut depth models and their formulations are given by Vuong (1992). Other previous works are described in Kenis et al. (1981),

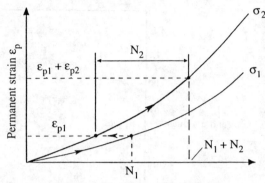

Figure 6 - Procedure for calculating cumulative loadings using the time hardening rule

Majidzadeh et al. (1976), Meyer (1976) and Monismith (1976).

The rut depth model adopts the deformation law observed in repeated load triaxial tests (as shown by Eqn (2)) to calculate surface permanent deformation. For simplification, the models also assume that surface deformation is the sum of the vertical plastic strain, by adopting the one-dimensional strain integration.

TABLE IV: MATERIAL PARAMETERS FOR RESILIENT MODULUS AND PERMANENT DEFORMATION USED IN THE ANALYSES

Load Stage	Material	Moisture Content (%)	E_1 or E_{max} (MPa)	n	p	E_{min} (MPa)	α	μ
1	Crushed rock	3.7	250	0.56	-	-	0.10	0.35
	Sandstone	7.0	450	-	1.20	120	0.07	0.09
	Siltstone	8.5	220	-	3.00	60	0.05	0.06
2	Crushed rock	3.7	300	0.56	-	-	0.10	0.35
	Sandstone	7.0	450	-	2.00	120	0.07	0.09
	Siltstone	8.5	220	-	4.00	60	0.05	0.06
3	Crushed rock	3.7	350	0.56	-	-	0.10	0.35
	Sandstone	7.0	450	-	2.20	120	0.07	0.09
	Siltstone	8.5	220	-	4.50	60	0.05	0.06
4	Crushed rock	3.7	350	0.56	-	-	0.10	0.35
	Sandstone	7.0	450	-	2.20	120	0.07	0.09
	Siltstone	8.5	220	-	4.50	60	0.05	0.06
5	Crushed rock	4.5	250	0.50	-	-	0.13	0.50
	Sandstone	7.0	450	-	2.50	120	0.07	0.09
	Siltstone	8.5	220	-	5.40	60	0.05	0.06
6	Crushed rock	2.5	400	0.56	-	-	0.08	0.20
	Sandstone	7.0	450	-	2.55	120	0.07	0.09
	Siltstone	8.5	220	-	5.50	60	0.05	0.0
7	Crushed rock	4.5	300	0.50	-	-	0.13	0.50
	Sandstone	7.0	450	-	2.60	120	0.07	0.09
	Siltstone	8.5	250	-	5.50	60	0.05	0.06
8	Crushed rock	2.5	450	0.56	-	-	0.07	0.20
	Sandstone	7.0	450	-	2.65	120	0.07	0.09
	Siltstone	8.5	250	-	5.60	60	0.05	0.06

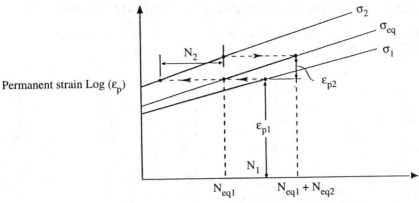

Permanent strain Log (ε_p)

Number of loading cycles, Log (N)

Figure 7 - Procedure for calculating equivalent wheel passes

Other assumptions used in the calculation of cumulative rut depth over time are:

(a) a time hardening rule in the calculation of cumulative deformation (see Figure 6), and

(b) a rule for equivalent wheel passes in the simulation of transverse loading distributions (see Figure 7).

These are discussed in detail by Vuong (1992).

Because the Benalla pavement test section was very short (about 10 m) and it was built under strict quality control, variability of the material properties with space was not considered in the analyses. However, changes of material properties with increasing loading cycles, as derived from the material data obtained from the repeated load triaxial tests (see Figures 4 and 5), were considered.

4.3 Input material parameters used in the analysis

Appropriate material constants of the models for elastic and plastic properties selected for each loading stage are tabulated in Table IV.

TABLE V: PAVEMENT CONFIGURATION ADOPTED IN THE ANALYSIS

Sub-layer No	Material	Thickness (mm)
1	Crushed rock	110
2	Crushed rock	110
3	Crushed rock	200
4	Sandstone 1	70
5	Imported Subgrade	300
6	Subgrade	300
7	Subgrade	4000
8	Subgrade	∞

4.4 Pavement configuration adopted in the analysis

In order to simulate the variation of moduli with depth, the top crushed rock layer of 420 mm was partitioned into 3 sub-layers and the subgrade was also partitioned into 3 sub-layers. Thicknesses of the sub-layers are given in Table V; whereas the final values of layer moduli for each loading stage are given in Table VI.

TABLE VI: LAYER MODULI OBTAINED FROM THE ANALYSIS

Layer	Thickness (mm)	Layer Modulus (MPa) Loading Stage							
		1	2	3	4	5	6	7	8
1	110	444	637	801	627	522	923	630	1044
2	110	232	309	388	298	286	429	335	474
3	200	167	215	251	212	196	268	223	291
4	170	350	257	214	298	184	200	186	201
5	300	166	133	111	150	94	101	95	102
6	300	188	165	148	177	133	138	134	138
7	4000	216	212	208	214	205	205	205	205
8	∞	220	220	220	220	220	220	220	220

TABLE VII: DEFLECTION BOWLS MEASURED FROM THE ALF LOADING TEST

Distance from Loading Centre (mm)	Surface Deflection (micrometer) for the Principal Load of 80 kN			
	Loading Stage			
	3	5	6	8
0	553	664	561	532
125	483	592	498	480
200	452	535	451	428
300	388	445	366	350
450	305	325	269	258
600	241	234	189	186
750	175	181	141	137
900	132	132	95	100
1000	116	103	80	85
1200	73	66	50	58

TABLE VIII: COMPARISON BETWEEN CALCULATED AND MEASURED
DEFLECTION BOWLS

Distance from Loading Centre (mm)	Percentage Difference in Surface Deflection (%)			
	Loading Stage			
	3	5	6	8
0	-4.7	-5.1	-6.0	-4.9
125	-3.9	-8.1	-6.0	-5.6
200	-11.1	-13.5	-9.0	-6.7
300	-15.5	-16.9	-7.4	-4.9
450	-19.0	-16.3	-3.7	-0.4
600	-19.9	-11.1	+7.9	+9.1
750	-10.9	-8.8	+16.3	+19.7
900	-3.0	+1.5	+42.1	+35.0
1000	-1.7	+14.6	+48.7	+41.1
1200	+26.0	+42.4	+88.2	+65.5

5. COMPARISON OF SURFACE DEFLECTIONS

The deflections measured with the Benkelman beam in the four principal loading stages 3, 5, 6 and 8 were reported in Kadar (1986) and are summarised in Table VII. The deflections measured in the shakedown periods (loading stages 1, 2 and 4) were not reported.

The differences between the calculated and measured deflections for these four principal loading stages are given in Table VIII.

The results indicated that the differences between the calculated and measured (at the loading centre) were about 4.7 to 6.0 per cent. The differences in deflection at offsets from the loading centre were much higher, possibly because their magnitudes were much lower. It should be noted that the solutions can also be improved by increasing the stiffness of the bottom subgrade stiffness (i.e. sub-layer 8), which strongly influences the deflections at large offsets (see Vuong et al 1988). However, this was not considered in this report as its influence on the pavement life (predicted based on surface permanent deformation or on the critical strain at the top of the subgrade) was insignificant.

6. PAVEMENT LIFE PREDICTED BY THE AUSTROADS SUBGRADE STRAIN CRITERION

Vertical strains at depths of 420 mm, 590 mm and 890 mm calculated for each loading stage are tabulated in Table IX. Considering the structure and materials of the Benalla pavement, the interface between the lower sub-base sandstone and siltstone subgrade (at a depth of 590 mm) was regarded as critical, so it was the vertical strains calculated at this depth that were used for estimating pavement life using the AUSTROADS subgrade strain criterion.

For the four principal loading stages, the differences between the calculated critical subgrade strains (at a depth of 590 mm) were also very high (up to 20%), and the corresponding differences between the estimated pavement lifes were more than 500% (because of 7th power). This indicates that the moisture condition strongly influences the pavement life. Under the worst condition (highest moisture condition), the estimated pavement life for the principal load of 80 kN was $4.74*10^7$, indicating a very long pavement life.

It should be noted that the AUSTROADS Design Guide

TABLE IX: CRITICAL SUBGRADE ELASTIC STRAIN AND THE PAVEMENT LIFE PREDICTED
BY AUSTROADS STRAIN CRITERION

Loading Stage	Load (kN)	Vertical strain (microstrain)			Estimated Life for 80 kN (x10^{7}cycles)
		At Depth of			
		420 mm	590 mm	890 mm	
1	40	202	250	112	542*
2	60	364	408	174	83*
3	80	513	575	238	22.7
4	40	219	252	111	512*
5	80	630	716	280	4.7
6	80	518	592	243	18.4
7	80	595	675	267	7.2
8	80	498	566	236	25.4

(*) estimated using the fourth power law

suggests the fourth power law for calculating the equivalent number of 80 kN axles which causes the same damage as the 40 or 60 kN axles, i.e.

$$N(i) = N(j)/[P(i)/P(j)]^4 \qquad (5)$$

where $N(i)$ and $N(j)$ are the equivalent numbers of loading applications for the two axle loads $P(i)$ and $P(j)$, respectively. However, the equivalent pavement lives for the 80 kN axle loads estimated from lives for lower axle loads (using the fourth power law) were much higher than those predicted for all the principal loading stages, indicating that the fourth power law may not be appropriate in this case.

7. COMPARISON OF RUT DEPTHS

The possible range of rut depths developed throughout the ALF loading test derived by Kadar (1986) is shown in Figure 8. The maximum deformation (at a distance of 190 mm from the centre of the load group) calculated with the rut depth model is also plotted in Figure 8 for comparison.

Although the variations in the measured rut depths were very large, it can be seen in Figure 8 that the rut depth model produced acceptable predictions of rut depth throughout the loading history.

Examination of the variation in both the measured and calculated rut depths with increasing loading cycles (see Figure 8) also showed that the rut depths developed in the initial loading stage (shakedown period) could contribute up to 50 per cent of the total final rut depth; and that, in the principal loading stages, the rut depths developed in the periods of dry and warm weather were insignificant compared to those developed in the earlier periods of high rainfall.

8. SUMMARY AND CONCLUSIONS

The program NONCIRL, which incorporates the AUSTROADS subgrade strain criterion and the rut depth model, has been used to predict the performance of the

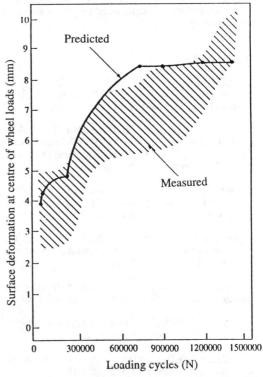

Figure 8 - Predicted versus measured deformations at various loading stages

Benalla granular pavement tested with ALF. The analysis has considered all the effects of the material non-linearity, loading configuration and distribution, environmental condition, pavement surface condition and loading history. The material data were obtained from repeated load triaxial tests for the loading and environmental conditions monitored throughout the ALF trial.

It was found that the calculated deflections at the loading

centre were reasonably close to the deflections measured with the Benkelman beam.

The calculated subgrade critical strains were found to be strongly influenced by the magnitude of the load and environmental conditions. To ignore these effects would introduce large errors into the estimation of pavement life using this parameter as a performance indicator. In this study, under the highest moisture condition, the AUSTROADS subgrade strain criterion still overestimated the pavement life, possibly because surface permanent deformation was not critical in the subgrade layers.

The rut depth model can take into account permanent deformation in all pavement layers and in the subgrade. In this study, validation of the rut depth models may be limited because the measured rut depths were comparatively small. However, comparisons of the output with the possible range of the measured rut depths indicated that the rut depth model could produce acceptable predictions and the results obtained may be regarded as most encouraging.

REFERENCES

Kadar, P. (1987). Test results and analysis of the ALF trial No. 2 at Benalla - Victoria, 1985/86. Australian Road Research Board Internal Report, AIR 430-1.

Kenis, W.J., Sherwood, J.A. and McMahon, T.F. (1981). Verification and application of the VESYS structural sub-system. Proc. 5th Int. Conf. on Structural Design of Asphalt Pavements.

Majidzadeh, K., Khedr, S. and Guirguis, H. (1976). Laboratory verification of a mechanistic subgrade rutting model. Trans. Res. Rec. 616, pp. 34-37.

Meyer, F., Haas, R. and Dharmawardene, M.W.W. (1976). Procedure for predicting rut depths in flexible pavements. Trans. Res. Rec. 616, pp. 38-40.

Monismith, C.L. (1976). Rutting prediction in asphalt concrete pavements. Trans. Res. Rec. 616, pp. 2-8.

AUSTROADS (1992). Pavement Design - A Guide to Structural Design of Road Pavements.

Saraf, C.L., Smith, W.S. and Finn, F.N. (1976). Rut depth prediction. Trans. Res. Rec. 616, pp. 9-14.

Sharp, K.G (1991). Australian experience in full-scale pavement testing using the Accelerated Loading Facility. Australian Road Research, 21(3), pp. 23-32.

Thrower, E.N. (1977). Methods of predicting deformations in road pavements. Proc. 4th Int. Conf. on Structural Design of Asphalt Pavements.

Van De Loo, P.J. (1976). Practical approach to the prediction of rutting in Asphalt Pavements: The Shell method. Trans. Res. Rec. 616, pp. 15-21.

Vuong, B. (1985-a). Non-linear finite element analysis of road pavements. Australian Road Research Board Internal Report, AIR 403-5.

Vuong, B. (1985-b). Permanent deformation and resilient behaviour of a Victorian crushed rock using the repeated load triaxial test. Australian Road Research Board Internal Report, AIR 403-6.

Vuong, B. (1986-a). Mechanical response properties of road materials obtained from a pavement test section at Rooty Hill, New South Wales. Australian Road Research Board Internal Report, AIR 403-7.

Vuong, B. (1986-b). Mechanical response properties of road materials obtained from the ALF pavement test section at Somersby, New South Wales. Australian Road Research Board Internal Report, AIR 403-8.

Vuong, B. (1987). Mechanical response properties of road materials obtained from the ALF pavement test section at Benalla, Victoria. Australian Road Research Board Internal Report, AIR 403-10.

Vuong, B. (1991). NONCIRL - A program for structural analysis of pavements having non-linear material characteristics - Technical Note. Australian Road Research Board Working Document, WD RI91-014.

Vuong, B. (1992). A discussion on the selection of rut depth models. Australian Road Research Board Working Document, WD RI92-000.

Vuong, B, Potter, D. and Kadar, P. (1988). Analyses of a heavy duty granular pavement using Finite Element method and linear elastic back-calculation models. Procs. 14th ARRB Coference, Part 8, pp. 284-297.

WARDLE, L.J., (1977). Program Circly Users Manual. CSIRO Division of Applied Geomechanics.

Prediction of behavior of reinforced embankments on Muar clay deposit

Dennes T. Bergado & A. S. Balasubramaniam
Asian Institute of Technology, Bangkok, Thailand

Jin Chun Chai
Kiso-Jiban Consultants Co. Ltd, Chiyoda-ku, Tokyo, Japan

ABSTRACT: The behavior of two Tensar polymer grid reinforced stage constructed embankments has been predicted by finite element method. In the finite element modelling, more accurate simulation of the actual construction process considering the soft ground permeability variations during the consolidation process was made. Predicted values have been compared with field data in terms of excess pore pressures, settlements, and lateral displacements. Fairly good agreement has been obtained between predicted values and field data up to the end of construction condition. Since the post-construction field data are not available, the predicted values after construction stage are presented alone with discussions and comments. Finally, the reinforcement tension force and soil/reinforcement shear stress distribution pattern from the finite element analysis are also presented and analyzed.

INTRODUCTION

Ground reinforcements have been widely used for embankment construction on soft soils. The reinforcements are usually placed at the base of the embankment to reduce the lateral spreading force from the embankment and increase the foundation bearing capacity. Since the finite element method has the ability to accommodate nonhomogeneous materials, nonlinear stress/strain behavior, and soil/reinforcement interaction properties, the behavior of base reinforced embankments on soft ground has been analyzed by several investigators using finite element method (e.g. Hird and Pyrah, 1990). Some important factors have been found such as the influence of the reinforcement stiffness on the soft foundation soil lateral displacements. However, the accuracy of finite element results depends mainly on both constitutive models and model parameters used. Although several case histories of comparing the finite element results with field data about reinforced embankment on soft ground have been recorded in literature, and the agreements are reasonably good, a more accurate simulation of the construction process has not been emphasized. Furthermore, the use of finite element method to predict the performance of stage constructed reinforced embankment on soft ground during the construction and consolidation processes, considering the variations of permeability values in the foundation subsoil, rarely appeared in literature.

The numerical procedure must simulate the actual construction process as closely as possible. For stage constructed embankments, the interval between different construction stages is long. In this case, considerable soft ground settlement will occur during construction period. However, the finite element mesh for the embankment system is usually drawn up at the beginning of the analysis. During the analysis, the incremental load is applied by assigning the gravity load of the embankment elements layer by layer. Therefore, it is necessary to consider the changes in the coordinates of the embankment elements above the current construction level. Ignoring the coordinate change of the elements above the current construction level will result in significant error because the applied fill thickness will be more than the actual value resulting in the foundation settlements during the construction period (Bergado et al, 1992).

For predicting the behavior of stage constructed embankment on soft ground, another key point is to simulate the consolidation process. The consolidation rate is mainly influenced by the foundation soil permeability. The behavior of embankment on soft ground has been systematically investigated by Tavenas et al (1980) through field observation. It has been found that for a soil element under embankment center line, before yield, it behaves close to drained condition, and after yield, close to undrained condition. This phenomenon indicates that the permeability of soft ground varied during the loading and consolidation process. However, most finite element model do not consider the significant change of the soft ground subsoil permeability before and after yield (Tavenas et al, 1980), and, therefore, cannot simulate the whole consolidation process well. In order to simulate the whole consolidation process, it is important to consider the foundation soil permeability variation during the construction and consolidation process.

In this paper, the finite element modelling is briefly described first. Then, two geogrid reinforced embankments, one with berm and another without berm, have been analyzed by the finite element method. The available field data are only up to the end of construction. The prediction has been made to 3 years after construction condition. Good agreement has been obtained between field data and predicted values in terms of excess pore pressures, settlements, and foundation lateral displacements up to the end of construction condition. The predicted reinforcement tension force and interface shear stress mobilization process and distribution pattern are also presented and discussed. The results of the analysis provide useful information regarding the use of finite element method to predict the behavior of stage constructed reinforced embankment on soft ground.

FINITE ELEMENT MODELLING

The reinforced embankment on soft ground system has been modelled by finite element method under plane strain condition. All the elements are formulated as isoparametric elements. Discrete material approach is used to model the reinforced embankment on soft ground system because the properties and responses of the soil/reinforcement interaction can be directly quantified. The bar elements and zero thickness interface elements are used to represent the reinforcement and the soil/reinforcement interface, respectively.

The behavior of the soft foundation soil is controlled by modified Cam clay model (Roscoe and Burland, 1968). The backfill soil is modelled by hyperbolic constitutive law (Duncan et al, 1980). The consolidation process of soft ground is simulated by coupled consolidation theory (Biot, 1941). Interface elements, above and below the reinforcement, work as pair elements. The interface properties are selected according to their relative shear displacement pattern (direct shear or pullout). Two different models are used to simulate the behavior of interface elements. The hyperbolic shear stress/shear displacement model (Clough and Duncan, 1971) is used to represent direct shear soil/reinforcement interaction mode. For pullout of grid reinforcement from soil, the resistance consists of skin friction from the

longitudinal members and bearing resistance from the transverse members. The skin friction is modelled by linear elastic-perfect plastic model and the pullout bearing resistance is simulated by a hyperbolic bearing resistance model which is only valid for grid reinforcements (Chai, 1992). It is assumed that the pullout resistance is uniformly distributed over the entire interface areas. For both direct shear and pullout interaction modes, when the normal stress at the interface is in tension, a very small normal and shear stiffness is assigned to allow the opening and slippage at the interface.

The technique for correcting the node coordinates considering the soft ground permeability variation during incremental analysis has been discussed by Bergado et al (1992). For the sake of completeness, a brief description is presented here. The node coordinates are updated to consider the large deformation phenomenon during incremental analysis. In order to simulate the actual construction procedure and ensure that the applied fill thickness is the same as the field value, the coordinates of the embankment elements above the current construction level are corrected based on the following assumptions: (a) the original vertical lines are kept at vertical direction, and the horizontal lines remain straight, (b) the incremental displacements of the nodes above current construction top surface are linearly interpolated from the incremental displacements of the two end nodes (left and right) of current construction top surface according to their x-coordinates (horizontal direction).

Two options are used to consider the foundation soil permeability variation. One is that the permeability is varied with the void ratio by Taylor's equation (Taylor, 1948) as follows:

$$k = k_o 10^{[\frac{-(e_o-e)}{c_k}]} \qquad (1)$$

where e_o is the initial void ratio; e is the void ratio at the condition considered; k is the permeability; k_o is the initial permeability; and c_k is constant, which is equal to 0.5 e_o (Tavenas et al, 1983). Another option is that the foundation soil permeability is drastically changed before and after soil yielding which is controlled by modified Cam clay soil model. In this case, the permeability is also varied with the void ratio before and after yielding by using Eq. 1.

MALAYSIAN REINFORCED TRIAL EMBANKMENTS

Two Tensar polymer grids reinforced embankments were constructed on soft Muar clay, about 50 km due east of Malacca on the Southeast coast of west Malaysia, with vertical drain (Desol) in the foundation. One of the embankments with berm (scheme 6/8) is referred as embankment A, and another one without berm (scheme 3/4) is called embankment B in later discussions. These two embankments were instrumented with piezometers, settlement gages, and inclinometers to monitor the field behavior of the embankments and the soft ground foundation, and reliable field data have been obtained (MHA, 1989).

The soil profile at the test site consisted of a weathered crust at the top 2.0 m which is underlain by about 5 m of very soft silty clay. Below this layer lies a 10 m thick layer of soft clay which in turn is underlain by 0.6 m of peat with high water contents. Then, a thick deposit of medium dense to dense clayey silty sand is found below the peat layer. Figure 1 shows the index properties, vane shear strength, and cone resistance of the foundation soil (Brand and Premchitt, 1989). The top 2 m weathered clay layer has an average over consolidation ratio (OCR) of 4 and the soil layers below are slightly overconsolidated with an average OCR of 1.1. Embankment A was constructed with a base width of 88 m, length 50 m, and initially to a fill thickness of 3.9 m. Then, a 15 m berm was left on both sides and the embankment was constructed to a final fill thickness of 8.5 m. The construction history is shown in Fig. 2. Two layers of Tensar SR110

geogrids were laid at the leveled ground surface in a 0.5 m thick sand blanket with 0.15 m vertical spacing between them. The vertical drains (Desol) were installed in a square pattern with 2.0 m spacing to 20 m depth (MHA, 1989).

The embankment B was constructed with a base width of 44 m, length 50 m and to a fill thickness of 6.07 m. Then, 0.8 m of the top layer was removed due to the occurrence of tension cracks. The construction history is also shown in Fig. 2. One layer of Tensar SR80 geogrids was laid at the ground surface in a 0.5 m sand blanket and vertical drains (Desol) were installed in the foundation soil similar to embankment A (MHA, 1989).

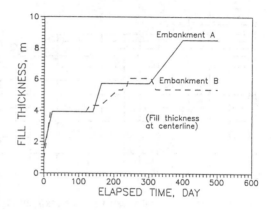

Fig. 2   Construction histories

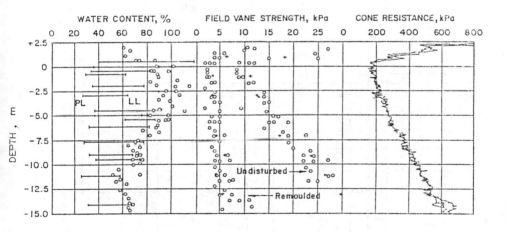

Fig. 1   Index properties and strength of the foundation soil at test site (after Brand and Premchitt, 1989).

147

FINITE ELEMENT ANALYSIS AND INPUT PARAMETERS

The finite element meshes together with the boundary conditions for analyzing the Malaysian reinforced trial embankments as shown in Figs. 3 and 4 for embankments A and B, respectively. For drawing up the finite element mesh, the horizontal boundaries are selected far enough (about 6 times of the fill thickness from the embankment toes) to ensure that boundary effect can be ignored. The bar elements representing the reinforcements are indicated by coarse solid line. However, their interface elements are not shown in the mesh for the sake of clarity. For embankment A, two layers of SR110 geogrids are simulated by a single layer of bar elements, because the distance between the two layers is very small, only 0.15 m.

The modified Cam clay model parameters for

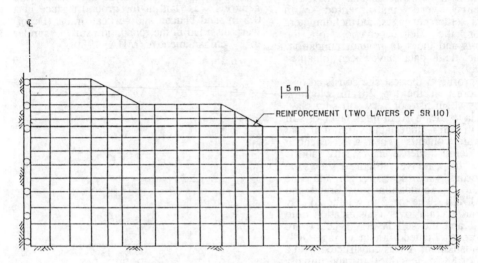

Fig. 3   Finite element mesh for embankment A.

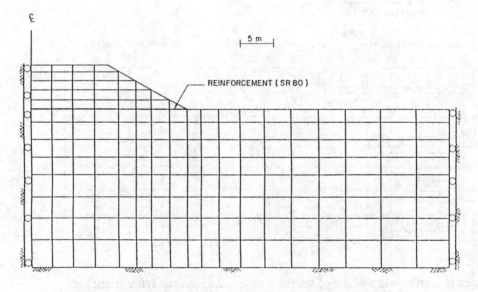

Fig. 4   Finite element mesh for embankment B.

foundation soil are listed in Table 1. The parameters, M, λ, and κ are obtained directly from test results (AIT, 1988; 1989). The Poisson's ratio for soft and stiff clay layers is 0.25, and for weathered clay layer and dense clayey silty sand layer is 0.2 (Balasubramaniam et al, 1989; Magnan, 1989).

Regarding the foundation permeability, existing test and analysis results show that the field permeability is 2 to 3 times the laboratory test value, and the horizontal permeability is 1.5 to 2 times of the vertical value (Poulos et al, 1989; Magnan, 1989). Concerning the effect of the vertical drain, comparing the performance of the trial embankments with and without vertical drain (Desol), it was found that the effect of the Desol drain was not significant and the replacement ratio is very small, only 0.01 % . Therefore, it is difficult to discretely model the vertical drains by finite element method. Based on the test data and the above considerations, two basic sets of the permeability values are selected for finite element analysis as shown in Table 1. The first set of parameters does not consider the effect of the vertical drain. while the second set of parameters considers the vertical drain effect as vertical seams which increased the vertical permeability and it is assumed that in the zone with vertical drain, the vertical permeability is twice as large as the value in the zone without vertical drains.

The backfill material is cohesive-frictional soil consisting of decomposed granite with consistency of sandy clay. The hyperbolic soil model parameters for backfill material and sand blanket are given in Table 2. These are taken from test results or selected from the parameters collected by Duncan, et al. (1980).

Tensar grid SR110 are spaced at 150 mm between transverse members and 22.7 mm between longitudinal members. The average cross section of transverse member is 5.7 mm in thickness and 16.0 mm in width, and the average cross section of longitudinal member is 2.1 mm in thickness and 10 mm in width. The Tensar SR80 are spaced at 160 mm between transverse members and 22.5 mm between longitudinal members. The average cross-section of transverse members has a thickness of 3.8 mm and width of 16 mm, and the average cross section of longitudinal member has a thickness of 1.4 mm and width of 10 mm. The stiffness of the polymer grids is influenced by the temperature and strain rate as well as the stress level (McGown et al, 1984). However, in the field, the temperature is varied and the strain rate is difficult to assess. Referring to the values used by Hird and Pyrah (1990), the constant stiffness of 450 kN/m and 650 kN/m were used in the analyses for Tensar SR80 and SR110, respectively.

The adopted interface hyperbolic direct shear model parameters were: interface frictional

Table 1 Soil parameters of Muar clay in Malaysia.

Parameter	Symbol		Soil Layer				
			1	2	3	4	5
	Depth, (m)		0-2	2-7	7-12	12-18	18-22
Kappa	κ		0.06	0.10	0.06	0.04	0.03
Lambda	λ		0.35	0.61	0.28	0.22	0.10
Slope	M		1.2	1.07	1.07	1.07	1.2
Gamma (P'=1 kPa)	Γ		4.16	5.5	3.74	3.45	2.16
Poisson's Ratio	ν		0.20	0.25	0.25	0.25	0.2
Unit Weight, (kN/m³)	γ		15.5	14.5	15	15.5	17.0
Horizontal Permeability	SET 1	k_h	2.78	1.40	1.04	0.70	14000
(m/sec), (10^{-8})	SET 2	k_h	2.78	1.40	1.04	0.70	14000
Vertical Permeability	SET 1	k_v	1.39	0.70	0.52	0.35	7000
(m/sec), (10^{-8}) Drain/No Drain	SET 2	k_v	2.78/ 1.39	1.40/ 0.70	1.04/ 0.52	0.70/ 0.35	14000/ 7000

Drain/No Drain: In the zone installed with vertical drain (Desol), the vertical permeability is equal to the horizontal value and two times of that the zone without vertical drain.

Table 2 Hyperbolic parameters used for backfill material and sand blanket of Malaysian embankment.

Parameter	Symbol	Backfill	Sand Blanket
Cohesion, (kPa)	C	19	0
Friction Angle, (°)	ϕ	26	38
Modulus Number	k	320	460
Modulus Exponent	n	0.29	0.50
Failure Ratio	R_f	0.85	0.85
Bulk Modulus Number	k_b	270	392
Bulk Modulus Exponent	m	0.29	0.50
Unit Weight, (kN/m³)	γ	20.5	20.5

Table 3 Summary of analyses for Malaysian test reinforced embankments.

Analysis No.	Analysis Type	Foundation Permeability	Reinforcement	Remarks
H1	C	Set 1 (Table 1)	SR110	Embankment A
H2	C	Set 2 (Table 1)	SR110	
H3	C	Variation I	SR110	
H4	C	Variation II	SR110	
H5	C	Variation III	No	
L1	C	Set 1 (Table 1)	SR80	Embankment B
L2	C	Variation I	SR80	
L3	C	Variation II	SR80	

C = consolidation analysis
Variation I: = initial values are Set 2 in Table 1.
Variation II: = initial values for after yield soils are Set 1 in Table 1.
Variation III: = initial values for after yield soils are Set 2 in Table 1.

angle, ϕ, of 35 degrees, cohesion, C, zero, shear stiffness number, k_j, 4,800, shear stiffness exponent, n_j, 0.51, and failure ratio, R_{fj}, of 0.86. The skin friction angle between Tensar grid plane surface and the sand was 10 degrees and adhesion was zero. The maximum relative displacement for mobilizing the peak skin friction was 2 mm.

Considering the uncertainties of the foundation soil permeability, a parametric study was carried out. The influence of the reinforcement on the performance of the embankment has been also investigated. Table 3 is the summary of the analyses. Permeability variation (1) means that the permeability was varied with the void ratio with initial value of high permeability (set 2 in Table 1). While, for permeability variation (II) and (III), before yield foundation soil permeability is five times the corresponding after yield values which were low and high permeability values in Table 1, respectively. All the analyses conducted are consolidation analyses.

PREDICTED VALUES AND COMPARING WITH THE FIELD DATA

Although the prediction is class C prediction (Lambe, 1973), the important thing is the analyses were conducted systematically and the input parameters are determined based on the test results. Comparing the predicted value with actual results, it was shown that using constant permeability values cannot simulate the field behavior well. Therefore, the results from varied permeability analysis (I) served as the main predicted values and the results from the varied permeability analyses (II) and (III) are included for discussions. Comparisons between predicted and field data are made in terms of excess pore pressures, settlements, and lateral displacements. The finite element results of reinforcement tension forces and interface shear stresses are also presented. The results of embankment A and embankment B are presented in a parallel manner.

Excess pore pressures

The typical variation of the excess pore pressure with elapsed time for a piezometer point 4.5 m below ground surface and on the embankment centerline of embankment A is shown in Fig. 5. Figure 6 shows the comparison of predicted and measured variation of excess pore pressures along the depth at different fill thickness on the embankment centerline. From both Figs. 5 and 6, it can be seen that the agreement between predicted and measured data up to end of construction is good. However, at early stage of construction, the predicted value is still higher than the measured one and the results, considering the drastic changes of the permeability before and after soil yield (variation (II) and (III), gave better prediction at early stage of contruction. This confirms the necessity of modelling the permeability change before and after the soil yield. It should be noted that, when using the option of drastic permeability changes before and after the soil yield, in the zones away from the loading area

the soil is always with higher permeability and excess pore pressure dissipation is quicker. As shown in Fig. 6, the option of permeability variation II yield lower excess pore pressure below the depth of 18 m. The predicted excess pore pressure at the piezometer point 4.5 m below ground surface and on embankment center line (embankment A) 3 years after construction (end of September, 1992) is 30 kPa. The actual value might be higher than this because of the possibility of stronger permeability variation (see Fig. 5).

Figures 7 and 8 show the excess pore pressure variation of embankment B at a point 4.5 m below ground surface and on the embankment center line and excess pore pressure profiles along the depth for different fill thickness, respectively. The general tendency is the same as for embankment A. However, the analysis that considers the drastic permeability variation before and after the soil yields seemed to give better prediction through the whole construction process. In Fig. 7, the predicted results show a sharp reduction due to

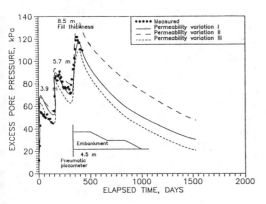

Fig. 5   Typical excess pore pressure versus elapsed time curve for embankment A

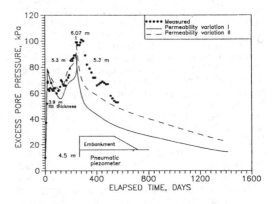

Fig. 7   Typical excess pore pressure versus elapsed time curve for embankment B

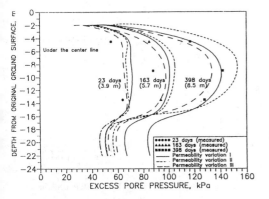

Fig. 6   Excess pore pressure profile on the centerline of embankment A

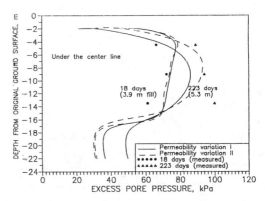

Fig. 8   Excess pore pressure profile on the centerline of embankment B

the unloading of the embankment fill from 6.07 m thickness to 5.3 m thickness. The field data did not show this tendency. The predicted excess pore pressure 3 years after construction (end of September, 1992) at the piezometer point 4.5 m below ground surface and on embankment centerline (embankment B) is 20 kPa which is lower than embankment A due to the lower embankment fill thickness and shorter horizontal drainage path (see Fig. 8).

Settlements

The comparison of surface settlement profiles for embankment A is shown in Fig. 9. Up to the end of construction, the agreement between the predicted and measured values is good. An interesting factor is at the early stage of the embankment construction (3.9 m fill thickness). At the zone near the embankment toe, both predicted and measured values show larger settlements than the center point of the embankment because this zone has high shear stress level. It can be also observed that using the options of drastically changing the permeability before and after the soil yield results in larger settlement at early stage of construction, but lower heave near the embankment toe due to the higher permeability of the soil outside of the embankment base. Although there is no measured data about heave, the information from a rapidly built to failure embankment on same site shows that using constant permeability the values of heave were overpredicted (Brand and Premchitt, 1989). Therefore, this trend might be more closer to the actual behavior. Figure 10 shows the comparison of settlement-time curves for the points on the embankment center line for embankment A. It can be seen that the predicted values also agreed well with the field data. It needs to be mentioned that at the beginning of construction, the predicted values were lower

than measured data which coincided with the higher predicted excess pore pressure. Using the permeability variation option (III) with high initial permeability values resulted in better surface settlements at early stage of construction. At later stages, the settlements were overpredicted (short dashed line in Fig. 10). The predicted surface settlement under embankment A at 3 years after construction is 2.9 m.

Figures 11 and 12 show the surface settlement profiles and typical settlement-time plots for embankment B, respectively. The tendency is the same as for embankment A and the agreement between predicted and measured values seems better than that of embankment A during construction period. Figure 12 also shows that during the unloading period, both predicted and measured data indicate that the settlements increased due to the foundation consolidation effect which was stronger than the rebound. The predicted surface settlement at 3 years after construction for embankment B is 2.0 m.

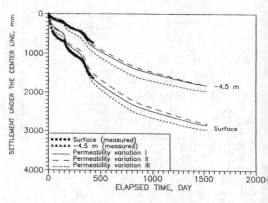

Fig. 10   Typical settlement versus elapsed time plots for embankment A

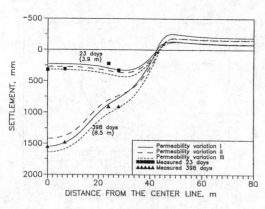

Fig. 9   Surface settlement profile for embankment A

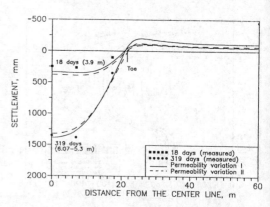

Fig. 11   Surface settlement profile for embankment B

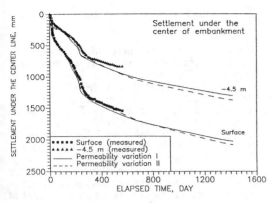

Fig. 12   Typical settlement versus elapsed time plots for embankment B

Lateral displacements

Figure 13 compared the predicted and measured lateral displacement profiles at inclinometer position of embankment A. It shows that the agreement between the predicted and the measured data is good. However, at the early stage of the construction, the predicted values considerably overestimated the lateral displacements, and at the end of construction, the predicted values slightly underestimated the lateral displacements. Using the permeability variation (II), only slightly better predictions were obtained, i.e. lower lateral displacement at early stage and larger values at later stage. Using the permeability variation option III with high initial values resulted in lower lateral displacement for all construction stages. This lateral displacement variation tendency coincided with the values of settlement, and it indicated that the consolidation process is still not simulated well by the finite element analysis.

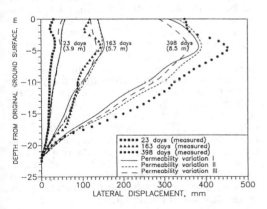

Fig. 13   Lateral displacement profile at inclinometer location of embankment A

Figure 14 shows the predicted and measured maximum lateral displacements at the location of the inclinometer casing of embankment A. It indicates that up to the end of construction, the agreement between predicted and measured data is fair, and most discrepancies mainly occurred during the consolidation period between the different construction stages.   One possible reason is the creep effect at high stress level zone, which the modified Cam clay model does not consider. The predicted maximum lateral displacement at inclinometer location is 660 mm at 3 years after the construction (end of September, 1992) as shown in Fig. 14. However, it seems certainly underpredicted.

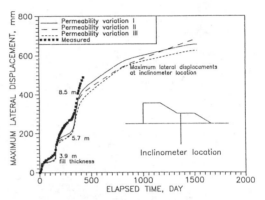

Fig. 14   Maximum lateral displacement versus time plot at inclinometer location of embankment A

For embankment B, the lateral displacement profiles and maximum lateral displacement plots at inclinometer location are indicated in Figures 15 and 16, respectively. Again the tendency is the same as for embankment A, but the discrepancy is larger than that of embankment A. During the construction, the stress level at the inclinometer location of embankment B is higher than that of embankment A maybe due to creep effects. It is obvious that the finite element analysis underpredicted the lateral displacements for embankment B (Fig. 16).

Both embankments A and B were first constructed to a fill thickness of 3.9 m. For embankment A, the inclinometer location was inside the embankment body. Therefore, at inclinometer location, both finite element results and measured lateral displacements were much smaller than those of embankment B. However, comparing the maximum lateral displacement which occurred under the toe by finite element analysis at the end of the first stage construction (3.9 m fill thickness), they are nearly the same, which is about 200 mm. For the late construction stages, since the two embankments have different reinforcements and different geometries, direct comparison is difficult to make. However, finite element results show that for fill thickness increase from 3.9 m to about 6 m, for embankment A with stronger reinforcement (two layers of Tensar SR110) and berm, the maximum lateral displacement increased from 200 mm to 325 mm. For embankment B with weaker reinforcement (one layer of Tensar SR80) and without berm, the maximum lateral displacement increased from 200 mm to 480 mm. Comparing the finite element results, with and without Tensar grid reinforcements for embankment A, it showed that the small lateral displacement increments mainly contributed to the presence of the berm.

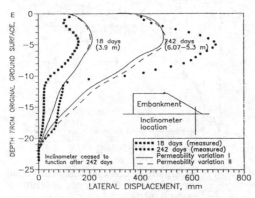

Fig. 15 Lateral displacement profile at inclinometer location of embankment B

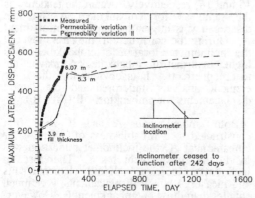

Fig. 16 Maximum lateral displacement versus time plot at inclinometer location of embankment B

Reinforcement tension forces and interface shear stresses

There are no measured data about the reinforcement tension forces and interface shear stresses. The results from finite element analyses are presented and the effect of the reinforcement on the performance of the embankment is discussed. The predicted reinforcement (Tensar SR110) tension force distributions of embankment A for different fill thickness are shown in Fig. 17. It shows that at the early stage, higher tension force developed near the embankment toe at the location of higher shear stress level zone. Later on, since the berms are placed on both sides of the embankment and also due to the soft ground consolidation effect, the tension force increased at the embankment center position and decreased under the berm. At the end of the construction, the maximum tension force in each SR110 geogrid is 13 kN/m, equivalent to 2% of axial strain in the reinforcement. For 3 years after construction condition, the maximum tension force in each SR110 greogrid is 20 kN/m or 40 kN/m in two layers of SR110 geogrids. This indicates that during the consolidation process, the maximum tension force in the reinforcement increased.

Figure 18 shows the shear stress distributions at soil/reinforcement upper and lower interfaces of embankment A at different fill thickness. The maximum interface shear stress immediately after the construction was 12 kPa and 15 kPa 3 years after construction. The sign convention is also shown in the figure by key sketch. It can be seen that for polymer grids, the signs of shear stresses at upper and lower interfaces are the same for most interface areas, i.e., the direct shear interaction mode is applicable for this case. In the zone near the toe of embankment and the intersection point between the berm and the main embankment, because of the free face of the embankment fill, the lateral displacement of the fill is large, and the interface shear stress has negative sign. In other zones, the lateral squeezing of the foundation soil causes the interface shear stresses to have positive sign. During the increase of the fill thickness, the maximum interface shear stress increased, and at the zone near the embankment centerline, the shear stress shifted from negative at the early stages of construction to positive at the end of construction. This indicates that at the zone near the embankment centerline, and at the early stage of construction, the lateral spreading of the fill material is larger than the lateral movement of the soft foundation soil. Later on the lateral movement of the foundation soil is larger than the fill material.

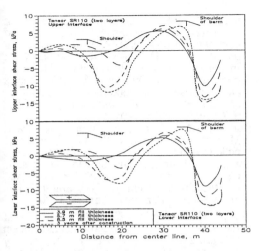

Fig. 17 Tension force in Tensar SR110 reinforcements in embankment A

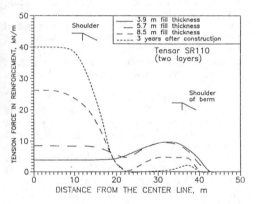

Fig. 18 Interface shear stress between soil and Tensar grid SR110 for embankment A

For embankment B, the tension forces increased with the increase of the fill thickness and during the consolidation of the foundation soils. The maximum tension force in the reinforcement (SR80) immediately after the construction is 12 kN/m, which is equivalent to 2.7% of axial strain in the reinforcement. The predicted maximum tension force is not in the center position of the embankment, but under the shoulder of the embankment, where the distortion stress was higher in the foundation soil. A similar shear stress distribution and variation tendency is obtained for embankment B as in embankment A. However, the variation of the shear stress is simpler than that of embankment A, and the shear stress only changes the sign approximately under the shoulder of the embankment. It indicates that at the zone near the embankment toe, the lateral

spreading of the fill material is larger than the lateral squeezing of the foundation soil. The maximum interface shear stress is about 15 kPa immediately after construction, and increased to about 25 kPa 3 years after the construction which are higher than those of embankment A. For both embankments A and B, the parametric study indicates that for high stiffness reinforcements, such as steel grids, the reinforcement tends to hold the lateral spreading forces both from embankment fill and foundation soil, and the pullout soil/reinforcement interaction mode governs the interface behavior, and for extensible reinforcements, such as polymer grids, the direct shear mechanism is applicable.

The effect of the reinforcement on the performance of the embankment is also investigated by finite element analysis. Comparing the results of with and without reinforcements, it was found that the polymer grid reinforcements placed at the base of the embankment have negligible effect on the foundation deformation pattern. For embankment A, two layers of SR110 polymer grids reduced the maximum foundation lateral displacement by about 5% (25 mm) at the inclinometer location for the end of construction condition. For embankment B, one layer of Tensar SR80 polymer grid only slightly reduced the foundation lateral displacement (less than 2%). Reinforcements can influence the foundation deformation pattern by reducing the undrained distortion of the soft foundation soil. For the cases analyzed, there was a 2 m thick weathered clay layer on top of the foundation which possesses higher shear strength and permeability. Therefore, the effect of the reinforcement was not significant. However, the mobilized reinforcement tension forces can increase the stability of the embankment. Especially for embankment B, during the construction, where cracks were observed on the embankment. This indicated that the embankment B was close to limit equilibrium condition, and the mobilized tension force of 12 kN/m in the reinforcement (one layer of Tensar SR80) benefitted the embankment stability.

CONCLUSION

The behavior of two polymer grids reinforced stage constructed embankments on Muar clay deposit have been predicted by plane strain finite element analyses. In the finite element modelling, the construction process was accurately simulated and the foundation soil permeability variation during the loading and consolidation process was considered. Up to the end of construction, fairly good agreement has been found between predicted values and field data. The comparison was made in terms of excess pore pressures, settlements, and lateral displacement. However, finite element method overpredicted the lateral displacement at early stage of construction and underpredicted at the later stages. Considering the permeability change before and after the soil yielding, the lateral displacement predictions were improved. Most of the discrepancy occurred during the consolidation process between different construction stages. The reasons cited were the creep effect under high shear stress level zone and inability of the modified Cam clay model to consider the creep effect. The field data available are only up to end of construction, but the predicted values given in this study continued to 3 years after construction stage.

Finite element results show that the polymer grid reinforcements have negligible influence on embankment deformation pattern. The reduction of the foundation lateral displacement at the end of construction condition is less than 5%. However, the mobilized reinforcement tension forces might have increased the embankments stability. The effect of the reinforcement on the performance of the embankments was not significant maybe because of the influence of the 2 m thick weathered clay crust on top of the ground.

REFERENCES

AIT, Asian Institute of Technology. 1988. *Laboratory tests data on soil samples from the Muar flats test embankment*. Johore, Malaysian. GTE Division, AIT, Bangkok, 95p.

AIT, Asian Institute of Technology. 1989. *Laboratory tests data on soil samples from the Muar flats test embankments*. Johore, Malaysian. Research Report, Phase II, GTE Division, AIT, Bangkok, 89p.

Balasubramaniam, A.S., Phien-Wej, N.N., Indraratna, B. and Bergado, D.T. 1989. Predicted behavior of the test embankment on a Malaysian marine clay. *Proc. of the Intl. Symp. on Trial Embankments on Malaysian Marine Clays*. Kuala Lumpur. Vol. 2, pp. 1/1-1/8.

Bergado, D.T., Chai, J.C. and Balasubramaniam, A.S. 1992. Prediction of performance of MSE embankment using steel grids on soft Bangkok clay. *Intl. Symp. on Prediction Versus Performance in Geotechnical Engineering*. Nov. 1992. Bangkok, Thailand.

Biot, M.A. 1941. General theory of three-dimensional consolidation. *J. of Applied Physics*. 12, pp. 155-164.

Brand, E.W. and Premchitt, J. 1989. Comparison of the predicted and observed performance of the test embankment. *Proc. of the Intl. Symp. on Trial Embankment on Malaysian Marine Clays*. Kuala Lumpur. Vol. 2, pp. 10/1-10/29.

Chai, J.C. 1992. *Interaction between grid reinforcement and cohesive-frictional soil and performance of reinforced wall/embankment on soft ground*. D. Eng'g. Dissertation. Asian Institute of Technology, Bangkok, Thailand.

Clough, G.W. and Duncan, J.M. 1971. Finite element analysis of retaining wall behavior. *J. of Soil Mech. and Found. Eng'g. Div., ASCE*. 97(12), 1657-1673.

Duncan, J.M., Byrne, P., Wong, K.S. and Mabry, P. 1980. Strength, stress-strain and bulk modulus parameters for finite element analysis of stresses and movements in soil. *Geotech, Eng'g. Research Report No. UCB/GT/80-01*. Dept. of Civil Eng'g. Univ. of California. Berkeley, August, 1980.

Hird, C.C. and Pyrah, I.C. 1990. Predictions of the behavior of a reinforced embankment on soft ground. *Proc. Symp. on Performance of Reinforced Soil Structures*. Thomas Telford. pp. 409-414.

Lambe, T.W. 1973. Predictions in Soil Engineering (Rankin Lecture). *Geotechnique*, 23(2), 149-202.

Magnan, Jean-Pierre 1989. Experience-based prediction of the performance of Muar flats trial embankment to failure. *Proc. of the Intl. Symp. on Trial Embankment on Malaysian Marine Clays*. Kuala Lumpur. Vol. 2, pp. 2/1-1/8.

McGown, A., Andrawes, K.Z., Yeo, K.C. and Dubois, D. 1984. The load-strain-time behavior of Tensar geogrids. *Proc. Symp. on Polymer Grid Reinforcement in Civil Eng'g*. Mar. London. pp. 11-17.

MHA, Malaysian Highway Authority 1989. *Proceedings of the International Symposium on Trial Embankment on Muar Clay*, Vol. 1.

Poulos, H.G., Lee, C.Y. and Small, J.C. 1989. Prediction of embankment performance on Malaysian marine clays. *Proc. of the Intl. Symp. on Trial Embankments on Malaysian Marine Clays*. Kuala Lumpur. Vol. 2, pp. 4/1-4/10.

Roscoe, K.H. and Burland, J.B. 1968. On the generalized stress-strain behavior of wet clays. *Proc. of Eng'g. Plasticity*. Cambridge. Cambridge Univ. Press. pp. 535-609.

Tavenas, F. and Leroueil, S. 1980. The behavior of embankments on clay foundations. *Can. Geotech. J.* Vol. 17, pp. 236-260.

Tavenas, F. Jean, P., Leblond, P., and Leroueil, S. 1983. The permeability of natural soft clays. Part II, permeability characteristics. *Can. Geotech. J.* Vol. 20, pp. 645-660.

Taylor, D.W. 1948. *Fundamentals of soil mechanics*. John Wiley & Sons Inc., New York.

Prediction versus performance on lateral displacements in soft Bangkok clay under embankment loading

W. Teparaksa
Chulalongkorn University, Bangkok, Thailand

ABSTRACT: The embankment of 1.8 to 2.2 m. high has been constructed on very soft Bangkok clay parallel to the existing Bang Na-Bang Prakong Highway. The Bangkok subsoils consist of about 20 m thick very soft to medium grey clay and the strength is in the order of 0.8 to 2.5 t/m^2. Geotechnical instruments, such as inclinometers, settlement plates, and piezometers were installed to measure the lateral and vertical soil displacement. The results indicate that the ratio between maximum lateral and vertical displacement (Rhv) have a good relationship with embankment slope stability safety factor and with the maximum final shear stress ratio (fmax). The prediction of Rhv-value based on the finite element method with undrained linear elastic soil model agrees well with measurement only when the maximum final shear stress ratio is larger than 1.0 or local yield failure is encountered.

1. Introduction

The lateral displacements in soft clay during and after construction of embankments have been a subject of numerous studies in recent years. Those studies include the numerical analysis of lateral deformation (Poulos, 1972) and the empirical approaches to the problem (Tavenas et al, 1979; and Tavenas and Leroueil, 1980). The study of soil behaviour during and after construction of embankment in soft Bangkok clay has been emphasized on the effect of vertical settlement to the stability of embankment or effect to the adjacent structures such as natural gas pipe line (Chulalongkorn Report, 1984). The lateral displacement of soft clay due to embankment loading in soft Bangkok clay is becoming more of interest since lateral displacement has detrimental effect on the behaviour of adjacent structures such as pile foundation, movement of bridge abutment, movement of water and gas pipe line.

In this paper, the lateral displacement of soft Bangkok clay foundation under embankment loading during construction of a new frontage Bang Na- Bang Pakong Highway is monitored by various geotechnical instruments such as inclinometer, settlement plate, and piezometer and sondex. The lateral displacement behaviour of soft clay is evaluated and discussed with the maximum final shear stress ratio against local yield and safety factor of embankment. The prediction of lateral displacement behaviour by using Finite Element Method with undrained linear elastic soil model is also compared with the field performance.

2 Subsoil conditions along the Bang Na-Bang Pakong Highway

The Bang Na-Bang Pakong Highway is the main highway from Bangkok, capital of Thailand, to the eastern seaboard. The highway passes over the flat deltaic plains of Thailand, where the subsoil is soft marine clay called Bangkok clay (Figure 1). The typical soil conditions at km 30 from Bang Na is presented in Figure 2. The thickness of soft Bangkok clay varies from 15 m at km 0 at Bang Na to about 25 m at km 28 from Bang Na (Figure 3). Stiff clay of about 5-10 m thick is

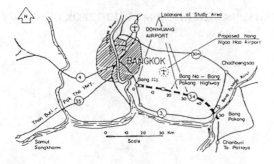

Figure 1   Location Map

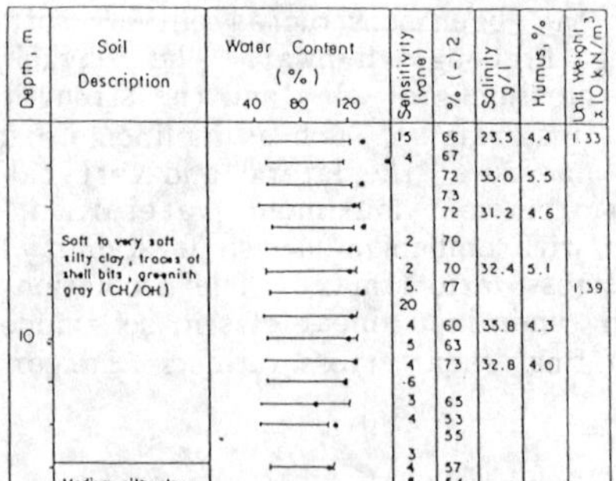

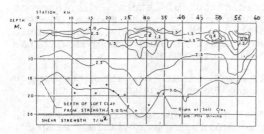

Figure 2   Typical soil condition at station 30

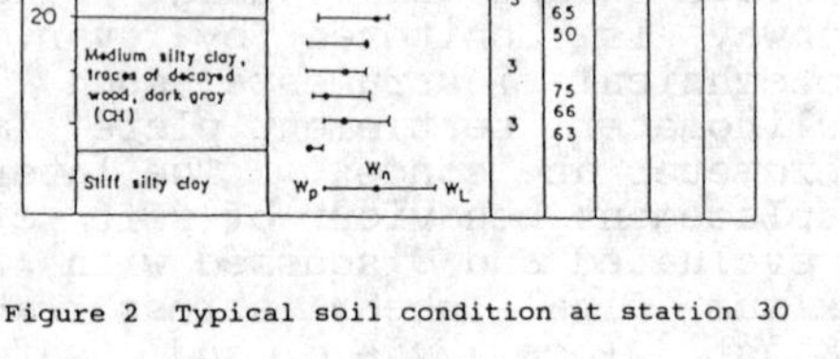

The undrained shear strength from field vane test is presented in Figure 3. It can be observed that two soft stretchs are encountered near km 30 and km 50. The natural water contents of soft Bangkok clay is in the order of 80-140% as shown in figure 4. The liquidity index profile shown in Figure 5 indicates that the moisture content exceeds the liquid limit in the first 5 m of soft clay. The compressibility index and apparent preconsolidation pressure of soft Bangkok clay along Bang Na- Bang Pakong Highway reported by Lea (1981) is presented in Figure 6.

The undrained elastic modulus, Eu is commonly expressed as a function of the undrained shear strength, S_u, i.e. $E_u = \alpha \ast S_u$. For Bangkok clay, the α-value lies between 70 to 250, however, Parnploy (1985) suggested α-value of 253 for weathered clay and 131 for soft clay layer ; based on the laboratory stress path test at km 2+899 of Bang Na- Bang Pakong Highway. The Poisson ratio in this study is in the range of 0.30-0.39.

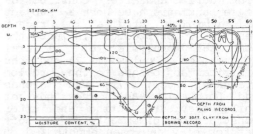

Figure 4 Moisture content contour of Bang Na-
Bang Pakong Highway (Lea, 1981)

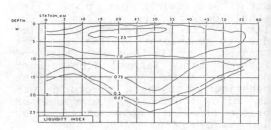

Figure 3 Shear strength contour of Bang Na-
Bang Pakong Highway (Lea, 1981)

Figure 5 Liquidity Index contour of Bang Na-
Bang Pakong Highway (Lea, 1981)

found below soft clay and followed by dense to very dense silty sand. The weathered crust of about 1-2 m thick is encountered at the topmost of soft clay layer.

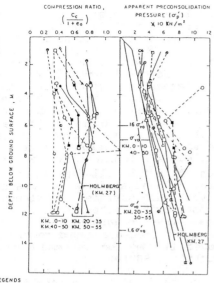

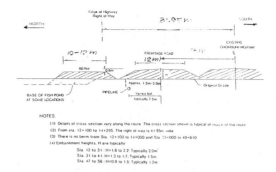

LEGENDS

DISTANCE FROM BANG NA IN KM. (KAMPSAX STATION)

○ 0+842(17+548)N   ▲ 28+783(45+013)N   ■ 42+776(59+020)N
⊙ 5+389(22+094)A   ▲ 28+785(45+013)N   □ 42+776(59+020)A
● 16+023(32+732)N   □ 33+930(50+160)N   ⊚ 53+678(81+735)N
△ 22+717(39+427)N   ⊚ 29+779(45+800)T   ○ 48+971(76+412)T
                    ▣ 16+284(33+250)T

σ_{vo}' = AVERAGE OVERBURDEN PRESSURE THROUGHOUT THE PROJECT ROUTE
A = BY ASIAN INSTITUTE OF TECHNOLOGY   - - - KM. 0 TO KM. 10, KM. 40 TO KM. 50
N = BY NORWEGIAN GEOTECHNICAL INSTITUTE ── KM. 20 TO KM. 35, KM. 50 TO KM. 55
T = BY THAI ENGINEERING CONSULTANTS

Figure 6   Compressibility properties of soil
along Bang Na-Bang Pakong Highway
(Lea, 1981)

3.   Embankment Construction

The embankment was constructed as frontage road (on the left side) of the existing Bang Na - Bang Pakong or Chonburi highway. The typical configuration of frontage road specified by the Department of Highways is shown in Figure 7. The frontage road is planned to be constructed into 2 typical categories; e.i. with and without berm depending on the soil conditions. For soft areas where existing fishpond is located very near to embankments, berm is provided to increase the stability of embankment, while in good soils (stiff soil), the embankment was constructed without berm.

The construction schedule was planned into 2 stages. The first comprised of the sand fill (embankment) compacted slightly higher than the subgrade elevation

Figure 7   Configuration of frontage road

(about 20-30cm above subgrade). This stage was kept for about 3-5 months for consolidation waiting period. The second stage comprised of the 0.5 m. thick pavement structure which includes 20 cm of laterite subbase, 20 cm of crushed gravel base, and 10 cm. of asphaltic concrete. The specifications are summarized below:

Item	Soft Soil	Stiff Soil
Location	Sta 14+000 to 35+000 Sta 47+120 to 55+670	Sta 12+100 to 14+000 Sta 35+000 to 40+680
Berm	10-12 m wide berm	No berm
Stage 1 Elevation	30 cms above subgrade	20 cms above subgrade
Waiting Period	5 months	3 months

The study areas for this research is located from station or KM. 15+670 to station 34+501 where is in the soft soil section area and the berm of 10-12 m. wide is provided.

4.   Geotechnical Instrumentations

The geotechnical instrumentations which include inclinometer, settlement plate, and piezometer were installed before the construction of embankment. Three typical instrumentation types (type A, B, and C) were provided to measure the ground response as shown in Figure 8. Type A instrumentation system is fixed at the very soft area, while the type B system is at a better ground condition area and type C system is at the general area. The location and number of instrumentation is summarized in Table 1. Total number of instrumentations is 19 for inclinometer, 50 for piezometer and 136 for settlement plate.

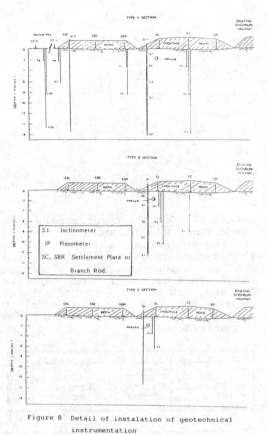

TYPE A SECTION

TYPE B SECTION

S.I.   Inclinometer

IP   Piezometer

SC, SBR Settlement Plate or

        Branch Rod.

TYPE C SECTION

Figure 8 Detail of instalation of geotechnical
          instrumentation

Table 1 Summary of geotechnical
         instrumentation

Station	Type	No. of Instrument		
		Inclinometers	Piezometers	Settlement Branch Rods
15+670	C	1	1	8
24+390	C	1	0	8
26+815	C	1	0	8
27+408	A	2	9	8
27+650	B	1	4	8
27+920	C	1	0	8
28+020	B	1	4	8
28+160	A	2	9	8
28+350	C	1	2	8
29+340	C	1	2	8
29+550	C	1	2	8
30+121	C	1	2	8
30+270	C	1	3	8
30+600	C	1	2	8
31+280	B	1	5	8
33+370	C	1	0	8
34+501	C	1	0	8
Total		19	50	136

An inclinometer consists of a 75 mm diameter pipe that is installed in a vertical borehole of about 14-16 m deep. Measurement by the inclinometer was taken at 0.6 m interval for full depth. Most of the piezometers used for this project is the pneumatic piezometer, however, 6 of stand pipe piezometers are provided as dummy which are installed far away from constructed embankment. Settlement plate or branch rod settlement is a steel rod driven into the fill embankment or natural ground surface to measure the vertical surface settlement. The monitoring schedule is from October 1990 to January 1992. The measurement procedures and recording followed the method recommended by Dunnicliff (1988).

The record data at station 27+408 through station 34+501 is the data recorded from those instruments installed before the construction of embankment. At station 15+670 to 26+815 the instruments were installed after the first stage embankment was filled for about 3-4 months. Therefore, at these stations consolidation and some lateral displacements had taken place before instrument installation.

5. Behaviour of Lateral Displacement under embankment loading

The measurement of lateral displacement and vertical settlement of soft clay foundation was recorded through out the construction period. The result of the measurement can be discussed as follows

5.1 Rate of lateral displacement. Rate of lateral movement is the ratio between the measured maximum lateral displacement at the toe of embankment and time (dYm/dt). The definition of geometry and deformation parameter is shown is Figure 9. The typical value of dYm/dt at first stage and second stage of construction at station 27+408 is presented in Figure 10. Rate of lateral displacement (dYm/dt) is plotted against the maximum final shear stress ratio (fmax) which is defined as the ratio between loading shear stress in the soil mass $[(\sigma_v - \sigma_h)/2]$ and the undrained shear strength of soils. The detail of estimated maximum final shear stress ratio and ,measured rate of lateral displacement is summarized in Table 2. Figure 11 presents the relationship between rate of lateral displacement and maximum final shear stress ratio. It can be seen that at fmax between 0.5-1.8, the lateral displacement is an exponential function with maximum final shear stress ratio as follows:

$$dYmax/dt = 0.007\ e^{2.807(fmax)}$$

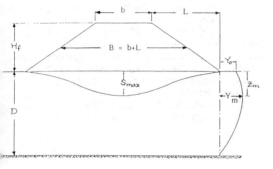

Figure 9   Geometry and deformation parameter

Table 2   Summary of rate of lateral movement and maximum final shear stress ratio

Station	Maximum Final Shear Stress Ratio (fmax)		Rate of Lateral Movement after Embankment Loading	
	1st Stage	2nd Stage	1st Stage (mm./day)	2nd Stage (mm./day)
15+670	NI	1.26	NI	0.11
24+390	NI	1.47	NI	0.21
26+815	NI	1.01	NI	0.15
27+408	1.31	1.56	0.29	0.61
27+650	1.00	1.58	0.26	0.23
27+920	0.95	1.10	0.18	0.32
28+020	0.94	NA	0.05	NA
28+160	0.82	NA	0.08	NA
28+350	0.53	NA	0.03	NA
29+340	1.39	1.64	0.33	0.35
29+550	1.31	NA	0.34	NA
30+121	1.39	NA	0.44	NA
30+270	1.34	1.67	0.32	0.56
30+600	1.06	1.21	0.24	0.26
31+280	0.60	NA	0.05	NA
33+370	0.94	NA	0.03	NA
34+501	NA	1.56	NA	0.34

Remark :   NI - Not Installed (slope indicators installed after 1st stage construction)

NA - Not Analysed (some locations still in 1st stage waiting period or not finish in 2nd stage construction)

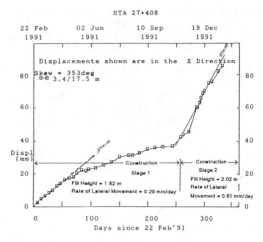

Figure 10   Typical rate of lateral movement at station 27+408

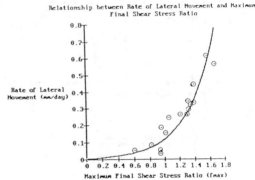

Figure 11   Relationship between rate of lateral movement and maximum final shear stress ratio

5.2 Relationship between lateral displacement and maximum vertical settlement (Rhv). The typical plot of maximum lateral displacement and maximum settlement measurement at station 27+408 at first stage and second stage of construction is shown in Figure 12. It can be seen that the ratio of maximum lateral displacement to the maximum vertical settlement (Rhv) at second stage is larger than at first stage. This is due to the higher filled embankment The field performance Rhv -value is plotted against the maximum final shear stress ratio (fmax) as shown in Figure 13. Table 3 summarizes the estimated maximum final shear

161

stress ratio (fmax) and measured Rhv -value at all stations both for first and second stages of construction. It can be seen from Figure 13 that the ratio between the maximum lateral displacement to the maximum vertical settlement (Rhv) is also an exponential function with maximum final shear stress ratio (fmax) for fmax between 0.5-1.8 as:

$$Rhv = dYm/dSm = 0.068e^{1.23(fmax)}$$

difference between frontage road and berm are also discussed.

The slope stability of embankment was analysed using Modified Bishop Method of stability analysis. The undrained shear strength of soft Bangkok clay is based on the data of field vane shear test with introduction of Bjerrum's correction factor, $\mu = 0.7$. The summary of safety factor of embankment and Rhv-value at all stations is presented in Table 4. Figure 14 presents the relationship of Rhv-value and factor of safety of embankment.

The Rhv-value is also plotted with the relative height difference in elevation between frontage road and berm as shown in Figure 15. It can be seen that the Rhv-value shows some scatter with safety factor and height differences than with the final maximum shear stress ratio. This might be due to the configuration of embankment, and is not due to the soil behaviour.

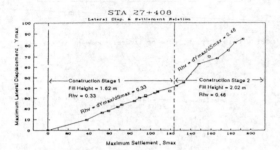

Figure 12 Typical estimation of ratio between maximum horizontal and vertical displacement (Rhv)

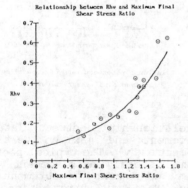

Figure 13 Relationship between Rhv and maximum final shear stress ratio

Apart from the relationship between Rhv and maximum final shear stress ratio (fmax) which is in the parameter of yield strength and embankment loading, the other relationship of Rhv and other parameteres such as safety factor of embankment and relative hight

Table 3 Summary of Rhv-value and maximum final shear stress ratio

Construction Stage No.	Station	Maximum Final Shear Stress Ratio (fmax)	Rhv
1	27+408	1.31	0.33
	27+650	1.30	0.25
	27+920	0.95	0.24
	28+020	0.94	0.17
	28+160	0.82	0.22
	28+350	0.53	0.15
	29+340	1.39	0.41
	29+550	1.31	0.34
	30+121	1.39	0.38
	30+270	1.34	0.38
	30+600	1.06	0.23
	31+280	0.60	0.13
	34+501	1.46	0.33
2	15+670	1.56	0.32
	24+390	0.74	0.19
	26+815	0.67	0.26
	27+408	1.56	0.46
	27+650	1.58	0.60
	27+920	1.10	0.66
	29+340	1.70	0.62
	30+270	1.67	0.45
	30+600	1.21	0.26
	34+501	1.56	0.42

Table 4   Summary of Rhv-value and estimation of slope stability

Construction Stage No.	Station	Slope Stability (FS.)	Rhv
1	27+408	1.394	0.33
	27+650	1.394	0.25
	27+920	1.576	0.24
	28+020	1.733	0.17
	28+160	1.898	0.22
	28+350	1.897	0.15
	29+340	1.393	0.41
	29+550	1.465	0.34
	30+121	1.293	0.38
	30+270	1.646	0.38
	30+600	1.744	0.23
	31+280	3.460	0.13
	33+370	2.832	0.09
	34+501	1.443	0.33
2	15+670	1.653	0.32
	24+390	1.826	0.19
	26+815	1.723	0.26
	27+408	1.156	0.46
	27+650	1.154	0.60
	27+920	1.218	0.66
	29+340	1.080	0.62
	30+270	1.010	0.45
	30+600	1.488	0.26
	34+501	1.299	0.42

both of the embankment and clay foundation is shown in Table 5.

Table 5   Soil parameter for FEM analysis

Parameters		Embankment Parameters	Clay Foundation Parameters
Unit Weight	KN/m3	20.1	13.0–13.6 (Increased with depth)
Undrained Modulus (Eu) KN/m2		11000	253·Su (For Weathered Clay Layer) 131·Su (For Soft Clay Layer)
Poisson's Ratio (ν)		0.3	0.5

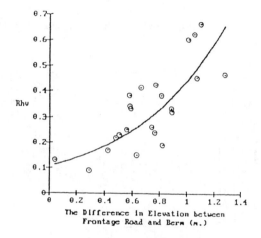

Figure 15   Relationship between Rhv and height difference (Δh) between frontage road and berm

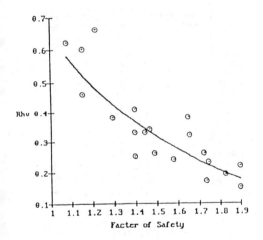

Figure 14   Relationship between Rhv and factor of safety of embankment

6. Comparison of prediction and field performance of lateral displacement

The prediction of lateral displacement under embankment loading is carried out using finite element method with undrained linear elastic soil model. The soil parameters of undrainded elastic modulus (Eu) and Poisson ratio (ϑ)

The boundary condition of the discretization system are that both vertical and horizontal movements at stiff clay layer are restrained, and the horizontal movement at the toe of frontage road at the existing highway side is restrained. The details of finite element mesh used for this analysis is shown in Figure 16. The result of the finite element analysis of Ymax and Smax is presented in Table 6. The predicted Rhv.-value based on finite element method of analysis using undrained linear elastic soil model is presented against the field performance as shown is Figure 17. The comparison of the estimated Rhv -value and field performance is summarized as follows :

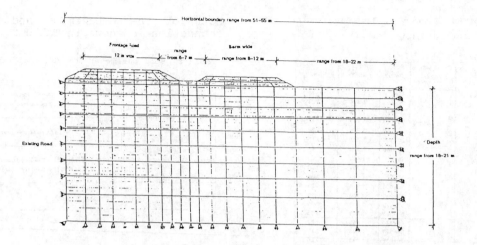

Figure 16   Detail of mesh for finite element method of analysis

Rhv Characteristic
FEM & Measured Data Comparison

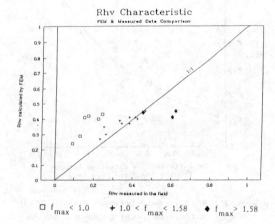

□ $f_{max} < 1.0$ + $1.0 < f_{max} < 1.58$ ◆ $f_{max} > 1.58$

Figure 17   Comparison of Rhv between measurement and
estimated value by FEM

Table 6   Summary of analysis of Ymax and
Smax by FEM

Station	Construction Stage No.	Results of Finite Element Analysis		
		Ymax (cm.)	Smax (cm.)	Rhv = Ymax/Smax
27+408	1	5.05	12.80	0.39
27+650		4.22	12.06	0.35
27+920		5.18	11.96	0.43
28+020		4.90	11.74	0.42
28+160		4.37	11.05	0.40
28+350		4.65	11.29	0.41
29+340		5.74	14.07	0.41
29+550		6.94	15.75	0.37
30+121		5.74	14.07	0.41
30+270		4.60	12.60	0.37
30+600		2.96	10.84	0.27
31+280		2.55	8.77	0.29
33+370		1.81	7.65	0.24
34+501		3.10	9.09	0.34
15+670	2	4.61	8.50	0.54
24+390		5.55	12.78	0.43
26+815		4.00	11.81	0.34
27+408		6.57	14.70	0.45
27+650		5.70	13.95	0.41
27+920		6.41	13.55	0.47
28+020		6.13	13.31	0.46
29+340		7.28	16.10	0.45
30+270		6.94	15.75	0.44
30+600		3.46	11.60	0.30
34+501		4.07	10.25	0.40

Remark : Ymax - maximum lateral displacements
Smax - maximum settlements

- At fmax < 1.0, the estimated Rhv-value by FEM is higher than the field measurement. This is due to the partially drain loading of overconsolidated clay at the first stage or begining of construction which represent the fmax less than 1.0 and Poisson ratio is much less than 0.5 (Tavenas, et al 1979). However, in this analysis, the Poisson ratio of 0.5 is assumed.

- At 1.0 < fmax < 1.58, the predicted Rhv-value by FEM agrees well with field performance, this is might be due to the reason stated by Tavenas, et val (1979) that at final stage of construction, soils under embankment corresponds to plastic flow of normally consolidated clay (fmax > 1.0) and Poisson ratio is about 0.5 as assume. Therefore, the predicted value agrees with the measurement.

Apart from estimation of Rhv-value, the shape of lateral displacement using FEM is also estimated as shown is Figure 18 with the measurement. Prediction of shape of lateral displacement using FEM by linear elastic soil model does not agree with the measurement as also stated by Poulos (1972) in Figure 19.

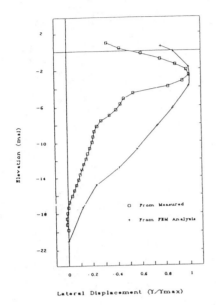

Figure 18   Prediction and measurement of shape of deformation

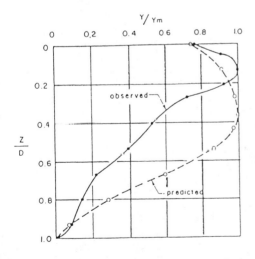

Figure 19   Prediction and observation of shape of deformation (After Poulos, 1972)

7.   Conclusion

Based on the measurement and prediction of lateral displacement of soft Bangkok Clay under embankment loading, it can be concluded that

1. Rate of maximum lateral movement shows a good relationship with the maximum final shear stress ratio (fmax)

2. The ratio between maximum lateral displacement and maximum vertical settlement (Rhv) shows a good relation with maximum final shear stress ratio

3. Prediction of Rhv-value based on finite element method with undrained linear elastic soil model agrees well with field performance only when the maximum final shear stress ratio is greater than one or when plastic flow is encountered.

REFERENCE

Chulalongkorn University (1984) : Technical report on impact to natural gas pipe line due to embankment, Civil Engineering Department, Submitted to Petrolium Authority of Thailand. (in Thai)

Dunnicliff, J (1988) : Geotechnical Instrumentation for monitoring filed performance, John wiley and sons.

Lea, N.D and associates, and TEC. (1981): Technical report submitted to Department of Highways,Thailand.

Parnploy, U. (1985) : Deformation analysis and settlement prediction of Bang Na-Bang Pakohg Highway, M.Eng. Thesis, AIT, Bangkok.

Poulos, H.G. (1972): Difficulties in prediction of horizontal deformations of foundations, ASCE, Jour. of SMFE., SM8.

Tavenas,F.,Mieussens,G.,and Bourges,F. (1979): Lateral displacements in clay foundations under embankments, Canadian Geotechnical Journal, Vol. 16., No. 3.

Tavenas, F, and Leroueil, S. (1980): The behaviour of embankments on clay foundation, Canadian Geotechnical Journal, Vol 17.

Embankment on sand columns: Comparison between model and field performance

Bujang B. K. Huat
Universiti Pertanian, Malaysia

Faisal Hj. Ali
Universiti Malaya, Malaysia

ABSTRACT: The paper describes the performance of a full scale embankment on sand columns which has been built and instrumented in Malaysia. Reference is made to a parallel design study carried out using centrifuge modelling technique.

1 INTRODUCTION

Extensive deposits of compressible and low strength soils are found worldwide, and the difficulties of supporting loads on such foundations have been widely reported. In Malaysia, quaternary erosion accentuated by climatic and sea level changes has produced widespread and thick alluvial deposits in the coastal areas and major river valleys. These alluvial clay formations vary in thickness from 5 m to 30 m and a review of the basic and engineering properties of some of these deposits has been published by Ting et al. (1987).

In recent years pile supported embankments have frequently been used in road construction but the high cost of this method of construction has been of concern to the Malaysian Highway Authority (LLM). At the same time specialist contractors and suppliers of proprietary products have claimed that their alternatives to piled embankments can result in cost saving. This has led to the construction of a trial embankment by the Highway Authority between the period of 1986 to 1989. The site chosen for the trial construction was at a lay-by on the Seremban-Ayer Hitam section of Malaysia's north-south expressway project. Approximately 20 km of this road cross areas of very soft clay deposits with thickness varying from 10 m to 20 m. The trial embankment itself was built on an area with approximately 20 m of soft to very soft marine clay underlain by loose to dense, medium to coarse sand.

The objectives of the trial were to verify the costs and effectiveness of different methods of soil improvement, and methods of design and analysis of settlement and stab-ility. The criteria for acceptable performance which LLM stipulated were that the pavement be constructed 15 months after start of fill placement and that the subsequent residual settlement after completion of paving work would be less than 100 mm over 2 years. Altogether there were 14 different trial sections, 11 of which were supported by different methods of soil improvement, 2 were control sections on untreated ground, whilst 1 section (also on an untreated foundation) was built to failure, Table 1. This paper describes the performance of scheme 6/5 where the trial embankment was supported on sand columns. Attempt has also been made to model the principal features of the field trial in a centrifuge for comparative studies.

The use of sand or stone columns offers one solution to the problem of stability and settlement posed by construction of road or highway embankments on soft ground. Under a wide-spread or embankment type of loading, for reasons of economy, granular support columns need to be widely spaced. The ultimate load capacity of the columns is insufficient to support the whole applied loading hence a significant proportion of the applied load will be carried by the ground between (Greenwood and Kirsch, 1984; Greenwood, 1990). This load sharing process between the columns and original ground will invariably influence the settlement behaviour of the treated foundation complex with simultaneous and interdependent changes of soil-column stress ratios, pore pressure and resulting stiffness in both soil and column, depending on several parameters including area ratio, loading rate and the group effect.

Table 1. Trial schemes

Section	Designer	Final Design Height of Embankment above Original Ground Level	Scheme No.
Control (no treatment)	Malaysian Highway Authority	3m & 6 m	3/2 & 6/6
Test Section to Failure	"	---	3/5
Electro-Chemical Injection	Process Kimia	3m & 6m	3/1 & 6/1
Sand Sandwich	Jurutera Konsultant	3m	3/3
Preloading, Geogrid & Vertical Drains	Zaidun Leeng	3m	3/4
Well Point Preload	Energoproject	6m	6/2
Electro-osmosis	Esa-Maunsell	6m	6/3
Concrete Piles	IJM-ICP	6m	6/4
Sand Columns	Fudo	6m	6/5
Vacuum Preloading & Vertical Drains	SSP	6m	6/7
Preloading, Geogrid & Vertical Drains	Pilecon	6m	6/8
Preloading & Vertical Drains	Moh & Associates	6m	6/8

2 FIELD STRUCTURE

Field information from the trial site included data from 2 deep boreholes, 4 piezo-
cone tests, 11 insitu vane profiles, and
piston sampling at 11 locations. Detailed
descriptions of the site, geology and subsoils are given in the investigation reports prepared by LLM (1987), and AIT (1989).
Within the area of the trial, the thickness
of the strata were known to vary a little
but the subsoil profile can be generalised.
The upper 17 m consist of very soft to soft
silty clay with natural water content of
50 - 120 %, liquid limit, W_L of 40 - 80 %,
plastic limit, W_P of 20 - 40 %, and over-
lain by a surface crust about 1 m thick.
Traces of sea shells indicate a marine
origin. Underlying this clay layer is a
layer of peat of about 0.5 m, followed by
some 2 m of sandy clay which is underlain
in turn by a thick deposit of loose to dense, medium to coarse sand with SPT values
ranging from 6 to 50. There is little
evidence of any drainage structure that
would permit rapid dissipation of excess
pore water pressure within the clay layer.
A summary of the geotechnical properties of

the clay is given in Figure 1. The undrained strengths obtained from the vane tests
show an almost linear increase below a surface crust with an average strength, S_u, of
9 kPa at 1 m, increasing to 36 kPa at depth
17 m, or 8 - 36 kPa if corrected with Bjerrum's correction factor for anistropy and
shear rate. The clays have a sensitivity
ratio in the range of 3 to 6. Results
obtained from the oedometer tests indicated
that they are slightly overconsolidated but
highly compressible. Values of C_v are low,
ranging from 1 - 10 m²/yr, and scattered.
The soil permeabilities ($k_v \leq k_h$) are generally less than 5×10^{-9} m/s, with clay
fraction of the order of 50 % and kaolinite
as the dominant mineral present.
Figure 2 shows cross section and instrumentation of the prototype sand column
scheme as built. 462 compacted sand columns 0.7 m diameter for the upper 10 m clay
foundation above sand drains (uncompacted
columns) 0.5 m clay foundation above sand
drains (uncompacted columns) 0.5 m diameter
for the lower 10 m, were installed vertically in a square grid at 2.2 m centres.
The area ratio, A_c/A was 0.08 at the top
of the column, where A is the total plan

area attributed to a column of cross section A_c. Instrumentation installed settlement gauges, piezometers, an inclinometer and land-slip meters.

A total fill thickness of 9.9 m inclusive of a 1 m sand banket was to be placed. Some 90 % of the primary consolidation settlement was expected by the end of the 15 month construction period, leaving 1.1 m of surcharge to be removed to give the desired embankment level. Subsequent settlement over 2 years was expected to be comfortably less than the specified limit of 100 mm. A minimum factor of safety against slope instability was computed greater than 1.2 for the embankment constructed without berms, as initially designed, assuming a gain in strength using the method proposed by Aboshi and Suematsu (1985). The actual performance of the field structure in terms of settlement, stability, generation and dissipation of excess pore water pressure is described below.

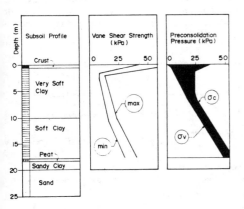

Fig. 1   Subsoil properties of trial site

3 CENTRIFUGE MODEL

A series of model test has been performed aboard the Manchester centrifuge, all nominally at 1 : 100 scale. The objectives were to simulate the essential features of the prototype structure and to provide a design study of the various parameters which influence the performance of such a structure. A description of the centrifuge used and discussion on the various considerations in its design and use are given by Craig and Rowe (1981). All tests were carried out inside a purposely made box of alluminium alloy plates 25 - 35 mm thick with internal dimensions 950 x 350 x 375 mm and an open top. Bolted construction allowed the sides to be removed for ease of model assembly and strip down. Greased rubber sheets were used to minimise boundary friction.

Figure 3 shows a typical cross section of the model embankment. Troll clay with W_L = 60 %, W_P = 30 % and C_v = 1 - 3 m²/yr that reasonably matched the field material was used to model the foundation. The clay was consolidated from slurry in a 1.0 x 1.0 m cell using the hydraulic gradient method (Zelikson, 1969), to a similar undrained strength profile as in the prototype (Figure 1). From each cell two clay beds 950 mm long, 350 mm wide and 200 mm deep (to simulate 20 m at 1 : 100 scale) were cut. A very sandy clay mixed with kaolin to a sand : clay ratio of 4 : 1 by weight, and which yielded compacted values of c' = 20 kPa, $\emptyset'$ = 30° and ρ_b = 2.09 Mg/m³ was used to model field fill. The sand columns were formed of fine Mersey River sand with D_{50} of 0.2 mm. Arguably this represents a 20 mm gravel at 1 : 100 scale, hence a stone rather than a sand column, but the intrinsic strength of the material would still be that of a sand.

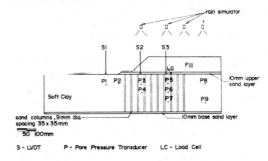

Fig. 3   Cross section of model embankment, SC 1

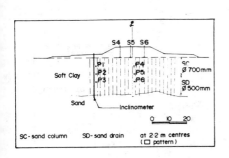

Fig. 2   Cross section of field trial embankment (section 6/5)

Sand columns of 9 mm diameter were installed to full depth of the clay foundation

169

200 mm, in a square grid spacing to the required area ratio. Ideally the EI of the column should be scaled down by 100^4 relative to the prototype, but since the insitu values are not exactly known, the sand columns were only modelled to approximately scaled dimensions and installed in a densest possible state to a similar area ratio as in the prototype. Modelling of the sand columns involved utilisation of a freezing technique, similar to those described by Masaaki and Masaki (1990) and Shinsha et al. (1991). Flexible plastic tubing, 9 mm internal diameter cut to 220 mm lengths (open at both ends) and with a slot cut down on one side were used as reusable formers. Dry sand was poured into each tube with a thin wire inserted at the centre to provide rigidity to the column whilst being handled. By tapping the sand was densified to an estimated relative density of 90 %. The assembly was slowly submerged in water until saturated, and placed inside a freezer for two hours. Once frozen, the tubes were recovered. The frozen columns were then withdrawn from the tube and inserted inside prebored vertical holes in the clay foundation. Once the ice has melted the wire insert was slowly withdrawn, tamping the sand column in the process.

Instrumentation installed included miniature load cells, Druck pore pressure transducers, LVDT and spaghetti displacement indicators.

The stage construction of the prototype was stimulated by spinning the model at different gravity levels, i.e. using the gravity turn-on-technique. During centrifuging a steady head of water in the base sand layer was achieved by linking via external piping to the upper surface of the model foundation. Climatic conditions around the model slope was simulated with a precipitation simulator system (Craig et al., 1991).

4 PROTOTYPE AND MODEL PERFORMANCE

The prototype at Section 6/5 was instrumented as shown in Figure 2. Installation of the columns took place in March and April 1987, after placement of a 1 m working platform of sand, and construction began in April 1988 and continued until June 1989, with intermittent rest periods, Figure 4. However in November 1988 (Days 223 - 225) the fill thickness was reduced from 8 m to 6.9 m after significant lateral movements and tension cracks were observed on Day 213 at both slope and crest of the embankment. Stabilising

berms (Figure 2) were constructed, before the fill was brought to final thickness, 9.2 m. Figure 4 also shows the field settlement and pore pressure records. Both rate of settlement and reduction in excess pore water pressure were significantly larger than those at a nearby control section without any foundation treatment (LLM, 1989). Unfortunately, all piezometers except P1 malfunctioned from Day 200 onwards, but during the construction pause periods (Days 100 - 170), some 43 % of excess pore water pressure had dissipated with larger dissipation shown at the shallowest piezometers, P1 and P4. Ignoring dissipation due to vertical drainage, coefficient of consolidation back calculated from Barron's equation was around 1 m²/yr, half the value assumed in the original design based on laboratory oedometer testing. This seems to suggest that installation of the column, in particular driving of the casing, caused the formation of a smear zone and reduced soil permeability in the vicinity of the column, though in a soil with little fabric the reduction is modest. The greater importance of this effect on other sites has

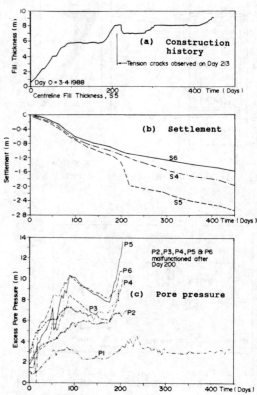

Fig. 4 Field results

been recognised (Hansbo, 1989; Greenwood, 1990) but owing to the close spacing of the columns rapid dissipation of excess pore pressure has still been achieved. A final settlement of 3.1 m was expected beneath the centre of the embankment, giving a settlement improvement n = 1.33 (defined as the ratio of untreated to treated ground settlement) for 4.1 m final untreated ground settlement calculated using conventional 1-D method. The end of construction (Day 450) settlement of the treated ground was still fairly large (2.6 m) which is not so surprising for the improvement area ratio was fairly small, but this was offset to some extent by means of preloading the soft clay. Immediate removal of 0.6 m surcharge as scheduled would leave the required 6 m high embankment, but the residual settlement (s_r) over the following 2 years period would exceed LLM's stringent performance criterion of s_r < 100 mm. The alternative of leaving the surcharge in place for several more months was adopted. The final 0.3 m surcharge was removed in November 1989, 5 months behind schedule. Figure 5 shows the plot of maximum lateral deformation (y) measured by the inclinometer versus centreline settlement (s). The initial ground deformation up to Day 180 under 6 m of fill was similar to that at the control section with $\Delta y/\Delta s = 0.27$. However from Day 180 to 200, when the fill thickness was raised to 8 m, $\Delta y/\Delta s$ increased sharply to 0.75. Unfortunately the inclinometer became inaccessible from Day 201 onwards but measurements from the landslip meter installed beneath the embankment toe indicated a substantial increase in movement between Day 200 – 220. This resulted in the formation of tension cracks described above, attributable to the instability of the adjacent pile scheme (Section 6/4) as well as to a period of heavy rainfall on the site. The calculated factor of safety against undrained failure was still greater than unity (Mizuno et al., 1989). Subsequent removal of 1.1 m of surcharge and construction of berms on both sides of the embankment appear to have been successful in containing

an overall collapse. The width of the crack ceased to increase on Day 225, and filling to 9.2 m was resumed on Day 285. Test SC1, Figure 3, will be used to give an indication of the data obtained from the models since this is the closest simulation of the prototype. Rapid dissipation of excess pore water pressure was in evidence beneath the embankment centre (Figure 6), and at locations closest to the drainage boundaries (transducers P5 & P7), when compared with the section without the sand columns (transducers P8 & P9). Average consolidation degree (U) in the foundation strata beneath the centre of the column section was 48 % at the end of the simulated construction, and after 2.08 hours (2.37 yrs) was 93 % compared with 30 % at the untreated section. No delayed pore pressure response after end of construction was observed beneath the toe and slope of the model embankment, although pore pressure stagnation did initially occur (transducers P2 & P3). This is consistent with the prototype behaviour. Transducer P1 located 60 mm (6 m) forward of the toe indicated no apparent pore pressure development beyond the extent of the loaded area, with pore pressure ratio ($\Delta u/\Delta \sigma_v$) remaining close to 0.8 at 100 g and 134 g. The rapid dissipation of excess pore water pressure resulted in rapid consolidation settlement (LVDT S2 & S3, Figure 6).

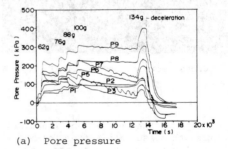

(a) Pore pressure

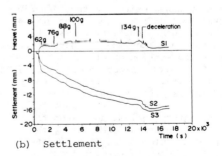

(b) Settlement

Fig. 6 Test results, SC 1

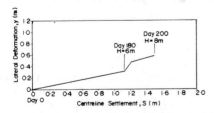

Fig. 5 Lateral displacement of field trial embankment (as function of settlement)

During simulation of the embankment cons-
ruction only a small heave was observed
at the toe, and this recompressed once
further consolidation was allowed after end
of construction. At the end of 2.08 hours
acceleration at 100 g (2.37 yrs) a total
settlement of 15.3 mm (1.53 mm) was measured beneath the embankment centre. For an
average U of 0.93, a final maximum consolidation of 11.6 mm (1.16 m) was expected,
giving a total settlement of the treated
foundation 16.1 mm (1.61 m) and a settlement improvement, n, of 1.36. This was
close to that of the prototype. End of
test observation of the model, consistent
with the above, indicated no apparent failure even after the acceleration was
brought to the maximum available, 134 g
(12.3 m fill). The spaghetti indicators
deformed uniformly with depth without any
sharp changes or breaks (Figure 7) and no
crack was observed at the embankment slope.
It can also be seen from Figure 7 that the
columns deformed with the surrounding soil,
both columns and clay appeared to have
settled together with equal strain.

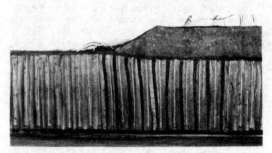

Fig. 7 Cross section showing final profile
of model SC 1

Figure 8 shows the initial and final clay
shear strength measurements from the model
foundation together with calculated strength gains obtained using a method described
by Barksdale (1987). Significant strength
increase through consolidation was observed
beneath the embankment centre but not under
the untreated section or at a location 200
mm (20m) forward of the toe where consolidation
had been minimal.   Based on these estimated and measured strength gains of the
foundation, calculations were made for
factor of safety against undrained failure.
Total stress Fellenius analysis by the computer program 'Slope' was used to determine
a possible slip surface. The frictional
strength of the columns was again accounted
for using the method described by Barksdale
(1987). Consistent with the observation,

the factors of safety were greater than
unity indicating no failures. Figure 9
shows the column load measurement and
stress concentration ratio, m (defined as
the ratio of vertical stress in column to
vertical stress in clay) of the model
embankment. In general as the soil consolidates, increases of soil strength (and
degree of confinement from the soil to the
column) enabled the column to sustain greater load. Although no measurements of
column loads had been made in the prototype, the results of the model were consistent with that observed at other field
sites on sand columns. For example Aboshi
et al. (1979) found an increase in m from 1
to 4 as the embankment was raised, and an
increase in m from 4 to 5 after end of
construction.

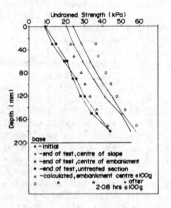

Fig. 8 Undrained strength profile, model
SC 1

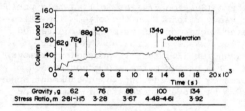

Fig. 9 Column loads, model SC 1

5 CONCLUSION

The centrifuge technique had enabled a comparative study of a controlled embankment
site to be undertaken, retaining essential
features of the prototype. The centrifuge
modelling permits monitoring of consolidation effects within a few hours.

Inclusion of the columns reduces soil drainage path lengths, and increases the rate of excess pore pressure dissipation. Installation of the columns by displacement techniques such as use of the casing driver for sand may reduce soil permeability due to formation of smear zones along the column boundary but since the columns are closely spaced, rapid dissipation of excess pore water pressure can still be achieved.

The behaviour of the composite soil/sand column foundation is complex with simultaneous and interdependent changes in stress ratios and resulting stiffness in both soil and column. In general as the soil consolidates, strength gain increases the degree of confinement afforded by the soil to the column, permitting greater stresses to be transferred from the settling ground to the column.

ACKNOWLEDGEMENT

The field data is made available by the Malaysian Highway Authority.

REFERENCES

Aboshi, H. & N. Suematsu 1985. The State of the Art on Sand Compaction Pile Method. Proceedings 3rd. NTI International Geotechnical Seminar, Singapore: 1 - 12.

Aboshi, H., E. Ichimoto, K. Harada & E. Emuki 1979. A Method to Improve Characteristics of Soft Clays on Inclusion of Large Diameter Sand Columns. Proceedings International Conference Soil Reinforcement - Reinforced Earth and Other techniques, Paris: 211 - 216.

AIT 1989. Laboratory Test Data on Samples from the Muar Flat Test Embankment. Asian Institute of Technology, Research Report Phase II, March.

Barksdale, R.D. 1987. State of the Art for the Design and Construction of Sand Compaction Piles. US Army Corps of Engineers, Report No. REMR-GT-4, Washington DC, 55.

Craig W.H. & P.W. Rowe 1981. Operation of Geotechnical Centrifuge from 1970 - 1979. Geotechnical Testing Journal, Vol. 4: 9 - 25.

Craig, W.H., Bujang B. K.H. & C.M. Merrifield 1991. Simulation of Climatic Condition in Centrifuge Model Test, ASTM Geotechnical Testing Journal, Vol. 14: 406 - 412.

Greenwood, D.A. 1990. Load Tests on Stone Columns. Proceedings ASTM Symposium on Design, Sand Columns and Other Related Techniques, Las Vegas, January.

Greenwood, D.A. & K. Kirsch 1984. Specialist Ground Treatment by Vibratory and Dynamic Method. Piling and Ground Treatment, Thomas Telford Limited, London: 17 - 45.

Hansbo, S. 1987. Design Aspects of Vertical Drains and Lime Column Installation. Proceedings 9th South East Asian Geotechnical Conference, Bangkok: 8.1 - 8.12.

LLM 1987. Express Way Lay - By in the Sungai Muar Plain. Assessment of Ground Improvement Methods, Factual Report on Geotechnical Investigation, Malaysia Highway Authority, Vol. Kuala Lumpur, June.

LLM 1989. Prediction and Performance. Proceedings of International Symposium on Trial Embankments on Malaysian Marine Clay, Malaysian Highway Authority, Kuala Lumpur, November 6 - 8.

Masaaki, T. & K. Masaki 1990. Sand Compaction Pile Methods for Soft Clay Improvement. Design and Actual Performance. Port and Harbour Research Institute Bulletin No. 68: 76 - 91.

Mizuno, Y., W. Shibata & Y. Kanda 1989. Trial Embankment with Sand Compaction Pile Method at Muar Flats. Proceedings International Symposium on Trial Embankments on Malaysian Marine Clays, Kuala Lumpur, Vol. 2: 2.53 - 2.66.

Shinsha, H.K. Takata, Y. Kurumada & N. Fujii 1991. Centrifuge Model Tests on Clay Partly Improved by Sand Compaction Piles. Proceedings Conference Centrifuge 91. (Editors) Ko and McLean. Balkema, Rotterdam: 311 - 318.

Ting W.H., T.F. Wong & C.T. Toh 1987. Design Parameters for Soft Ground in Malaysia. Proceedings 9th. South East Asian Geotechnical Conference, Bangkok: 5.45 - 5.60.

Zelikson, A. 1969. Geotechnical Models using the Hydraulic Consolidation Similarity Method, Geotechnique Vol. 19: 495 - 508.

Settlement prediction at Muar Flats using Asaoka's method

Saaidin Abu Bakar
Public Works Institute, Malaysia (IKRAM)

ABSTRACT: Asaoka's method with different time intervals and varying the time settlement data is use to predict the settlement performances of 3m and 6m embankment at Muar Trial. Varying time interval between 30 and 100 days does not significantly affect the primary settlement prediction and using 1 to 3.5 years settlement data gives a 10% final settlement prediction accuracy. The back calculated field c_v from Asoaka method agrees with those using Terzaghi curve fitting and square time method.

1 INTRODUCTION

There are numerous methods of predicting the magnitude of consolidation settlement, these includes the conventional Terzaghi one dimensional theory, two dimensional Rendulic theory, Skempton-Bjerrum 3 dimensional, stress path, elastic theory, finite element etc.

Inspite of all these advancement prediction of settlement has remain more of an art than of an established procedure. It remain a fact that a soil is a variable medium influence by deposition and other nature's element. It is therefore not surprising that Terzaghi consolidation theory couple with local experience is still widely used by practising engineer. Therefore using settlement data can be a useful tool to predict the final settlement, since settlement plate is cheap and easily install.

2 PREDICTION METHODS

There are a few methods that used settlement data to predict settlement e.g (Tan, 1977), (Asaoka, 1978), (Magnan, 1980) etc. By using the observed settlement data many uncertainties regarding the variability of soil, magnitude and distribution of load can be minimize. Therefore using the settlement data gives a good estimation of final settlement for practicing engineer.

(Magnan, 1980) shows that good settlement prediction using Asaoka method is possible after 60% consolidation is achieved and the accuracy is between 10% to 20%. (Hudson, 1990) using Tan's method indicate that extrapolation using early stage data can lead to unreliable prediction for more than 3 years.

In this paper Asaoka method is used to predict the primary and secondary settlement using the settlement data from the 3m and 6m control embankments at Muar Trial. Back calculated c_v value using curve fitting method proposed by Terzaghi and square root time is also compare with those obtain from Asaoka method.

3 DESCRIPTION OF EMBANKMENTS

The trial embankments were constructed in 1987 by Malaysian Highway Authority (MHA) to assess the effectiveness of various soil improvement methods used in the construction of embankment on Malaysian soft marine clay. Prior to construction the site were cultivated with rubber trees. It is located adjacent to the existing

DEPTH (m) +2.5mRL		SOIL DESCRIPTION	DOMINANT MINERALS	GRAIN SIZE (%)				k_h (m/sec)	$\frac{Cc}{1+e_0}$	Pc (kPa)
				Clay	Silt	Sand	Gravel			
+0.5	CRUST	Yellowish brown mottled red CLAY with roots, root holes and laterite concretions	—	62	35	3	0	—	0·3	110
-5.6	UPPER CLAY	Light greenish grey CLAY with a few shells, very thin discontinuous sand partings, occasional near vertical roots and some decaying organic matter (less than 2%)	Kaolinite Montmorillonite Illite Quartz	45	52	3	0	4×10^{-9}	0·5	40
-15.3	LOWER CLAY	Grey CLAY with some shells, very thin discontinuous sand partings and some decaying organic matter (less than 2%)	Kaolinite Montmorillonite Illite Quartz	50	47	3	0	1×10^{-9}	0·3	60
	PEAT	Dark brown PEAT with no smell (carbon dated to 10,000 years BP)								
-15.9 -19.9	SANDY CLAY	Greyish brown sandy CLAY with a little decaying organic matter	—	20	36	44	0	2×10^{-7}	0·1	60
	SAND	Dark grey very silty medium to coarse SAND (SPT greater than 20)	—	4	20	71	5	—	—	—

Fig: 1 General sub soil profile at Muar Flats.

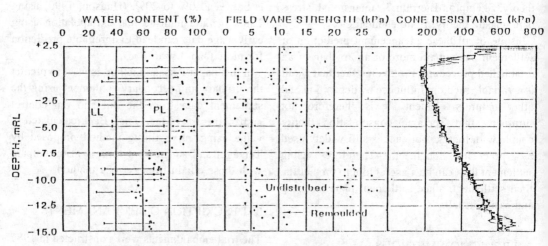

Fig: 2 Atterberg limit, undrained strength and cone resistance for Muar Flats.

Table: 1 Predicted primary consolidation settlement using Asaoka method with different Δt and period of data used.

Period of data used for prediction	Δt day	0.5 year	1.0 year	1.5 year	2.0 year	2.5 year	3.0 year	3.5 year
Predicted primary consolidation settlement using different Δt for 3m embankment (m)	15	0.84	0.98	1.03	1.07	1.11	1.15	1.18
	30	1.00	1.03	1.06	1.09	1.13	1.16	1.20
	60	-	1.00	1.05	1.08	1.07	1.15	1.19
	100	-	1.06	1.09	1.09	1.12	1.15	1.19
Predicted primary consolidation settlement using different Δt for 6m embankment (m)	15	0.87	1.34	1.40	1.47	1.54	1.59	1.62
	30	-	1.29	1.35	1.42	1.48	1.54	1.56
	60	-	1.30	1.35	1.42	1.48	1.62	1.57
	100	-	1.33	1.39	1.47	1.54	1.59	1.62

North South Highway about 1km from one of the meandering of Muar river.

The fill material is lateritic soil with plasticity index under 45% and its compacted bulk density of 1.85 Mg/m^3. The existing ground were not disturbed except for the removal of undergrowth and roots. The subsoil condition profile and strength properties are shown in Figs: 1 and 2. For further details regarding the sub soil, please refer to the Proceeding of the "International Symposium On Trial Embankment On Malaysian Marine Clay", by MHA. Settlement plate data located at the centre of the 3m and 6m

Table: 2 Comparisons between predicted settlement using Asaoka method and observed performance for 3m embankment.

| Δt(day) | Predicted settlement using Asaoka method for 3m embankment | | | Actual settlement recorded (m) |
	Primary settlement (m)	Secondary settlement (m)	Predicted total settlement (m)	
15	1.18	0.30	1.48	1.25
30	1.20	0.15	1.35	1.25
60	1.19	0.04	1.14	1.25
100	1.19	0.01	1.20	1.25

embankment is used for this study. These are control embankment and no improvement is done to the subsoil.

Performance of the 3m and 6m control embankments in term of plots of height of fill, settlement, and excess pore water pressure and lateral movement against time is shown in Figs: 3 and 4. Settlement monitoring of the embankment is on going. The excess pore water pressure is not dissipating but the embankment is still settling.

4 PREDICTIONS USING SETTLEMENT DATA

Asaoka method is use to predict the primary and secondary consolidation settlement using the procedure outlined by Magnan in (Brand 1980). Back calculated field c_v is also computed using Terzaghi curve fitting, square root time and Asaoka method.

In Asaoka method different time interval of 15, 30, 60, and 100 days with varying duration of settlement data of 0.5, 1.0, 1.5, 2.0, 2.5, 3.0 and 3.5 years are use tp predict the primary consolidation settlement and is summarized in Table: 1. Varying time interval between 30 and 100 days with settlement data exceeding 1 years does not significantly change the prediction. Varying the period of settlement data used between 1.0 and 3.5 years gives 10% accuracy primary settlement prediction. Therefore time interval exceeding 30 days with 1 year settlement data gives a reasonable good primary settlement prediction.

Predicted secondary settlement is also computed using Asaoka method as outline by Magnan in (Brand 1980). Comparison between total settlement predicted using 15, 30, 60, 100 days time interval with 3.5 years settlement data and actual settlement recorded is shown in Table:2 and 3 for the 3m and 6m embankment respectively. For the 3m embankment the 100 days interval give a good prediction. For the 6m embankment the settlement is over predict by 30%. This is due to large lateral movement

177

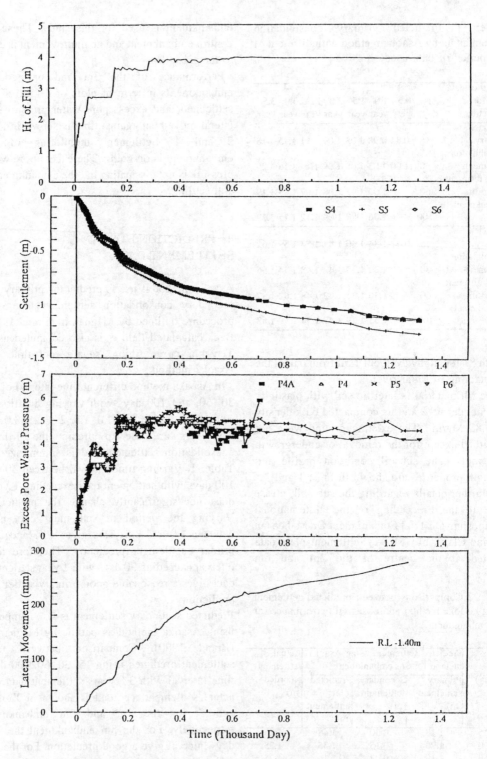

Fig: 3 Actual performance of 3m embankment.

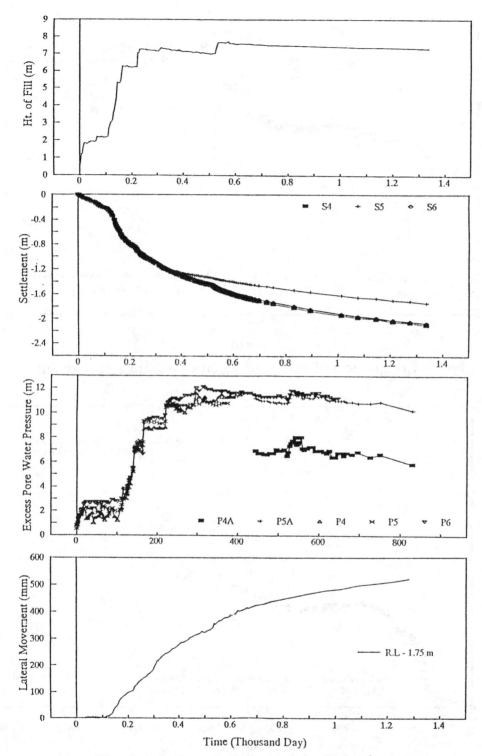

Fig: 4 Actual performance of 6m embankment.

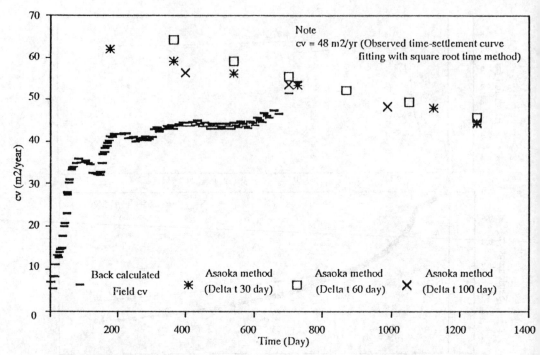

Fig: 5 Comparison of back calculated c_v for 3m embankment.

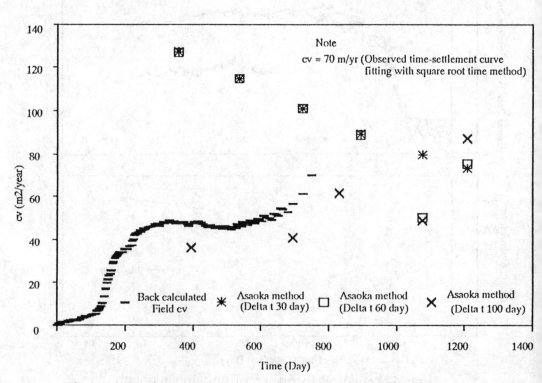

Fig: 6 Comparison of back calculated c_v for 6m embankment

Table: 3 Comparisons between predicted settlement using Asaoka method and observed performance for 3m embankment

Δt(day)	Predicted settlement using Asaoka method for 6m embankment			Actual settlement recorded (m)
	Primary settlemen t (m)	Secondary settlement (m)	Predicted total settlement (m)	
30	1.56	0.69	2.25	1.71
60	1.56	0.74	2.30	1.71
100	1.64	0.68	2.32	1.71

experience by the 6m embankment.

Back calculated c_v using Asaoka method ssuming 18m thick, two way drainage is compare with those using Terzaghi heoretical curve fitting and square root time method is shown in Figs: 5 and 6. The three method used give similar c_v for the 3m embankment and is approximately equal to 50 m^2/yr. For the 6m embankment the square root time method gives a slightly higher c_v when compare to those computed from Asaoka and curve fitting method. Therefore the c_v value computed from the Asoaka and curve fitting method for 3m embankment gives a reasonably good field c_v.

From Figs: 5 and 6 the back calculted c_v for Muar Flats is 50 m^2/yr and is approximately 15 times the laboratory c_v.

5 CONCLUSIONS

Asaoka method gives a reasonably good settlement prediction when the time interval exceeding 30 days and settlement data of 3 years is used. The back calculated field c_v from Asaoka plot is similar to those obtained from Terzaghi method for embankment less than 3m height.

REFERENCES

Brand, E.W. & Brenner, R.P. 1980.*Soft clay engineering*, Elsevier Scientific Publishing Company, Amsterdam, Oxford, New York.
Hudson, R.R. 1990, Low embankment on soft ground, Seminar on geotechnical aspects of north south highway, PLUS.
Malaysian Highway Authority. 1989 Symposium on trial embankment on malaysian clays, vol 1 and 2.

Prediction versus Performance in Geotechnical Engineering, Balasubramaniam et al. (eds)
© *1994 Balkema, Rotterdam, ISBN 90 5410 355 8*

Effect of embankment stiffness on deformations of soft clay foundation

Madhira R. Madhav
Institute of Lowland Technology, Saga University, Japan & I.I.T., Kanpur, India

Norihiko Miura
Institute of Lowland Technology, Saga University, Japan

ABSTRACT: An approach considering the geometry and relatively high shear stiffness of the embankment, and the compressibility and shear stiffness of the soft soil, is presented to analyze the interaction between the two. The analysis, with and without shear stiffness of the soft soil, indicates that the embankment stiffness has more significant effect in the former case. A new parameter and a criterion for identifying and quantifying the relative embankment rigidity is proposed. A parametric study reports the influences of the various parameters considered. Application to a case study is illustrated.

INTRODUCTION

In the analysis of stability of embankments on soft clays, the strength of the embankment is usually taken into account. However, the stiffness of the embankment is rarely considered in the prediction of settlements of embankments on the same clay foundations. Often, the solution obtained by Jurgenson (1937) or the influence chart developed by Osterberg (1957) are used to compute the vertical stresses in the soil due to embankment type of load on a semi-infinite medium. The above approach implies that the embankment has only weight and no stiffness nor rigidity as a structural element. Alternatively, solutions are available (Perloff et al 1976) for the distribution of vertical stresses within and below long elastic embankments which are continuous with the underlying material, based on the premise that the embankment and the foundation soil have the same elastic properties. The solutions obtained indicate that the vertical stresses beneath the center of the embankment are smaller than those computed by neglecting embankment stiffness, i.e. stresses based on

embankment type of loading. Thus it may be noted that consideration of even a small amount of stiffness of the embankment equal to that of the soft foundation soil reduces the vertical stresses and consequently the settlements of the soil below the center of the embankment. In practice, embankments built by heavy compaction are often very stiff and stronger than the soil on which they are founded.

As an extreme measure, some of the analyses for stability of embankments on soft soils consider the embankment to be rigid. It is particularly true and may probably be appropriate in case of reinforced embankments. Jenner et al. (1988) use slip line fields to assess the improvement in bearing capacity of an embankment with a cellular mattress at its base. The mattress is assumed to be rigid and the theory of plasticity for metal pressing and extrusion applied for the prediction of the improved bearing capacity of soft cohesive soils under stiff rough embankment base. In their analysis of the collapse of reinforced embankments on soft ground, Hird et al. (1990) also treat the reinforced embankment as rigid, and evaluate the foundation stability as a bearing capacity problem.

For embankments built on soft soils, the stiffness of the built up embankment is considerably different from that of the in situ soil. Often, the strength and the stiffness of the compacted embankment is one or more orders of magnitude higher than those of the soil on which it is founded. Because of its stiffness, the embankment could act as a semi-rigid or a rigid structural member. As a result of the interaction between the stiff embankment and the soft soil, embankment gets to redistribute the load and tends to settle more uniformly.

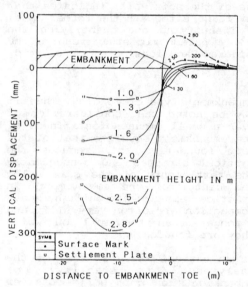

Fig.1 Vertical displacement profiles
(after Romalho et al, 1983)

Settlements from carefully conducted measurements (Figure 1) from a test embankment (Ramalho et al. 1983) indicate uniform settlements along the whole width of the embankment possibly as a consequence of redistribution of embankment load due to its stiffness. The embankment failed at a height of 2.8m. But up to a height of 2.0m, the settlements were uniform beneath the embankment. Also the displacements of the ground outside of the embankment, were very small up to an embankment height of

2.0m, beyond which there was considerable heave. Absence of displacements outside of the embankment width indicates a Winkler type of response of the soft clay. The differences in the settlements reported by Crawford et al. (1992) can be reasoned to be a result of embankment stiffness particularly because they are relatively high, 12.4m and 11.4m. In this paper, a simple approach is presented to account for embankment stiffness on deformational response of the embankment-soft soil system.

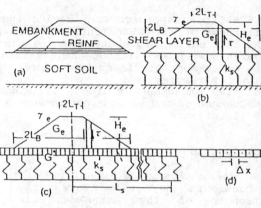

Fig.2 Definition sketch

PROBLEM DEFINITION

An embankment of height, H_e, base width, L_B, top width, L_T, unit weight, γ_e, resting on a soft soil is considered (Figure 2). Since embankments are constructed by heavy compaction, their compressibility may be neglected in comparison to that of soft foundation below, unless they are very high. The embankment is assumed to act as a shear layer and is characterized by its shear modulus, G_e. The foundation soil is usually a highly compressible soft clay deposit and can be characterized by its deformation modulus, E_s, or the subgrade modulus, K_s. These soft soils may have some small but finite shear stiffness. In the analysis presented two cases are considered:

I Soft Soil with no Shear Stiffness; and

II Soft Soil with Shear Stiffness.

The compressibility of the soil is represented by the modulus of subgrade reaction, K_s, and the shear stiffness by the shear parameter, G_s^* ($=G_s*H_s$, where G_s is the shear modulus of the soil and H_s- a length parameter). Using Vlasov and Leontev's theory, (Scott 1981) K_s and G_s^* can be related to the modulus of deformation and thickness of the soil. Or they can be estimated from Fletcher and Herrmann (1971) who compared the results from one and two parameter foundation models with those from a semi-infinite medium.

I STIFF EMBANKMENT ON SOIL WITH NO SHEAR STIFFNESS

The model in this case (Figure 2b), consists of an embankment shaped trapezoidal shear layer representing the embankment, resting on a Winkler medium representing the soft soil. The equations for the response of the model are

For $0 < x < L_T$

$$\gamma_e H_e = K_s.w - G_e.H_e.\frac{d^2w}{dx^2} \quad (1)$$

and for $L_T < x < L_B$

$$\gamma_e H_e(x) = K_s.w - \frac{d}{dx}(G_e.H_e(x)\frac{dw}{dx}) \quad (2)$$

where w is the displacement, x - the distance from the center of the embankment, $H_e(x)=(L_B-x)/(L_B-L_T)$ and K_s - the modulus of subgrade reaction of the soft clay foundation. The weight and the stiffness of the embankment vary linearly with distance in the sloping zone. Equations (1) and (2) are normalized and rewritten as

For $0 <= X <= \alpha$

$$1 = W - \beta_e \frac{d^2W}{dX^2} \quad (3)$$

and for $\alpha <= X <= 1$

$$\frac{(1-X)}{(1-\alpha)} = W - \{\beta_e \frac{(1-X)}{(1-\alpha)} \frac{d^2W}{dX^2}\}\frac{}{} + \frac{\beta_e}{(1-\alpha)} \frac{dW}{dX} \quad (4)$$

where $X=x/L_B$, $\alpha=L_T/L_B$, $W=w/w_0$, $w_0=\gamma_e H_e/K_s$, and $\beta_e=G_e.H_e/K_s.L_B^2$. The

boundary and continuity conditions are

$$X = 0; \quad \frac{dW|1}{dX} = 0 \quad (5)$$

$$X = \alpha; \quad W|1 = W|2 \text{ and } \frac{dW}{dX}|1 = \frac{dW}{dX}|2 \quad (6)$$

and $X = 1; \quad W|2 = 0 \quad (7)$

where 1 and 2 refer to solutions for the flat and the sloping portions of the embankment, respectively.

CASE II STIFF EMBANKMENT ON SOIL WITH SHEAR STIFFNESS

The embankment resting on a soil with small but finite shear stiffness is modelled (Figure 2c) as a shear layer and loading of trapezoidal shape overlying a Pasternak foundation defined by the parameters, K_s, and G_s^*. The equations governing the response of the embankment are

For $0 <= x <= L_T$

$$\gamma_e H_e = K_s.w - (G_s^*+G_e.H_e)\frac{d^2w}{dx^2} \quad (8)$$

for $L_T <= x <= L_B$

$$\gamma_e H_e(x)=K_s w-\{G_s^*\frac{d^2w}{dx^2} - \frac{d}{dx}[G_e.H_e(x)\frac{dw}{dx}]\}(9)$$

and for $L_B < x$

$$0 = K_s.w - G_s^* \frac{d^2w}{dx^2} \quad (10)$$

Once again normalizing and simplifying Equations (8) through (10), one gets

For $0 <= X <= \alpha$

$$1 = W - (\beta_s + \beta_e)\frac{d^2W}{dX^2} \quad (11)$$

for $\alpha <= X <= 1$

$$\frac{(1-X)}{(1-\alpha)}=W-\{\beta_s+\beta_e\frac{(1-X)}{(1-\alpha)}\frac{d^2W}{dX^2}\}\frac{}{}+\frac{\beta_e}{(1-\alpha)}\frac{dW}{dX}(12)$$

and for $1 <= X$

185

$$0 = W - \beta_s \frac{d^2W}{dX^2} \qquad (13)$$

with $\beta_s = G_s^*/K_s.L_B^2$. It may be noted that case I is a particular case of Case II with $\beta_s = 0$. Equations (11) through (13) in finite difference form are

for $i \leq \alpha N$

$$1 = W_i - (\beta_e + \beta_s)\{W_{i-1} - 2W_i + W_{i+1}\}/(\Delta X)^2 \quad (14)$$

for $\alpha N < i < N$

$$(1-X_i)/(1-\alpha) = W_i - [\beta_s + \beta_e(1-X_i)/(1-\alpha)].$$

$$\{W_{i-1} - 2W_i + W_{i+1}\}/(\Delta X)^2 + [\beta_e/2(1-\alpha)][W_{i+1} -$$

$$W_{i-1}]/\Delta X \qquad (15)$$

and for $N < i < 4N$

$$0 = W_i - \beta_s\{W_{i-1} - 2W_i + W_{i+1}\}/(\Delta X)^2 \quad (16)$$

where N is the number of elements into which the embankment base width is divided into, so that $\Delta X = 1/N$, and $X_i = (i-0.5)\Delta X$ - the distance to the center of the ith element. The displacements are verified to be zero beyond a distance of $L_s = 4L_B$. Hence, the extent of the foundation soil is limited to $4L_B$. Equations (14) through (16) are evaluated for the settlements or displacements along and outside the base of the embankment.

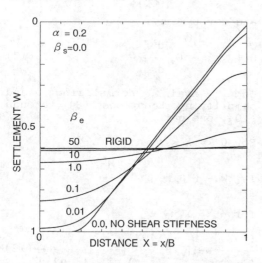

Fig.3 Settlement profile: Soft soil with no shear stiffness

RESULTS

Equations (14) through (16) are solved by choosing N = 50 and iteratively with a convergence criterion

$$\{W_{new}(i) - W_{old}(i)\}/W_{new}(i) < 10^{-7} \quad (17)$$

is satisfied. Increase in the value of N beyond 50 did not improve the accuracy of the results any further. A parametric study has been conducted for the following ranges of the parameters:

$\alpha = 0.1$ to 0.9; $\beta_e = 0$ to 10 and $\beta_s = 0$ to 0.1

The parameter α signifies the shape of the embankment while the parameters β_e and β_s represent the shear stiffnesses of the embankment and soft soil, respectively. The variation of displacements with distance for embankments with relatively large base width ($\alpha=0.2$) resting on soft soil with no shear stiffness ($\beta_s = 0$), are depicted in Figure 3 for different values of embankment stiffnesses (β_e). If the stiffness of the embankment is small or is neglected as is often done in practice, i.e. $\beta_e = 0$, the settlement profile is an image of the embankment shape. In this case, the settlement is uniform over the crest width and decreases linearly along the slope and is nearly zero at the toe or edge of the embankment. With increasing values of embankment stiffness, β_e, the central settlements decrease and the edge settlements increase and a tendency for uniform settlement pattern can be noticed. For β_e values of 0.01, 0.1, 1.0, and 10.0, the central settlements are 0.98, 0.85, 0.67, and 0.62 while the toe or edge settlements are 0.06, 0.24, 0.52, and 0.59, respectively. As the stiffness of the embankment increases, it acts more and more as a rigid body and when $\beta_e > 50$, the embankment acts as a perfectly rigid block with a uniform settlement of 0.6. In practice, if the relative stiffness of the embankment is more than 10, it can be considered to be rigid if resting on a soft soil with no shear stiffness.

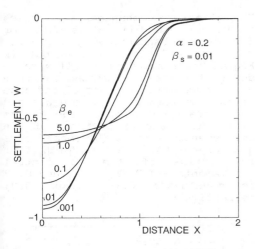

Fig.4 Settlement profile:Soft soil
with shear stiffness

If the embankment rests on soft soil
with shear stiffnesses of β_s equal to
0.01 and 0.1, its settlement
profiles are as presented in Figures
4 and 5, respectively. These results
are very similar to those presented
in Figure 3 in that the settlements
decrease with distance from the
center and tend to become uniform
with increasing values of β_e. The
response of the embankment on soil
with shear stiffness differs with
that of an embankment on soil with
no shear stiffness in two aspects.

Firstly soil with shear stiffness (β_s
>0) can transmit stresses to outside
and beyond the loaded width. For
the cases studied, the width over
which the settlements are observed,
extended up to a distance of three
times of embankment width (L_B) beyond
its toe. Pasternak foundation model
can account for this continuity of
the foundation due to its shear
stiffness and the spread of the load
over greater widths. Consequently,
the settlements beneath the crest of
the embankment are smaller and those
of the toe are more than the
corresponding settlements of
embankments on soil with no shear
stiffness (β_s=0). Secondly, the
effect of embankment stiffness in
equalizing the central and toe
settlements is less significant. For
an increase of β_e from 0 to 1.0, the
central settlements decreased from
0.95 to 0.62 for β_s=0.01 and 0.81
to 0.66 for β_s=0.1.

The effect of embankment shape as
described by the parameter α,
($=L_T/L_B$), on embankment deformations
is shown in Figure 6. If α is
smaller, the embankment is
relatively flatter and has a larger
slope width than when α is higher.
As the top width of the embankment
increases relative to the base
width, it is steeper and has more
self weight. If the stiffness of the
embankment is neglected, the central
settlement increases from 0.81 to

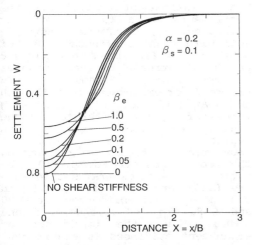

Fig.5 Settlement profile: Soft soil
with shear stiffness

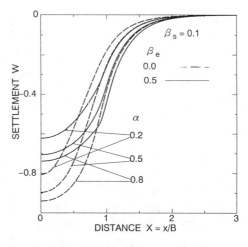

Fig.6 Effect of embankment shape

187

0.94 for α increasing from 0.2 to 0.8. If the embankment is five times stiffer than the soil (β_e=0.5 and β_s=0.1), the central settlements reduce to 0.62 and 0.74 respectively for α values of 0.2 and 0.8. The percentage reductions, 76.5% and 78.4%, are nearly the same in both cases   Thus curves for α=0.2 can be used for other values of α, if percentage changes in settlements are being considered.

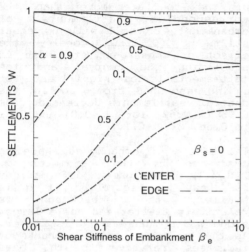

Fig.7 Central and toe settlements
     Soil with no shear stiffness

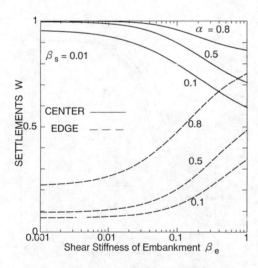

Fig.8 Central and toe settlements
     Soil with shear stiffness

The effect of embankment stiffness on center and toe settlements for different embankment shapes (α=0.1, 0.5, and 0.9) is brought out in Figure 7 for foundation soil with no shear stiffness. As stated earlier, as β_e increases, the central and edge or toe settlements tend to equal and the embankment tends to become rigid. The rate of increase of toe settlements is rapid for embankments with higher values of α. For β_e=1.0, the ratios of toe to central settlements are 0.75, 0.86, and 0.97 for α values of 0.1, 0.5, and 0.9, respectively.

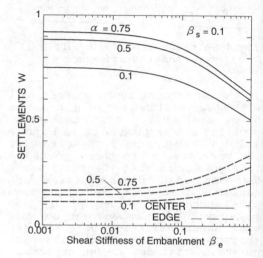

Fig.9 Central and toe settlements
     Soil with shear stiffness

The variations of central and toe settlements as a function of its stiffness, β_e, for embankments on soil with shear stiffnesses, β_s, of 0.01 and 0.1 are shown in Figures 8 and 9. The trends in these figures are very similar to those noted in Figure 7 for embankment on soil with no shear stiffness. With increasing values of β_s, the difference in the central and toe settlements is smaller, and the effect of embankment stiffness in equalizing the two is lesser. For β_e=1.0, the ratios of toe to central settlements are 0.576 and 0.883 for α=0.1 and 0.8 and β_s=0.01; while they are 0.39 and 0.54 for α=0.1 and 0.75 and β_s=0.1. Thus, the importance of embankment stiffness is significant if they are founded on softer soils than on stiffer ones.

SETTLEMENT OF RIGID EMBANKMENTS

If an embankment is reinforced at its base or throughout its height, and acts a rigid shear layer, its uniform settlement, W_r, can be predicted (Scott 1981), as

$$W_r = (1 + \alpha)/\{2(1 + 2\ \beta_s)\} \qquad (18)$$

Table 1 tabulates selected values of W_r.

Table 1. Settlements of rigid embankments

β_s	α				
	0.1	0.25	0.5	0.75	1.0
0	.550	.625	.750	.875	1.0
0.0001	.539	.613	.735	.858	.980
0.001	.517	.588	.705	.822	.940
0.01	.458	.521	.624	.728	.833
0.1	.337	.383	.459	.535	.613

APPLICATION.

Crawford et al. (1992) describe a case history of settlements of two test embankments (Figure 10), one founded on vertical drain treated ground and the other on untreated one. Even though the heights (12.4m and 11.4m) and the base widths (83m and 84m) are comparable, the settlements of the two embankments were very different. The West Abutment test fill settled by 2.5m to 3.25m, while the Waterline test fill settled by 0.4m to 0.65m. The authors postulate that the difference in the settlements is most likely due to sample disturbance or to spreading of the load in the upper regions of the soil. While the heights and the base widths in the cross-section of the two embankments are nearly the same, the overall shapes of the test fills, i.e. the heights of the

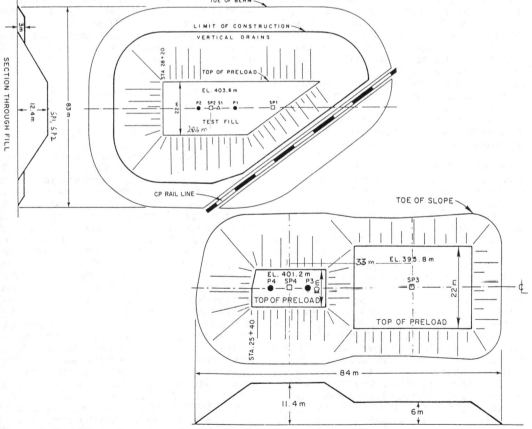

Fig.10 Details of test embankment
(after Crawford et al, 1992)

189

Table 2. Comparison of predicted and measured settlements

FILL	L_B m	H_e m	p_0 kPa	K_s kN/m^3	α	β_e	w_0 m	W_r	SETTLEMENT Pred. m	Meas. m	POINT
WEST ABUTMENT											
CS	62.5	12.4	252	72	0.35	5.3	3.5	.675	2.36	2.56	SP_1
LS	103.0	12.4	252	44	0.50	3.2	5.7	.75	4.27	3.28	SP_2
WATERLINE											
CS1	51.0	11.4	226	88	0.38	6.0	2.6	.69	1.86	0.67	SP_4
CS2	52.0	6.0	120	86	0.6	3.1	1.4	.8	1.12	0.4	SP_3
LS1	40.0	11.4	226	112	0.25	7.6	2.0	.63	1.24	0.67	SP_4
LS2	38.0	6.0	120	118	0.58	4.2	1.0	.79	0.81	0.4	SP_3

CS - Cross Section; LS - Longitudinal Section.

berms, 3m and 6m, and the lengths in the longitudinal direction, 103m and 40m, respectively, are different. The fill material, consisting of clean coarse sand and gravel, has been compacted to an average unit weight of 20.3 kN/m^3. Following Scott (1981), its modulus of deformation can be of the order of 300 MN/m^2. The shear modulus is taken to be 120 MN/m^2. The modulus of subgrade reaction for the in situ soil with C_u equal to about 50 kPa, is taken to be 1500 kN/m^3 for 0.3m wide plate. The embankments transfer stresses of intensity, 252 kPa, 226 kPa and 120 kPa. Table 2 summarizes and compares the relevant data for the two fills. It should be noted that while the settlement plate, SP_2, is beneath the center of the embankment in both directions, SP_1 is below the top edge of the West Abutment in the longitudinal direction. SP_3 and SP_4 are below 6m and 11.4m high embankments, respectively. The settlements are predicted based on the premise that both the embankments are rigid and the shear stiffness of the in situ soil can be neglected. The predicted settlements compare reasonably with the measured ones. It should be noted that SP_3 and SP_4 are on untreated ground and so would continue to settle for a long period than just during the period of observation reported (365 days). If the stiffness of the in situ soil (β_s) is taken as equal to 0.01, the settlements reduce by a factor of 1.2 and get closer to the observed

ones and enhance the value of the predictions. The analysis presented in this paper accounting for the shear stiffness of the embankment explains partly at least, the differences in settlements reported, and corroborates the postulate and quantifies the effect of load redistribution along the base of the embankment.

CONCLUSIONS

Embankments on soft clays possess significant shear stiffness, as they are often built by heavy compaction. Presently, settlements of soft soil underlying the embankment are predicted neglecting the shear stiffness of the latter. An approach considering the same is proposed. The embankment is modelled as a shear layer while the foundation soil is considered to be compressible, and without or with shear stiffness. Settlements of the embankment - soft soil system are evaluated using an extended Pasternak type model. The newly proposed relative embankment stiffness parameter β_e, is shown to affect the displacements, particularly if the soft soil has no shear stiffness, $\beta_s=0$. In that case, the embankment acts as a rigid block, if β_e is more than 50. The contribution of embankment stiffness in equalizing settlements decreases with increasing shear stiffness of the soft soil. A parametric study highlights the contributions of the effects due to embankment geometry and stiffness and the stiffness of the soil. An example illustrates the application of the approach to a practical case.

ACKNOWLEDGEMENT: The help of Mr.Y-M. Park in drafting the figures is gratefully appreciated.

REFERENCES

Crawford, C.B., Fannin, R.J., de Boer, L.J., and Kern, C.B. 1992. Experiences with prefabricated vertical (wick) drains at Vernon, B.C., Can. Geotech. J., 29:67-79.

Fletcher, D.Q., and Herrmann, L.R. 1971. Elastic foundation representation of continuum, J. EM. Div. ASCE, 97:95-107.

Hird, C.C., Pyrah, I.C., and Russell, D 1990. Finite element analysis of the collapse of reinforced embankment on soft ground, Geotechnique, 40:633-640.

Jarquio, R., and Jarquio, V. 1984. Vertical stress formulas for triangular loading, J. GT. Div., ASCE, .110:73-78.

Jenner, C.G., Bassett, R.H., and Bush, D.I. 1988. The use of slip line fields to assess the improvement in bearing capacity of soft ground given by a cellular foundation mattress installed at the base of the embankment, Theory & Practice of Earth Reinforcement, IS KYUSHU, p.209-214.

Jurgenson, L. 1937. The application of theories of elasticity and plasticity to foundation problems, J. Boston Soc. of Civil Engr.s.

Osterberg, J.O. 1957. Influence chart for vertical stresses in a scmi-infinite mass due to an embankment loading, Proc. 4th ICOSMFE, 1:393-398.

Perloff, W.H., Baladi, G.Y., and Harr, M.E. 1976. Stress distribution within and under long elastic embankments, TRB Publ.

Ramalho-Ortigao, J.A., Werneck, M.L.G., and Lacerda, W.A. 1983. Embankment failure on clay near Rio de Janeiro, J. GT. Div. ASCE, 109:1460-1479.

Scott, R.F. 1981. Foundation analysis, Prentice-Hall Inc., Englewood Cliffs.

Prediction versus Performance in Geotechnical Engineering, Balasubramaniam et al. (eds)
© *1994 Balkema, Rotterdam, ISBN 90 5410 355 8*

Prediction of performance of MSE embankment using steel grids on soft Bangkok clay

Dennes T. Bergado & A.S. Balasubramaniam
Asian Institute of Technology, Bangkok, Thailand

Jin-Chun Chai
Kiso-Jiban Consultants Co. Ltd, Chiyoda-ku, Tokyo, Japan

ABSTRACT: The behavior of a full scale test reinforced wall/embankment on Bangkok clay has been predicted by finite element method. In the numerical modelling, two aspects have been emphasized: (1) selecting proper soil/reinforcement interface properties according to the relative displacement pattern (direct shear or pullout) of the upper and lower interface elements between soil and reinforcement; and (2) simulating the actual construction process by updating the node coordinates including those of the wall or embankment elements above the current construction level which ensures that the applied fill thickness simulate the actual field value. Finite element results were compared with the field data in terms of excess pore pressures, settlements, lateral displacements, tension forces in the reinforcements, and base pressures. It was found that the foundation settlements and the wall face lateral displacements were predicted reasonably well. In addition, the predicted pore pressures, tension forces in reinforcements, foundation lateral displacements, and embankment base pressures agreed fairly with measured values. Several factors, such as permeability variation of the foundation soil and compaction effects of embankment fill have also been investigated. Furthermore, some of the deficiencies in finite element modelling are also discussed.

INTRODUCTION

For the last two decades, extensive researches and investigations have been carried out to understand the behavior and mechanisms of the reinforced earth structures. The behavior of the reinforced walls and embankments on soft ground have been analyzed by several investigators using finite element methods (e.g. Hird and Pyrah, 1990; Adib et al, 1990). However, the accuracy of the finite element analysis depends mainly on the accuracy of the models used and the correctness of the parameters inputted into the models. Although different soil models, such as nonlinear hyperbolic and modified Cam clay, have been used to represent the stress/strain behavior of the soils, the soil/reinforcement interface properties and actual construction process are not properly simulated. All these factors influence the ability of using finite element method to predict the behavior of reinforced wall/embankment on soft ground.

The most important parameters controlling the performance of the reinforced earth structure, among others, are the soil/reinforcement interface properties. The interface properties are usually determined by direct shear or pullout tests. However, for grid reinforcements, the different soil/reinforcement interaction modes (direct shear or pullout) yield different interface properties. In order to properly simulate the soil/reinforcement interaction behavior, it should be considered in the numerical modelling to use different properties for different interaction mode (Rowe and Mylleville, 1988).

In finite element analysis, the wall or embankment load is applied by one of the following methods: (1) by increasing the gravity of the whole or part of wall or embankment elements; and (2) by placing a new layer of wall or embankment elements. In the case of embankment construction, if the incremental load is applied by increasing the gravity of the whole embankment, in the beginning, the center part of the embankment is loaded more, while the area near the toe is loaded less than the actual value. Furthermore, the stiffness of the embankment may not be modelled properly. However, if the incremental load is applied by applying a new layer of elements, the node coordinates of the embankment elements above the current construction level must be updated to account for the deformation during the construction process. Otherwise, the applied total fill thickness will be more than the actual value especially in the case of embankment on soft ground. Most computer programs used for analyzing the behavior of embankment on soft ground do not treat this factor well. The original CRISP program (Britto and Gunn, 1987) did not consider this factor, and the CESAR program (Magnan and Kattan, 1989) assumed that the mechanical properties of the embankment fill elements exist from the beginning even when the element gravity load is not applied.

In this paper, the concepts of considering the different soil/reinforcement interaction modes and simulating the actual construction process are presented first. A full scale test reinforced embankment is then analyzed by the proposed finite element method. Consequently, the finite element results are compared with the field data. Finally, some of the difficulties in finite element modelling are discussed.

NUMERICAL MODELLING

General aspects

The reinforced wall/embankment on soft ground system has been modelled by finite element method under plane strain condition. All the elements are formulated as isoparameteric elements. Discrete material approach is used to model the reinforced wall/embankment on soft ground system so that the properties and responses of the soil/reinforcement interaction can be directly quantified. The face panel of reinforced wall is modelled by 3-node beam elements with axial, shear and bending stiffness. The reinforcement is modelled by 3-node bar elements. The soil/reinforcement and soil/wall face interfaces are modelled by 6-node zero thickness joint elements. The soil elements are modelled by 8 or 6-node solid elements with or without pore pressure degree of freedoms. Finally, nodal links are used at the free end of reinforcement to allow realistic vertical stress condition (Collin, 1986).

Several soil behavior models are employed in the analysis. The behavior of the soft foundation soil is controlled by modified Cam clay model (Roscoe and Burland, 1968). The linear elastic/perfect plastic model is used to model the heavily overconsolidated clay, and the yielding is controlled by Mohr-Coulomb criterion. The backfill soil is modelled by hyperbolic constitutive model (Duncan et al, 1980). The compaction operation is modelled by bi-linear hysteretic loading/unloading model (Duncan and Seed, 1986). The consolidation process of soft ground is simulated by coupled consolidation theory (Biot, 1941).

In modelling the soil/reinforcement interface behavior, two interaction modes are considered, namely: pullout and direct shear modes. The hyperbolic shear stress/shear displacement model (Clough and Duncan, 1971) is used to represent direct shear interaction mode. Pullout resistance of grid reinforcement consists of skin friction in the longitudinal members and bearing resistance in the transverse members. The skin friction is modelled by linear elastic-perfect plastic model and the pullout bearing resistance is simulated by a hyperbolic bearing resistance model which is only valid for grid reinforcements (Chai, 1992). The techniques of selecting proper interface properties and simulating the actual construction procedure are briefly discussed below.

Modelling different soil/reinforcement interaction modes

Soil/reinforcement interaction mode can either be direct shear or pullout. For grid reinforcement, these two different interaction modes will yield different interface strength and deformation parameters. The interface elements above and below reinforcement work as pair elements and the direct shear (the same sign of shear stresses) and pullout (different sign of shear stresses) soil/reinforcement interaction modes are automatically adopted according to their relative shear displacement pattern.

Pullout of reinforcement especially the grid reinforcement from the soil is a truly three-dimensional problem and it can only be approximately modelled in a two-dimensional analysis. It is assumed that the pullout resistance is uniformly distributed over the entire interface areas. Pullout interface shear stiffness consists of stiffness from skin friction resistance, k_{sf}, stiffness from passive bearing resistance, k_{sp}, respectively. The total equivalent tangential shear stiffness k_s is the sum of k_{sf} and k_{sp}.

$$k_s = k_{sf} + k_{sp} \qquad (1)$$

For both direct shear and pullout interaction modes, when the normal stress at the interface is in tension, a very small (e.g. 100 kN/m³) normal and shear stiffness is assigned to allow the opening and slippage at interface.

Simulating the actual construction process

The actual embankment construction is carried out by placing and compacting the fill material layer by layer. Therefore, in finite element analysis, the incremental load should be applied by placing the embankment elements one layer after another. In analyzing the problems, such as embankment on soft ground, the large deformation phenomenon can be considered by updating the node coordinates during the incremental analysis. In this case where considerable deformation occurs during the construction process, the coordinates of the wall or embankment elements above current construction levels are also corrected based on the following assumptions: (a) the original vertical lines are kept at vertical direction, and the horizontal lines remained straight; (b) the incremental displacement of the nodes above current construction top surface are linearly interpolated from the displacements of the two end nodes of current construction top surface according to their x-coordinates (horizontal direction). This operation ensures that the applied fill thickness is the same as the actual value and, thus, the actual construction process is most closely simulated.

TEST REINFORCED EMBANKMENT AND INPUT PARAMETERS

Test reinforced embankment

The full scale welded steel grid reinforced test embankment was constructed at the campus of Asian Institute of Technology (AIT). The original embankment was 5.8 m (19.5 feet) above the existing ground surface with about 26.0 m (87 feet) base length. It has three sloping faces with 1:1 slope and one vertical front face (wall). The welded wire mats used in the test wall/embankment system consisted of W4.5 (6.1 mm) x W3.5 (5.4 mm) galvanized welded steel wire mesh with 152 mm x 228 mm (6 x 9 inches) grid openings in the longitudinal and transverse directions, respectively. The length of reinforcement was 5 m and the vertical spacing between the reinforcements was 0.45 m. The cross-sectional view of the embankment is shown in Fig. 1a,b. The subsoil profile at the site consists of the topmost 2.0 m thick layer of dark-brown weathered clay overlying a blackish-grey soft clay layer which extends to a depth of about 8 m below the existing ground. The soft clay layer is underlain by a stiff clay layer. A typical subsoil

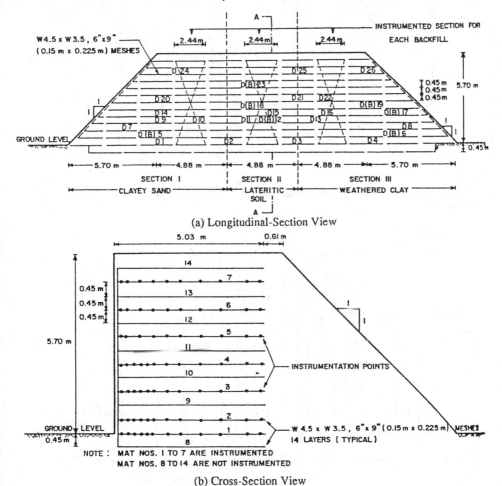

(a) Longitudinal-Section View

(b) Cross-Section View

Fig. 1   Cross-sectional view of the test reinforced wall/embankment.

profile together with the general soil properties at this site are depicted in Fig. 2. The finite element mesh and the boundary conditions are shown in Fig. 3. The bar and beam elements are indicated by darker solid line. For clarity, the interface elements are not shown in the mesh. The horizontal boundaries were selected far enough from the reinforced embankment so that the influence on the structure response will be negligible. The vertical fixed boundary was selected at 12 m depth because as can be seen in Fig. 2, at this depth, the subsoil is very stiff.

Model parameters

The linear elastic-perfectly plastic model parameters for topmost 1.0 m thick weathered clay layer and modified Cam clay parameters for soft to medium stiff clay layers are shown in Table 1. The parameters were determined based on actual test data (Balasubramaniam et al, 1978; Asakami, 1989) and some of them are shown in Fig. 4. Since there is uncertainty of the permeability of the foundation soil, 3 sets of permeabilities, namely: high, middle, and low

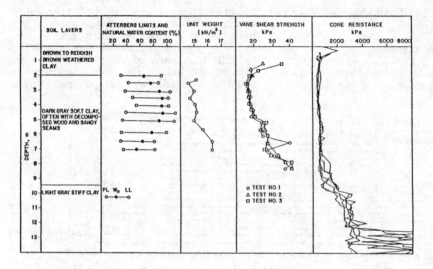

Fig. 2 Index properties, vane shear strengths and Dutch cone resistances of the subsoil at test site.

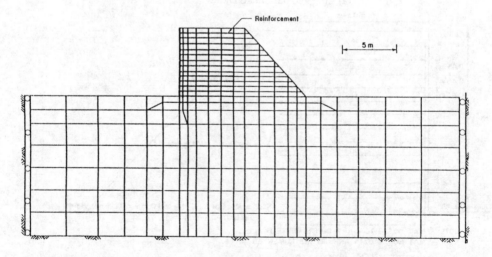

Fig. 3 Finite element mesh used for AIT test reinforced wall/embankment.

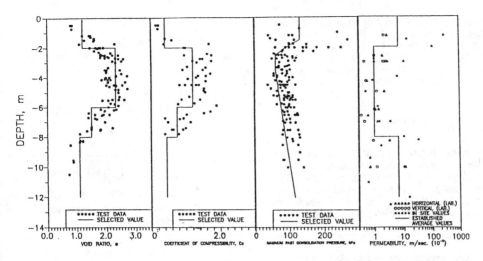

Fig. 4   Void ratiö, compressibility index, past maximum vertical pressure, and laboratory permeability of the Bangkok clay.

permeabilities, were determined based on existing information (Ahmed, 1977; Bergado, 1990) and indicated also in Table 1. The top 2 m weathered clay is overconsolidated with an average overconsolidation ratio (OCR) of 5 and the underlying soil layers are slightly overconsolidated with an average OCR of 1.2.

The hyperbolic, non-linear elastic soil model parameters for compacted lateritic fill material (middle section of the embankment) are tabulated in Table 2. These were determined based on triaxial unconsolidated undrained (UU) test results (Bergado et al, 1988) following the technique established by Duncan et al (1980).

The interface hyperbolic direct shear model parameters were determined from direct shear test results of the fill material (Macatol, 1990). The parameters adopted were: friction angle, ϕ, of 32.5 degrees, cohesion, C, 60 kPa, shear stiffness number, k_1, 10,500, shear stiffness exponent, n_1, 0.72, and failure ratio, R_{f1}, of 0.85. The skin friction parameters between reinforcement frictional surface and lateritic soil were determined from test results of Shivashankar (1991) with adhesion of 50 kPa, skin friction angle of 9 degrees. The spacing between the grid reinforcement bearing member was 225 mm and the diameter of the bearing member was 5.4 mm. For both direct shear and pullout models, the normal stiffness of the interface was defined as 10^7 kN/m³ for compression case and 10^2 kN/m³ for tension case.

For welded wire reinforcement including the wall face, the Young's modulus was $2.0 \cdot 10^8$ kPa and the cross-sectional area of longitudinal bar per meter width was 180 mm². For the reinforcement, the yielding stress was $6.0 \cdot 10^5$.

For the wall face, the shear modulus was $8.3 \cdot 10^7$ kPa, and the moment of inertia of cross sectional area was 45 mm⁴ which was the sum of the moment of inertia of individual bars within 1.0 m width. The shear and normal stiffness for nodal link were assigned as $1.5 \cdot 10^4$ kN/m and $5.0 \cdot 10^6$ kN/m, respectively, for the current problem.

The parameters adopted for hysteretic compaction model (Duncan and Seed, 1986) were: at-rest lateral earth pressure coefficient, k_o, of 0.55, friction component of limiting coefficient of at-rest lateral earth pressure, $k_{1,\phi}$, of 2.21, cohesion under dynamic load, C_d, 50 kPa, at-rest lateral earth pressure coefficient for unloading and reloading, k_2, 0.15, and softening depth of 0.4 m. Considering light compactor (i.e. Ingersollrand, D 23) and the factor that there was about 0.3 m gap between wall face and the soil being compacted which was later filled up during the placement of next reinforcement layer, the peak compaction induced lateral stress profiles used in the analysis are as shown in Fig. 5. The effect of the compaction operation on soft ground soil was ignored.

Finite element analysis

The wall/embankment above the ground surface was simulated by 13 incremental layers. For each layer, the gravity force was applied by two increments. A total of 6 analyses have been conducted. The first three analyses were using 3 sets of the permeabilities listed in Table 1, respectively. Analysis number 4 used middle permeability but with compaction effect. Both analyses numbers 5 and 6 were conducted with two different options of varying the permeability during the loading and consolidation process. For analysis number 5 (varied I), the permeability was varied from the formula proposed by Taylor (1948) and verified by Tavenas et al (1983) as follows:

$$k = k_o 10^{[\frac{-(e_o-e)}{c_k}]} \quad (2)$$

where e_o is the initial void ratio; e is the void ratio at the condition under consideration; k is the permeability; k_o is the initial permeability; and c_k is constant, which is equal to 0.5 e_o (Tavenas et al, 1983). The initial value of permeability was the middle permeability.

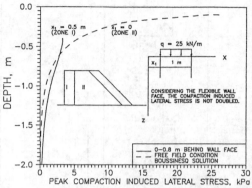

Fig. 5 Maximum compaction induced lateral earth pressures

Table 1 Soil parameters of Bangkok clay

Parameter	Symbol		Soil Layer			
		1	2	3	4	5
	Depth, (m)	0-1	1-2	2-6	6-8	8-12
Kappa	κ		0.04	0.11	0.07	0.04
Lambda	λ		0.18	0.51	0.31	0.18
Slope	M		1.1	0.9	0.95	1.1
Gamma (P'=1 kPa)	Γ		3.0	5.12	4.0	2.9
Poisson's Ratio	ν	0.25	0.25	0.30	0.30	0.25
Modulus, (kPa)	E	4000				
Friction Angle, (°)	φ'	29.0				
Cohesion, (kPa)	C'	29.0				
Unit Weight, (kN/m³)	γ	17.5	17.5	15	16.5	17.5
Horizontal	High k_h	69.4	69.4	10.4	10.4	69.4
Permeability	Middle k_h	34.7	34.7	5.2	5.2	34.7
(m/sec), (10^{-8})	Low k_h	13.9	13.9	2.1	2.1	13.9
Vertical	High k_v	34.7	34.7	5.2	5.2	34.7
Permeability	Middle k_v	17.4	17.4	2.6	2.6	17.4
(m/sec), (10^{-8})	Low k_v	6.9	6.9	1.0	1.0	6.9

NOTE: High: k_v = 50 times of estimated average test value;
      Middle: k_v = 25 times of estimated average test value;
      Low: k_v = 10 times of estimated average test value.

Horizontal permeability is always 2 times of the vertical value.

Table 2 Hyperbolic soil parameters used for lateritic backfill material.

Parameter	Cohesion	Friction Angle	Modulus Number	Modulus Exponent	Failure Ratio	Bulk Modulus Number	Bulk Modulus Exponent	Unit Weight
	C, (kPa)	φ, (°)	k	n	R_f	k_b	m	γ,(kN/m³)
Value	60	32.5	1078	0.24	0.96	1050	0.24	20.0

For analysis number 6 (varied II), the permeability variation was also controlled by Eq. 2. However, the values of permeability of the soft soil were different before and after yield with much higher value before yield (Tavenas and Leroueil, 1980). In this analysis, before the soil yield (yielding is controlled by modified Cam clay model), the high permeability values (Table 1) were used and after the soil yielded, the permeability values of 1/5 of the values before the yield were adopted, i.e. low permeability in Table 1. A computer program named CRISP-AIT which was developed by modifying the CRISP computer program (Britto and Gunn, 1987) was used for the analyses. All the analyses were consolidation analyses.

PREDICTED RESULTS AND COMPARISON WITH FIELD DATA

Since the prediction is class C type prediction (Lambe, 1973), the predicted results are presented together with the field data. The data included excess pore pressures, vertical settlements, wall face and subsoil lateral displacements, tension forces in the reinforcements, and the wall/embankment base pressures. The finite element results obtained by using middle permeability with compaction effect are mainly used as predicted values. The results of using high and low permeabilities (Table 1) did not predict the field data well. For the sake of clarity they are omitted from the presentation. However, some of the results from varied permeability analyses and using the middle permeability without compaction effect are also included for discussion.

Excess pore pressures

For predicting the excess pore pressure, the key parameter is the permeability of the subsoil. Figures 6 and 7 show the typical predicted excess pore pressure variations with different assumptions of the foundation permeability together with the field data at piezometer points 4 m and 7 m below the ground surface. The agreement between predicted and measured data is fairly good. From the figures, it can be seen that all the analyses overpredicted the excess pore pressure at the end of construction. However, the varied permeability analyses predict the excess pore pressure dissipation process better. For analysis 6 (varied II), the soil elements under the embankment yielded at an early stage of construction (OCR = 1.2) and after yielding, the soil permeability was varied with the initial value of low permeability (Table 1). However, the soil elements away from the embankment may not yield in the whole construction process, and thus, still possess with high value of permeability. The overall effect

is that the predicted excess pore pressures are higher than those obtained using middle permeability (Table 1) at end of construction and closer to the value of using middle permeability at later stages of consolidation.

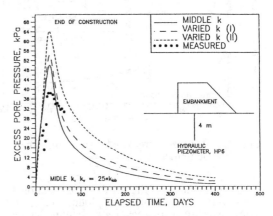

Fig. 6   Predicted and measured excess pore pressure variation at piezometer point HP6

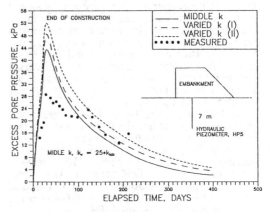

Fig. 7   Predicted and measured excess pore pressure variation at piezometer point HP5

Settlements

The predicted and measured surface settlements under the center point of reinforced mass are compared in Fig. 8. The locations of settlement plates are also shown by the key sketch in the figure. It can be seen that the predicted values have remarkable agreement with measured data. However, the varied permeability analyses yielded higher settlement rate at the early stage of construction and lower settlement rate during the consolidation process.

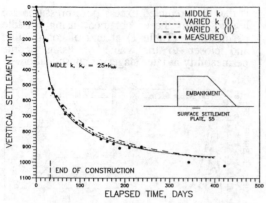

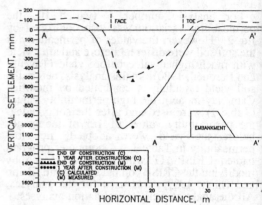

Fig. 8   Typical predicted and measured settlement curves

Fig. 9   Predicted and measured surface settlement profiles

The settlement profile on the cross-sectional lines on the ground surface is plotted in Fig. 9. The comparisons are given for both immediately after construction and one year after construction conditions. From the figure, it can be seen that the agreement between the predicted and measured data are reasonably good. Unfortunately, there are no measured data for foundation heave. The predicted maximum foundation heave in front of the wall face is 125 mm immediately after construction. One year after construction, it reduced to 62 mm. As mentioned previously, the varied permeability analysis (varied II) allows the variation of the permeability in the horizontal direction with high value for the soil elements away from the embankment loading area, and the subsequent predicted value of heave is less than that of the constant permeability analysis. For example, at end of construction, the maximum heave is 110 mm. This tendency seems more closer to field behavior, since most finite element analyses using constant permeability overpredicted the foundation heave (e.g. Magnan and Kattan, 1989).

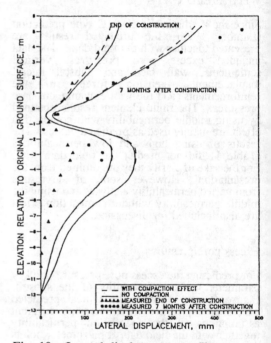

Fig. 10   Lateral displacement profiles

Lateral displacements

Lateral displacement is one of the most difficult items to predict. Figure 10 is the comparison of predicted and measured lateral displacement profiles for both end of construction and 7 months after construction cases. For lateral displacements in the foundation soils, the measured data up to 7 months after construction only reach down to 3 m depth because the inclinometer probe could not be inserted into the deformed casing below 3 m depth. At the end of construction, the predicted wall face lateral displacements agreed well with the measured data. However, the predicted subsoil lateral displacements are twice as large as that of

measured data. At 7 months after construction, the predicted subsoil and wall face lateral displacements reasonably agreed with the measured values. However, at the top of the wall face, the predicted values are less than the measured ones and the predicted maximum subsoil lateral displacements are still larger than the field data. It can also be seen that compaction effect increased wall face lateral displacement by about 10% at end of construction even with the light compactor. This effect became less significant at one year after construction.

200

The time-lateral displacement relationship is shown in Fig. 11 for two points, namely: (a) top of the wall face, and, (b) 3 m below the original ground surface where maximum lateral displacement occurred in the foundation. For the top of the wall face, the discrepancy between the predicted and the measured values appears at 3 months after the construction (August 20, 1989). At that time, the measured data showed an increased rate of lateral displacements. It was probably due to the occurrence of heavy rainfall because there was a sudden ground water level increase at that time (Bergado et al, 1991). For the point under the wall face and 3 m below original ground surface, the discrepancy between the predicted and the measured lateral displacements mostly occurred during the construction period. After construction, both the predicted and the measured lateral displacements show small increment rate. There are two reasons for the differences obtained between the measured lateral displacements and those predicted by the finite element analyses, namely: (1) the deficiency of the analytical method (Poulos, 1972); and (2) the influence of inclinometer casing stiffness which may result in relative displacements between the soil and the casing because it is difficult for casing to freely follow the "S" shape deformation pattern (see Fig. 10).

Tension forces in reinforcements

The predicted maximum tension forces in reinforcements at immediately after construction and one year after construction are shown in Fig. 12, together with the measured data at immediately after construction. Also shown are the active and at-rest earth pressure lines without considering the cohesion in drawing the active earth pressure line. The agreement between predicted and field data for immediately after construction case is quite good. The data are presented in terms of per meter width and per reinforcement layer (0.45 m). The measured data one year after construction was not included because of too much scatter. Both the predicted and measured data showed that at the end of construction, the maximum tension forces in the reinforcements at the top half of the wall are much larger than k_o line. At the middle wall height, the data are closer to the k_o line. At the bottom of the wall, the data are much higher than k_o line again. For reinforced wall on soft ground, under the wall loading, the soft soil tends to squeeze out of the wall/embankment base which causes large relative movement between the reinforcement and the soil. Therefore, large tension force can be developed in the reinforcements. The maximum tension forces in the reinforcements increased during the foundation soil consolidation process. At the top half of the

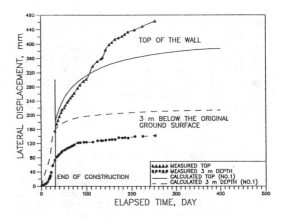

Fig. 11   Maximum lateral displacement curves

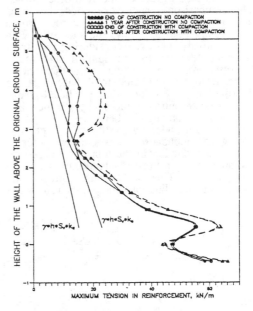

Fig. 12   Maximum reinforcement tension force profiles

wall, the maximum reinforcement tension forces after one year of construction are twice as large as those immediately after construction due to the large lateral displacement of the wall face. The figure also shows that the compaction effect may cause the tension force in reinforcement at the top of the wall to increase significantly.

The tension force distributions along the reinforcements for instrumented mats are shown in Fig. 13. Generally speaking, the agreement between the predicted and measured data for immediately after construction case is fair. The reinforcement tension force distribution pattern was strongly influenced by the interaction between the reinforced wall/embankment mass and soft foundation soil. The foundation differential settlement causes a bending effect on the reinforced mass, i.e. top in compression and bottom in tension, so that the length of the reinforcement in tension is short in the upper part of the wall. The figure also shows that the location of maximum tension force is very close to the wall face (less than 1.0 m) except at the base reinforcement layer.

Wall/embankment base pressures

The base pressure is an important item for design of reinforced wall and embankment on soft ground because it controls the safety factor of bearing capacity of the foundation. Figure 14 shows the predicted and measured total earth pressures at the base of the reinforced mass. The comparison is not very good, but it is qualitatively sufficient. From vertical force equilibrium point of view, the measured data was relatively low even considering the stress spreading due to the foundation settlement. Nevertheless, there are two points that can be made from the predicted total base pressure distribution. Firstly, the predicted earth pressure distribution is more likely trapezoidal pattern eventhough there is a stress concentration under the wall face. Secondly, during the foundation consolidation process, there is a reduction in the total base pressure and an increase in stress concentration under the wall face. This is because during the foundation consolidation process, the reinforced

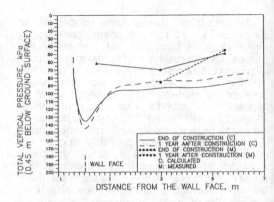

Fig. 14 Predicted and measured total embankment base pressure

mass sinks into the ground, and the vertical load is distributed into a larger horizontal area. At the same time, the overturning movement of the reinforced mass also increased due to the differential settlement. The overall effect is the increased stress concentration under the wall face and the reduction of the base pressure at other locations.

DIFFICULTIES IN PREDICTION

Difficulties related to determine the input parameters

As mentioned previously, the correctness of finite element results depends largely on both the constitutive model and the value of model parameters used. Although most of the model parameters can be determined from high quality laboratory test results with confidence, some parameters, such as soft soil permeability, are very difficult to determine. The laboratory permeability test can be subjected to error resulting from the size of sample, temperature, and the large difference between the hydraulic gradient in the field and in the laboratory. For Bangkok clay, as reported by Bergado et al (1990), the laboratory test values underestimated the field permeability significantly. Field permeability measurements such as the piezometer method can be affected by the clogging of the filters and disturbance of the soil during the equipment installation.

The values derived from back analysis of existing case histories are of great help for determining the actual permeability values. For Bangkok clay, the values of field permeability are about 10 to 30 times of the corresponding laboratory test data. Another point is the variation of the permeability. This factor has been noticed long time back (e.g. Taylor, 1948). In order to precisely predict the behavior of the embankment on soft ground, it

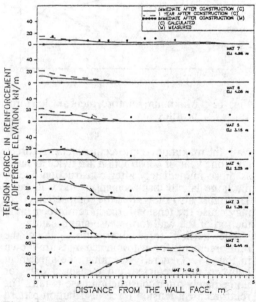

Fig. 13 Reinforcement tension force distributions

is necessary to consider the variation of the permeability with the yielding and the change of the void ratio of the soil.

Difficulties in modelling

The stress/strain behavior of the soil is influenced by several factors, such as elastoplastic behavior, nonhomogeniety, anisotropy, and structure of the soil, the stress path followed by the soil, etc. Even the sophisticated soil model, such as the modified Cam clay model cannot consider all these factors. At present, the settlement can be predicted reasonably well, but the agreement between predicted and measured lateral displacements is fair to poor. This phenomenon had been discussed by Poulos (1972) who cited such reasons as: the effect of Poisson's ratio, anisotropy, and nonhomogeniety which appeared to be most significant factors. Normally, the maximum lateral displacement occurs at the vicinity of the embankment toe, where considerable principal stress rotation occurs during the embankment construction. Most of the above mentioned factors cannot be properly modelled and some of them are still not well understood, e.g. the effect of the principal stress rotation on the behavior of soil.

CONCLUSIONS

The finite element modelling techniques presented in this paper improved the ability of finite element method to predict the behavior of the reinforced earth structure on soft ground. The modelling demonstrates that the soil/reinforcement interaction properties can be properly selected according to the relative displacement pattern between soil and reinforcement (direct shear or pullout), and the construction process can be most closely simulated.

Comparison of predicted and measured data indicates that the performance of the reinforced embankment on soft ground can be predicted by finite element method by selecting the foundation permeability based on back analyzed values from case histories and by considering the compaction effect on the embankment fill. It has been found that the predicted foundation settlements and wall face lateral displacements agreed reasonably well with the field data, and the agreement between predicted excess pore pressures, tension force in reinforcements, foundation lateral displacements, and embankment base pressures is fair. It was also observed that using constant permeability cannot precisely simulate the excess pore pressure variation. Regarding the discrepancy between the predicted and measured foundation lateral displacement, two reasons are cited as follows:

(1) the deficiency of the analytical method, and (2) the measured data might be influenced by inclinometer casing stiffness.

REFERENCES

Adib, M., Mitchell, J.K. and Christopher, B. 1990. Finite element modelling of reinforced soil walls and embankments, design and performance of earth retaining structure. *ASCE Geotech. Special Publication*. No. 25, 409-423.

Ahmed, M.M. 1977. *Determination of permeability profile of soft Rangsit clay by field and laboratory tests*. M. Eng'g. Thesis, No. 1002. Asian Institute of Technology, Bangkok, Thailand.

Asakami, H. 1989. *The smear effect of vertical band drains*. M. Eng'g. Thesis, No. GT-88-8, Asian Institute of Technology, Bangkok, Thailand.

Balasubramaniam, A.S., Hwang, Z.M., Vddin, W., Chaudhry, A.R. and Li, Y.G. 1978. Critical state parameters and peak stress envelopes for Bangkok clays. *Q. J. Eng'g. Geol.*, 11, 219-232.

Bergado, D.T., Sampaco, C.L., and Alfaro, M.C. Balasubramaniam, A.S. 1988. Welded-wire reinforced earth (mechanically stabilized embankments) with cohesive backfill on soft clay. *Second Progress Report Submitted to USAID Bangkok Agency*.

Bergado, D.T., Ahmed, S., Sampaco, C.L., and Balasubramaniam, A.S. 1990. Settlements of Bangna-Bangpakong Highway on soft Bangkok clay. *J. of Geotech. Eng'g. Div.*, *ASCE*, 116 (1), 136-155.

Biot, M.A. 1941. General Theory of Three-Dimensional Consolidation, *J. of Applied Physics*, 12, 155-164.

Britto, A.M. and Gunn, M.J. 1987. *Critical state soil mechanics via finite elements*. Ellis Horwood.

Chai, J.C. 1992. *Interaction behavior between grid reinforcement and cohesive-frictional soil and performance of reinforcement of reinforced wall/embankment on soft ground*. D. Eng'g. Dissertation, Asian Institute of Technology, Bangkok, Thailand.

Clough, G.W. and Duncan, J.M. 1971. Finite element analysis of retaining wall behavior. *J. of Soil Mech. and Found. Eng'g. Div.*, *ASCE*, 97(12), 1657-1673.

Collin, J.G. 1986. *Earth wall design*. Ph.D. Thesis, Univ. of California, Berkeley.

Duncan, J.M., Byrne, P., Wong, K.S. and Mabry, P. 1980. Strength, stress-strain and bulk modulus parameters for finite element analysis of stresses and movements in soil. *Geotech. Eng'g. Research Report No. UCB/GT/80-01*. Dept. of Civil Eng'g. Univ. of California, Berkeley, August, 1980.

Duncan, J.M. and Seed, R.B. 1986. Compaction-induced earth pressure under K_o-condition. *J. of Geotech. Eng'g. Div., ASCE.* 112(1),1-22.

Hird, C.C. and Pyrah, I.C. 1990. Predictions of the behavior of a reinforced embankment on soft ground. *Proc. Symp. on Performance of Reinforced Soil Structures.* Thomas Telford. 409-414.

Lambe, T.W. 1973. Predictions in soil engineering (Rankine Lecture). *Geotechnique.* 23(2), 149-202.

Macatol, K.C. 1990. *Interaction of lateritic backfill and steel grid reinforcements at high vertical stress using pullout test.* M. Eng'g. Thesis. GT-89-12, Asian Institute of Technology, Bangkok, Thailand.

Magnan, Jean-Pierre and Kattan, A. 1989. Additional analysis and comments on the performance of Muar flats trial embankment to failure. *Proc. of the Intl. Symp. on Trial Embankments on Malaysian Marine Clays.* Vol. 2, 11/1-11/7.

Poulos, H.G. 1972. Difficulties in prediction of horizontal deformations in foundations. *J. of Soil Mech. and Found. Eng'g. Div., ASCE,* 98(8), 843-848.

Roscoe, K.H. and Burland, J.B. 1968. On the generalized stress-strain behavior of wet clays. *Proc. of Eng'g. Plasticity.* Cambridge, Cambridge Univ. Press, 535-609.

Rowe, R.K. and Mylleville, B.L.J. 1988. The analysis of steel reinforced embankment on soft clay foundation. *6th Intl. Conf. on Numerical Methods in Geotechnics.* Innsbruck. 1273-1278.

Shivasankar, R. 1991. *Behavior of a mechanically stabilized earth (MSE) embankment with poor quality backfills on soft clay deposits, including a study of the pullout resistances.* D. Eng'g. Dissertation, Asian Institute of Technology, Bangkok, Thailand.

Tavenas, F. and Leroueil, S. 1980. The behavior of embankments on clay foundations, *Can. Geotech. J.* 17,236-260.

Tavenas, F., Jean, P., Leblond, P., and Lerouil, S. 1983. The permeability of natural soft clays, part II, permeability characteristics. *Can. Geotech. J.* 20, 645-660.

Taylor, D.W. 1948. *Fundamentals of soil mechanics.* John Wiley & Sons Inc. New York.

4 Earth structures, mines and slopes

Prediction versus Performance in Geotechnical Engineering, Balasubramaniam et al. (eds)
© *1994 Balkema, Rotterdam, ISBN 90 5410 355 8*

Prediction v. performance for rockfill dams and other structures using the Velocity Method

A.K.Parkin
Monash University, Clayton, Vic., Australia

ABSTRACT: The Velocity Method, wherein settlement rate is plotted against time to log-log scales, has been applied to the analysis of settlement behaviour of rockfill dams, large cast-in-situ test piles and road embankments. In this format, it is shown that the settlement record often displays discontinuities that may add considerably to an understanding of structural performance but which are incapable of being differentiated from a traditional settlement-time record.

1 INTRODUCTION

Long continuing settlements are frequently a feature of engineering structures, in most cases occurring as a result of consolidation. Where this is so, appropriate analytical procedures are well established, but should the settlement regime be one, or predominantly one, of creep, then analysis and prediction can become shrouded in some mystery.

Such creep settlement, when it manifests, will often follow a period of hydrodynamic consolidation (which may then serve to make the point of initiation vague), but can also initiate without the precedence of any observable phase of consolidation, as with rockfills and gravels. In either case, the traditional engineering approach is to graph settlement against log time, whereupon the resulting relationship is frequently observed to be linear, with a slope designated by C_α, the Coefficient of Secondary Compression (noting, however, that C_α has been variously applied to graphs of both settlement and void ratio, eg. Wahls, 1962).

For many purposes, the creep (or secondary compression) component of settlement may not be of particular consequence. In the design of rockfill dams, for example, 24 hour settlements in a stage load oedometer test are normally considered adequate for in situ predictions, with an arbitrary allowance of perhaps 5% for creep. However, there are cases where, either because the creep component is large in comparison with consolidation (or immediate settlement), or because the structure is particularly sensitive, more careful predictions of creep settlement are required. The available methods for synthesising such field projections from laboratory data are uncertain at best, firstly because they must be made from a very short time base (as there is no realistic method of accelerating creep in the laboratory), and secondly because creep curves are frequently found to be subject to erratic deviations (eg, Meigh, 1976).

With the advent of the Velocity Method, as first applied to the determination of c_v from the oedometer consolidation test (Parkin, 1978), it soon became apparent that creep curves, analysed by this procedure, can reveal a significant amount of data that is otherwise unsuspected (on the basis of the usual settlement-log time presentation). It is therefore in the nature of creep (and contrasting with consolidation) that the Velocity Plot has an important diagnostic function, as anticipated by Pusch and Feltham (1980), and one purpose of this paper is to show the importance of this function in enhancing the reliability of any projections of creep settlement.

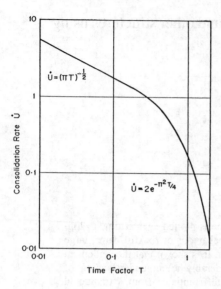
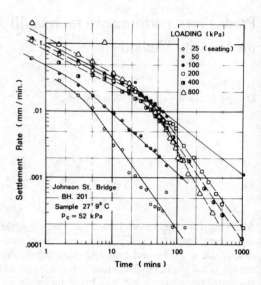

Fig. 1: The velocity method, theory and practice (Parkin, 1991)

2 THE VELOCITY METHOD

From the time of Terzaghi's earliest investigations of consolidation in the 1920's, settlement analysis has almost invariably focussed on the settlement-time relationship, usually as plotted to some convenient mathematical scale. It has, however, been shown that a velocity-time graph, produced by differencing the raw settlement observations, is not only sufficient for the evaluation of c_v (ie, no essential data has been lost in differentiation), but that it also yields a more accurate value through the elimination of the irrelevant data which gives rise to the familiar constructions in the t_{50} and t_{90} methods (Parkin, 1978). The Velocity Method then allows an experimental curve (plotted to log-log scales) to be compared directly with the theoretical solution by overlaying, thence leading to a value of c_v.

The Velocity Method has particular relevance to creep settlement. Where this consists of a secondary compression, following a period of consolidation, the velocity plot becomes linear (rather than following the continuously curved consolidation solution) at a slope C_β, which varies with load (Fig. 1). The decay rate, as represented by C_β, is seen to pass through a minimum for a load increment spanning the pre-consolidation pressure p_c, corresponding with the increase in C_α

reported at this stage by Mesri and Godlewski (1977). The reason for the variation in C_β is not clear, but in this writers view it is likely to relate to differences in the time of initiation of creep. Such uncertainty will not be present for rockfills and gravels (where any consolidation phase is immediate), and the velocity plot then consists of a straight line of unit slope commencing from the moment of loading.

A unit slope on the velocity plot, which in fact characterises transient creep in a diversity of materials, clearly translates into a constant value of C_α, as used in the traditional approach to the analysis of creep (eg, Poulos et al., 1976). However, by contrast, C_α is a material property (dependent also on effective stress) which must be determined experimentally. The degree of uncertainty in C_α is illustrated by a set of settlement records for rockfill dams (Fig. 2) assembled by Sowers et al. (1965), who also note that C_α values cannot be readily predicted from laboratory tests (as low as half the field value). Furthermore, it is a frequently observed feature of creep that C_α can change unpredictably in time, as in the case of some major foundations on a mudstone reported by Meigh (1976) (Fig.3), and as noted also by Mesri and Godlewski (1977). It will be shown that the Velocity Method offers a means to explore this uncertainty and to enhance the reliability of predictions that can be made.

3 ROCKFILL CREEP IN OEDOMETER TESTING

Rockfill is a free-draining material wherein effectively all the time dependent deformation are the result of creep, this being, to a large degree, attributable to the crushing of particle asperities. In an oedometer test, the load application is quick, so that, in contrast with many field situations, the time of origin for creep t_o is clearly defined, which ensures that the velocity plot shows a slope of unity.

For any load stage, a velocity plot can be prepared by differencing the raw settlement readings. Almost invariably, this shows a slope of -1, but not infrequently it may also show a sudden discontinuity, as in Fig. 4. Should this occur, then a conventional presentation (as in Figs. 2 and 3) must clearly have a discontinuity of slope. However, very significantly, the slope of the velocity plot is preserved across the discontinuity.

Whilst features of this type might normally be regarded as being avoidable under good laboratory technique, they in fact prove to be an intrinsic component of creep, almost wherever it occurs. Similar effects have been observed on clays, rockfills and soft rocks and described as "limited instabilities" by Bishop (1974), based on the work of Bishop and Lovenbury (1969) and Pigeon (1969), with Bishop noting that they could not be eliminated with even the most meticulous care. Their origin would appear to lie in the fundamental "stick-slip" nature of creep, wherein deformation under load leads to the formation of local stress concentrations, followed by yield or fracture in diminishing steps (and contrasting with the necessary continuity of consolidation under a hydraulic gradient). The discontinuity in Fig. 4 represents a locking-up of the system, perhaps due to stress generated on the oedometer wall.

4 PROBLEMS IN THE INTERPRETATION OF FIELD SETTLEMENT DATA

If the Velocity Method is to be applied to the analysis of field settlement records, as will be done subsequently, there are two problems that may arise outside the laboratory environment. Firstly, the time of origin of creep t_o may not be identifiable (perhaps due to the construction sequence), and secondly the settlement data may contain an appreciable component of "noise" from extraneous sources.

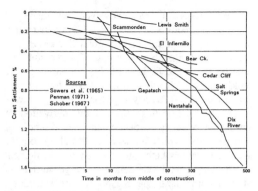

Fig. 2: Crest settlement of some rockfill dams (Sowers et al., 1965)

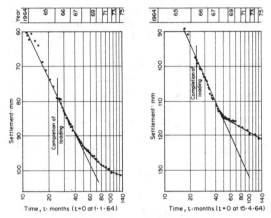

Fig. 3: Creep settlement of large pad foundations on a soft rock (from Meigh, 1976)

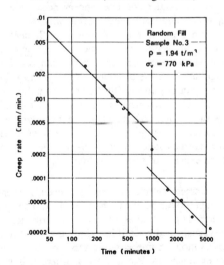

Fig. 4: Discontinuity in rockfill oedometer test (Parkin, 1985, 1991)

An estimate of t_o can, however, be made by plotting the reciprocal creep rate $1/S$ against time, whereupon, the if sufficiently good line results, t_o can be read as the time intercept. The method will here be applied to Cedar Cliff Dam, a sloping core rockfill dam in North Carolina, USA, for which a settlement record (as S v log t) is given by Sowers et al. (1965). For this purpose, the derivation of settlement velocity depends, in the absence of original data (which is clearly preferable), on the accuracy with which the published settlement graph can be read (the relatively large settlement of this dam being the reason for its selection). Whilst the data from which Fig. 5 was prepared are said to be referenced to the midpoint of construction (Sowers et al.), no date for this is quoted. However, on the presumption that this would be near April 1952 (the time of initial topping out, according to Growden, 1958), this analysis would suggest an effective time of origin some four months later, coinciding with first filling (soon after completion, August 1952), possibly because of a significant contribution to creep coming from the suddenly imposed water loading on the acutely-inclined core. With this value of t_o, the gradient of the velocity plot is then close to -1 (as against some 25% steeper on the original time base).

Should the field data contain a measure of scatter, as can normally be expected, then that scatter will be considerably amplified on the derived velocity graph. It is important then to realise that this enhanced scatter, despite appearances, does not represent any introduced error whatsoever, and that any projections of creep settlement will remain at least as accurate as by any other method. On occasions, the scatter can be of such magnitude that useful interpretation of the velocity plot is not possible, with the scatter tending to be biased below the mean creep line (by virtue of the log scale) and perhaps including negative values. However, a useful trend can still emerge if creep rates are calculated over a larger time increment, at the expense of some sensitivity.

If the creep rate plot, as then produced, is found to contain discontinuities, these will, if capable of interpretation, represent new information not previously discernible in the settlement observations. Some possible types of discontinuity are examined by Parkin (1971), where it is shown that a conventional settlement-time plot then consists of two quite unrelated segments. Any

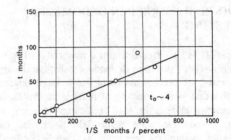

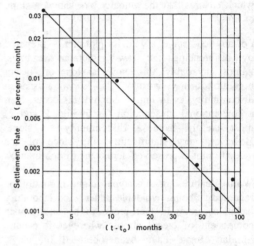

Fig. 5: Velocity Method applied to a sloping core rockfill dam (Parkin, 1985, 1991)

attempt to predict creep settlement based on a single continuous line cannot then be relied upon. However, because the slope of the creep line is not a material property (for the appropriate value of t_o), it follows that it cannot change across a discontinuity, as seen in Fig. 4. This has considerable significance, as a discontinuity can occur at any stage, and may not always have a sufficient span of data following to define a slope on which to base a settlement projection. Even if that necessary data exists, it is often broadly scattered, being derived from increasingly small measurements.

In making projections of creep settlement, it is always possible (though rare) that more than one discontinuity may be present in the data. However, unless the system is approaching collapse, the successive discontinuities will be of diminishing magnitude as the system moves toward an optimal distribution of stress. Even so, the identification of just one discontinuity will enhance the accuracy of a creep projection by an order of magnitude.

5 CREST SETTLEMENT BEHAVIOUR OF EARTH AND ROCKFILL DAMS

Whilst settlement-time records for very many structures are presented in literature, there is often not a lot that can be concluded from them. The Velocity Method, as shown below, may change this considerably, although the evaluation of dam settlement records is frequently limited on account of observations not being commenced soon enough. This happens because the installation of settlement monuments is often delayed until the completion of crest operations which may cause their loss. This is indeed important with respect to total settlement, but mishaps of this sort are of no consequence to settlement rates, so that the merits of installing temporary markers at an early stage are strongly commended.

The Velocity Method is here applied to several rockfill dams under the control of the Hydro-Electric Commission of Tasmania (HEC), as made possible by having access to original survey files, and essentially as in previous reports by Parkin (1985) and Parkin (1990). The five dams listed in Table 1 were selected for analysis ahead of others for which the time base was too short, or the movements too small, or where unusual events caused substantial interference (eg, Cole and Fone, 1979). In all cases, the results relate to precise levelling in the vicinity of the maximum section. The velocity diagrams are, of course, dependent on the chosen time of origin, generally taken at either the midpoint of construction or end of construction (EOC), as will be done here. However, all construction programmes differ, so that there may not be an obvious choice for t_o, and creep lines at a slope of other than unity are therefore possible.

For a time origin at EOC, crest settlements rates are generally found to decay along a line of near unit slope. This is true of Serpentine Dam (Fig. 6), which comprises an embankment of rolled weathered quartzite and schist over some 8m of in-situ river gravels (HEC, 1975). For this, EOC has been taken as September 1971, when the embankment was complete except for a final 1.2m of rockfill added after completion of the concrete facing. The significant feature of this graph is that the application of water load during reservoir filling (which extended over several years because of the very large capacity) has had no apparent effect on vertical settlement, a characteristic that generally applies to faced rockfills. It is also noticeable that that the scatter of points increases

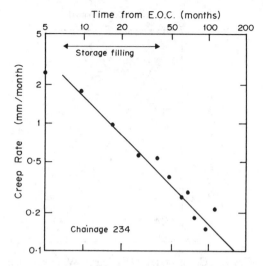

Fig. 6: Creep rate behaviour, Serpentine Dam, Tasmania (from Parkin, 1985, 1991)

with time as the measured movements become very small.

Wilmot Dam, the first faced rockfill to be built by the HEC (Cole, 1971), is ostensibly similar to Serpentine with respect to its creep behaviour, but highlights the effect of a somewhat abnormal construction routine. Here the rockfill is a hard greywacke, placement of which was completed in November 1968, but Wilmot is unique among this group of case studies in showing a gradient steeper than, -1 suggesting that this EOC date could be inappropriate (Fig. 7). A closer scrutiny of operations in the 16 month period from EOC up to the commencement of crest settlement readings shows that face construction took a further 10 months, during which a settlement of 10mm was recorded from hydrostatic settlement gauges.

These then showed negligible settlement in the succeeding 4 months up to the commencement of filling, indicating behaviour not at all consistent with the line on Fig. 7. The indications are therefore that some new phase of settlement has been initiated during filling, and taking t_o at this point does indeed lead to a line of unit slope.

By comparison, earth cored dams show notable differences in their settlement rate behaviour. In the case of Parangana Dam (Mitchell et al., 1968), the central core (containing the crest marks) consists of weathered granodiorite supported by rockfill shoulders of quartzite and schist. Taking

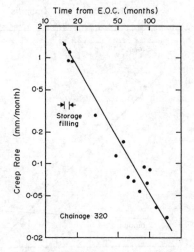

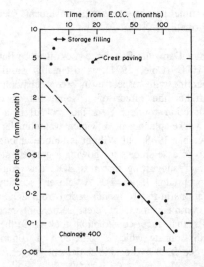

Fig. 7: Creep rate behaviour, Wilmot Dam, Tasmania (from Parkin, 1985, 1991)

Fig. 8: Creep rate behaviour, Parangana Dam, Tasmania (from Parkin, 1985, 1991)

EOC at June 1968, the creep rates define a 45° line over the greater part of the time spanned, except for some higher rates in the initial stages (Fig.8). These are evidently attributable, as indicated, to reservoir filling and settlement associated with the penetration of a wetting front, and, at a later stage, to either a 5 month period of reservoir drawdown or (more probably) to road paving operations on the crest. Such deviations

appear to be superimposed on an otherwise global trend for the main body of the dam.

On occasions, the plotted creep rates may be found to show such a level of scatter as to make any interpretation difficult, but even this scatter must exist for a reason. Such is the case of Rowallan Dam, an earth-cored dam with a central concrete spillway, wherein the core consists of a well-graded till, derived from quartzite and dolerite, between rockfill shoulders of quartzite and schist (Mitchell et al., 1968). Taking EOC at January 1967, the creep rates show a clear global trend at unit slope (Fig. 9), with a substantial superimposed scatter which is too great to be explained by rounding errors in the surveyed levels. Filling was gradual, with the final third occurring during months 7 and 8, where the creep rates reflect the passage of the wetting front. A low point around month 23 may be connected with a reservoir drawdown caused by piping problems (Mitchell et al., 1979), but later fluctuations are for reasons that can not be ascertained by the Author. A correlation of creep rates against rainfall was also attempted, and this gave some grounds for regarding the scatter as a seasonal shrink-swell phenomenon in the upper levels of the sometimes rather plastic core.

Cethana Dam proved to be another instance wherein crest settlement rates showed substantial scatter, whilst still maintaining a global trend at unit slope (Fig. 10). This structure, which is a

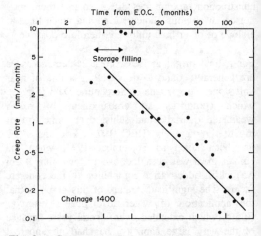

Fig. 9: Creep rate behaviour, Rowallan dam, Tasmania (from Parkin, 1985, 1991)

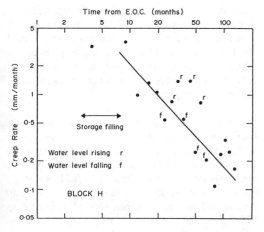

Fig. 10: Creep rate behaviour, Cethana Dam, Tasmania (from Parkin, 1985, 1991)

rather high rockfill in a narrow valley, consists of a well-rolled quartzite fill, placement of which was essentially complete by November 1969, except for a further 11.5m added in October 1970 (adopted EOC) after costruction of the concrete face (Fitzpatrick et al., 1973). A possible complication with Cethana is that the partly-built reinforced rockfill was overtopped in November 1968 at a height of 15m, incurring a substantial

washout (Hydro-Electric Commission, 1969), but this has had no discernible effect on the velocity plot, probably because it relates to a fairly early stage of construction. With Cethana, however, the crest level marks are located in the parapet wall, so that the recordings thereon are influenced by the movement of the facing slab. This is reflected dramatically in the high settlement rates on first filling, rising to a peak of 20 mm/month (off scale) in April 1971 as the lake rose 30m. This is clearly concrete shrinkage in contact with stored water at a temperature typically around 6°C. Thereafter, high and low creep rates can mostly be associated with a rise or fall in reservoir level during the corresponding time interval, while the underlying rockfill continues to creep independently, dominated by the final lift and such that only a resolved component is recorded on the parapet wall.

6 LOAD TESTS ON LARGE BORED PILES IN SOFT ROCK

Some additional case histories have been drawn from an investigation by Williams (1980) into the behaviour of bored piles socketted into soft rock. In this study, a number of   test piles, ranging up

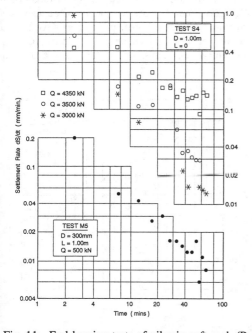 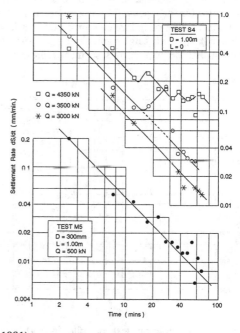

Fig. 11: End bearing tests of piles in soft rock (Parkin, 1991)

to 1m in diameter, was constructed in a highly to moderately weathered siltstone at two sites, one a motorway cut of some 5m depth (S) and the other a brick pit of about 25m depth (M). These locations, both close to Monash University, allowed the tests to be taken to failure and then excavated for examination.

In some cases, piles were provided with a collapsible base to ensure that they operate in side friction only, while others were designed to operate in end bearing only. Load was applied in increments, generally of one hour duration, up to failure, with sufficient data being available for the construction of velocity plots in many cases.

Two end-bearing pile tests were selected for analysis, with creep rates being calculated from the raw settlement observations (Fig. 11). In one of these examples (Test M5), the load is well below failure (F=2.8) and the creep is seen to be of a routine character showing unit slope. In the other example (Test S4), the rock is somewhat softer and more complex behaviour is apparent, at least in the left hand diagram as shown without an interpretation superimposed. However, if a 45° set square is used to pick out segments of lines, as in the right hand diagram, a coherent picture then emerges, showing development from a stable transient creep at a load of 3000kN into a rapidly deteriorating condition at 4350kN, evidently with the formation of cracks. The test report indicates that some tilting of this pile commenced at a load of 2550kN, increasing significantly at 3500kN (noting that the blip on this segment represents a non-creep deformation of about 1.5mm), and then further increasing to 1 in 27 tilt at maximum load, at which point the pile was deemed to have failed (load increment 3900kN not plotted). Being a failure situation, progressing towards eventual collapse, these successsive slips show no tendency to a diminishing magnitude.

A somewhat different picture emerges for a side friction pile, as in Test S3, for which the data is again presented both with and without a superimposed interpretation (Fig. 12). Here, load was increased at approximately 500kN per stage up to 4900kN (as plotted), where creep settlements became substantial. The load was taken briefly to 5100kN, but because of the excessive deformations the pile was then deemed to have failed. Again, by fitting 45° linear segments, a rather formless sequence of points can be shown to consist of a series of slips, in this case of diminishing magnitude. Final excavation showed that shearing had occurred through the roots of the deliberately formed helical asperities (of about 12mm depth) in the socket wall, creating a shear zone up to 100mm thick.

7 EMBANKMENT SETTLEMENTS ON SOFT SOILS

The South Melbourne section of Melbourne's Westgate Freeway consists mainly of an elevated structure on deep bored piles, but also has a number of low embankments on the approaches and entry/exit ramps, up to 2m high and 60m

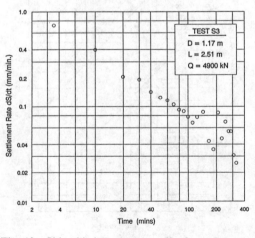 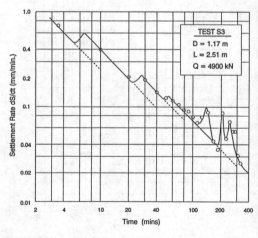

Fig. 12: Side friction test on pile in soft rock (Parkin, 1991)

wide, as described by McDonald and Cimino (1984). These embankments rest on a mixed alluvium, featuring 12 to 18m of soft clayey silt (Coode Island Silt) in the uppermost sequence. This silt classifies as highly plastic, with PI around 40+, and gives a ratio p_c/p_0' of around 2.5.

Because of the anticipated settlements, it was deemed appropriate to construct the embankments well in advance of the elevated structure, with surcharges taking the maximum embankment height to about 3.7m. This represents foundation loadings of the order of 50 to 75kPa, which puts much of the underlying silt up to a stress equal to, or exceeding, p_c. In this condition, it is commonly found (eg, Mitchell, 1986) that pore pressure dissipation is extremely slow as a result of excess water being liberated during internal structural breakdown. In addition, most of the embankment sites lay on resumed industrial land with a variable and unknown history of loading (from stored containers, timber &c.).

As settlements were found to be continuing at a rate of around 0.3mm/day after 4 years or so, attention (mid 1982) turned to the possible use of sand drains to hasten stabilisation, raising the question as to whether settlement was still pore pressure controlled and whether there was any real benefit to be gained this way. Some indication of this is possible by applying the Velocity Method to the observed settlement records, wherein, at least in principle, a line of slope 0.5 indicates hydrodynamic consolidation whilst a slope of 1.0 indicates creep (for which sand drains would be ineffective). This, of course, can be somewhat more complex in practice due to the time of origin being poorly defined, coupled with the unknown

stress history and the variable relationship between embankment loading and p_c.

In the region of Gittus Street, surface settlement markers were placed at intervals of about 5m across the embankment and adjacent land (designated 35L to 35R), and two typical records are given in Fig. 13. Outside the embankment area (which extends from 25L to 25R), the settlement rates define a line of near unit slope, as would be expected for creep, with settlements being generally small (<0.1mm/day), and indicating some 34mm additional settlement to occur over the next log cycle of time (25 years). Towards the edge of the embankment, the situation became less clear with wide fluctuations and high peaks in the latter stages, but underneath the embankment there was an immediate and sudden change to a slope of 0.5 to 0.6, with an order of magnitude increase in rate. Notably, the degree of scatter is also much reduced, with all these characteristics together being indicative of continuing consolidation. The conclusions to be drawn from this are that (i) there is a reasonable prospect of sand drains being beneficial and (ii) this graph cannot be extrapolated to estimate long term settlement (which indicates a further 800mm over 25 years). A trial sand drain installation did in fact achieve around 50% reduction in excess pore pressure within a month, as reported by McDonald and Cimino (1984) and again by McDonald (1988).

Other locations were analysed in a similar manner, but much less conclusively, although mostly with the appearance of creep. This is probably the result of too many unknowns, but the Velocity Method has, at least in the case of Gittus Street, provided insight which could not have been

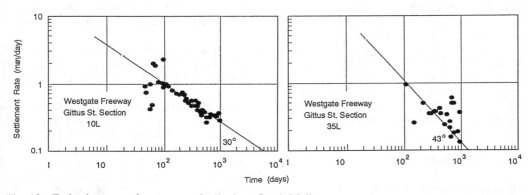

Fig. 13: Embankment settlement on soft alluvium, South Melbourne

possible by traditional methods. In particular, it shows that there is no point trying to apply consolidation theory in situations where the slope of the velocity plot is -1.

8 CONCLUSIONS

For situations in which creep is the predominant mechanism of deformation, it has been shown that traditional methods of settlement-time plotting cannot be relied upon for making settlement predictions. This is because of the intrinsically discontinuous nature of creep, arising from local stress concentrations and yield, contrasting with the essentially continuous nature of consolidation under a hydraulic gradient. It is therefore characteristic for plotted creep rates to be subject to sudden discontinuities, whose occurrence and magnitude cannot be satisfactorily predicted, at least at this time.

Under the Velocity Method, settlement rates (preferably derived from raw settlement readings) are plotted against time on log-log scales, whereupon creep will generally manifest as a line of slope -1. It is shown that other slopes may be attributable to an unknown time of origin, and that the frequent appearance of discontinuities can be expected to have an explanation in terms of loading history. These effects become submerged and cannot be viewed in isolation under traditional analysis, giving the Velocity Method an important diagnostic function, as anticipated by Pusch and Feltham (1980). In the matter of settlement prediction, the identification and isolation of just one creep discontinuity can increase predictive accuracy by an order of magnitude.

Case histories have been presented for 5 rockfill dams, load tests on large bored piles in soft rock and embankment settlements on soft clay. The success of the interpretations depends on the ability to define a time of origin, the availability of necessary details on construction or loading history and having not too many such complicating details. In such circumstances, the method can lead to a much increased understanding of structural performance.

ACKNOWLEDGMENT

The case histories on dam settlement have been made possible through the kindness of the Hydro-Electric Commission of Tasmania, in allowing access to original survey records. Permission of Melbourne Water to include Figs. 1b and 4 is acknowledged and appreciation is expressed to Dr. A. F. Williams for access to data from his Doctoral investigation of large bored piles. Data on embankment settlements is included by permission of the Roads Corporation of Victoria, with Fig. 13 being prepared by D. J. Cimino of RCV. In all cases, however, the interpretations, which are necessarily speculative to some degree, are those of the Writer.

TABLE 1: HEC Dams in Tasmania

Dam	Location	Height (m)	Type
Wilmot	Wilmot R.	30	Faced
Parangana	Mersey R.	55	Zoned
Rowallan	Mersey R.	42	Zoned
Cethana	Forth R.	110	Faced
Serpentine	Serpentine R.	30	Faced

REFERENCES

Bishop, A.W. (1974). The strength of crustal materials. *Engineering Geology* (Special Issue), 8:139-153.

Bishop, A.W. and H.T. Lovenbury. (1969). Creep characteristics of two undisturbed clays. *Proc. 7th Int. Conf. on Soil Mech. and Found. Engg.*, Mexico City, 1:29-37.

Cole, B.A. (1971). Wilmot rockfill dam - concrete face deflections. *ANCOLD Bulletin* No. 33, 19-26.

Cole, B.A. and P.J.E. Fone (1979). Repair of Scotts Peak Dam, Tasmania. *13th Congr. Int. Comm. on Large Dams*, New Delhi,2:211-231.

Fitzpatrick, M.D., T.B.Liggins, L.S.Lack and B.P.Knoop (1973). Instrumentation and performance of Cethana Dam. *11th Congr. Int. Comm. on Large Dams*, Madrid, 3:145-164.

Growden, J.P. (1958). Rockfill dams with sloping clay cores. *Proc. ASCE*, V. 84, No. PO4, Paper

1743. See also: Paper 1744, Rockfill dams: Performance of seven sloping core dams.

Hydro-Electric Commission of Tasmania (1969). Cethana Dam - Flood breach of partly completed rockfill dam. *ANCOLD Bulletin* No. 28, 23-36.

Hydro-Electric Commission of Tasmania (1975). Gordon River Power Development (Part II). *ANCOLD Bulletin* No. 41, 22-26.

McDonald, P. (1988). The real world of embankment settlement. *Proc. 5th Australia-NZ Conf. on Geomechanics*, Sydney, 110-117.

McDonald, P. and D.J.Cimino (1984). Settlement of low embankments on thick compressible soils. *Proc. 4th Australia-NZ Conf. on Geomechanics*, Perth, 1:310-315.

Meigh, A.C. (1976). The triassic rocks, with particular reference to predicted and observed performance of some major foundations. 16th Rankine Lecture. *Geotechnique* 26:3,391-452.

Mesri, G. and P.M. Godlewski (1977). Time- and stress-compressibility inter-relationship. *Proc. ASCE, V.* 103, No. GT5, 417-430, Paper 12910.

Mitchell, J.K. (1986). Practical problems from surprising soil behaviour. 20th Terzaghi Lecture. *Proc. ASCE, V.* 112, No. GT3, 255-289.

Mitchell, W.R., J.Fidler and M.D.Fitzpatrick (1968). Rowallan and Parangana rockfill dams. *Journal, Inst. of Engrs. Aust.* 40:239-249.

Mitchell, W.R. and M.D.Fitzpatrick (1979). An incident at Rowallan Dam. *13th Congr. Int. Comm. on Large Dams*, New Delhi, 2:195-210.

Parkin, A.K. (1971). Application of rate analysis to some settlement problems involving creep. *Proc. 1st Australia-NZ Conf. on Geomechanics*, Melbourne, 1:138-143.

Parkin, A.K. (1978). Coefficient of consolidation by the Velocity Method. *Geotechnique* 28:4,472-474.

Parkin, A.K. (1985). Settlement rate behaviour of some fill dams in Australia. *Proc. 11th Int. Conf. Soil Mech. and Found. Engg.*, San Francisco, 4:2007-2010.

Parkin, A.K. (1991). Creep of rockfill. Chap. 9 in *"Advances in Rockfill Structures"* (Ed. E. Maranha das Neves), NATO ASI Series V. 200 (Kluwer).

Pigeon, Y. (1969). The compressibility of rockfill, *Ph.D. Thesis*, University of London.

Poulos, H.G., L.P.de Ambrosis and E.H.Davis (1976). Method of calculating long term creep settlements. *Proc. ASCE,* V.102, No.GT7, 787-804, Paper 12273.

Pusch, R. and P. Feltham (1980). A stochastic model for the creep of soils. *Geotechnique* 30:4, 497-506.

Sowers, G.F., R.C.Williams and T.S.Wallace (1965). Compressibility of broken rock and the settlement of rockfills. *Proc. 6th Int. Conf. on Soil Mech. and Found. Engg.*, Montreal, 2:561-565.

Wahls, H.E. (1962). Analysis of primary and secondary consolidation. *Proc. ASCE,* V. 88, No. SM6, 207-231, Paper 3373.

Williams, A.F. (1980). The design and performance of piles socketed into weak rock. *Ph.D. Thesis*, Monash University, Australia.

See also: Williams, A.F. (1980). Proceedings, *3rd Australia-NZ Conf. on Geomechanics*, Wellington NZ, 1:87-94.

See also: Williams, A.F., I.W.Johnston and I.B.Donald (1980). *Proc. Int. Conf. on Struct. Foundns. on Rock*, Sydney, 1:327-347.

Prediction versus Performance in Geotechnical Engineering, Balasubramaniam et al. (eds)
© *1994 Balkema, Rotterdam, ISBN 90 5410 355 8*

Seepage analysis of Tarbela dam (Pakistan) using finite element method

Yusuke Honjo & Pham Huy Giao
Asian Institute of Technology, Bangkok, Thailand

Parvez Akhtar Naushahi
Pakistan Engineering Services, Lahore, Pakistan

ABSTRACT : The performance data on seepage control of Tarbela Dam , which was completed in 1975 in Pakistan, is analyzed using Saturated-Unsaturated Seepage Flow Finite Element analysis based on the Invariant Mesh technique. Computer code UNSAF2D has been used for the analysis. The monitoring data is collected to compare with the theoretically calculated values. The analysis is done for the various stages of reservoir filling, depleting and various conditions of sedimentation in the reservoir. The comparison, generally, shows a good agreement between calculated and observed values. One of the pronounced feature of the Tarbela dam is its very wide upstream impervious blanket. The purpose of the construction of this blanket was to control the seepage and the uplift pressure of the dam built on the very thick alluvium deposit of the Indus river. The calculated results explain the function of this upstream blanket, and also its change of permeability characteristics with time due to accumulation of large amount of sediment on it.

1 INTRODUCTION

For the design of any Earth Dam that retains a body of water it is very important to find out the expected amount of seepage, distribution of fluid pressure, the exit gradient and the zero effective stress conditions. Various methods and procedures have been used to find out these quantities.The recent trend is to use computer aided methods to handle these tasks. In this study, the Finite Element Method has been introduced to analyze the seepage through Tarbela Dam in Pakistan.

Tarbela Dam, which is the largest in terms of embankment volume in the world, is a heavily instrumented dam. All kinds of monitoring data like deformation, settlement, seismic records, or slope movement and seepage records are available with the Dams Safety Organization in Pakistan (Szalay and Marino, 1981). The situation, thus, provides an opportunity to study the behavior of this earth fill in many aspects. In this paper, a study on seepage behavior of main dam and its impervious blanket using

Saturated-Unsaturated Finite Element Seepage Analysis Program has been carried out.

Tarbela Dam was designed in late fifties to early sixties. At that time computer modelling of the seepage problems was not commonly available. The seepage design was done using conventional assumptions of steady state conditions at maximum reservoir level. Other assumptions on the boundary conditions were also much simplified compared to the actual situation.

The Finite Element Method provides a very flexible model to analyze the problem (Neumann, 1973). This technique is to simulate the steady state conditions as well as the unsteady state conditions of seepage during reservoir filling and depleting. The computer code UNSAF2D has been used. It considers saturated-unsaturated flow by using invariant mesh for analysis (Nishigaki and Takeshita, 1987).

The calculated results are compared with the actual observations. Special interests are evaluation of the performance of the exceptionally large upstream blanket and effect of the material deposition in later years of the operation.

2 TARBELA DAM PROJECT

The Tarbela Dam Project is on the Indus river in the North West Frontier Province of Pakistan, about 64 km northwest of Islamabad. The global location of the project is $72^0 42'E$ and $34^0 05'N$. The location map of the project is shown in Fig. 1. The Tarbela Reservoir conserves surplus water for the benefit of the vast irrigation system of Pakistan.

2.1 Geology of the area

Tarbela Dam is located in the Hazara Hills, which are part of the mountain group known as the Lesser Himalayas. The Hazara Hills are composed of crystalline and metamorphic rocks with some non-fossiliferous deposits and gabbroic intrusions, all ranging in age from Precambrian to Permian.

The major rock types which occur at Tarbela are quartzite, limestone, phyllite, carbonaceous and graphitic schists, chloritic schists as well as some basic igneous rocks. A geological section across Indus river valley is shown in Fig. 2 (a).

There are three distinct geological formations at the Tarbela dam site : the Salkhala formation, forming the right bank; the Hazara formation, forming the bedrock base of the Indus valley; and the Kingriali formation, forming the left bank. The Salkhala formation consists of quartz, marble, graphitic schists and quartzofeldspatic gneiss, belonging to Precambrian age and being the oldest known rocks in the region. The Hazara formation is mainly formed of quartzite and phyllite with some chloritic schists exposed on the right bank. The Kingriali formation consists, mainly, of limestone, dolomite, quartzite and phyllite with some basic igneous rocks. The Quarternary deposits mostly are loess and stream deposits, they are found in the terraces along the main river. The general orientation of bedding indicates that the banks of the river are the limbs of an anticline, the axis of which has been by the Indus river.

The bedrock surface below the Indus flood plain varies across the river and exceeds more than 213 m (700 ft) at the deepest point. The bedrock profile forms ridges and valleys running parallel with the river.

The right and left banks of the Indus river are separated by a flood plain of 1.8 km wide. The river flow passed the site in a braided stream pattern on alluvial deposits at about El. 338 m (1110 ft). At one time, alluvium filled the valley up to El. 488 m (1688 ft). Above the flood plain the hillsides are generally steep with slopes often controlled by the dip of bedding or joint systems. Topographic relief in the dam site area is about 460 m (1500 ft).

Sections through the alluvium at the dam site are shown in Fig. 2 (b).The depth of alluvium is greatest in the dam foundation area and becomes shallower upstream; the width of alluvium is narrowest at the dam and widens in the upstream direction. The rock of the right bank forms a bench at the location of Main Embankment Dam (MED) of approximately 610 m (2000 ft) wide, which is covered by shallow alluvial deposits. Generally speaking, in the project area there are three main types of alluvial deposits :

1. Dense rounded to sub-rounded boulder gravel predominates at the site. The voids are completely or almost completely filled with medium to fine sand. Coarse sand and fine gravel are almost missing from this gap-graded material.

2. Open voided, highly pervious rounded boulder gravel, termed "openwork", having the same coarse component as the material described above. However, the voids between boulders are not filled or partially filled with the sand. The openwork layers in the dam foundation area are between 24 and 61 m deep and range in thickness between 15-24 m on the right site of the valley, and 6 to 9 m on the left site.

3. Sand, gravelly sand, silty sand and silt layers, are found throughout the foundation. The largest sand layers are located in the central portion of the main embankment foundation where their thickness is as much as 15 to 30 m.

2.2 Principal components of the project

The principal elements of the Tarbela project are shown on general layout plan of the site in Fig. 3. Some of the particulars of the Project are listed in Table 1. The main features of the Project are as follows :

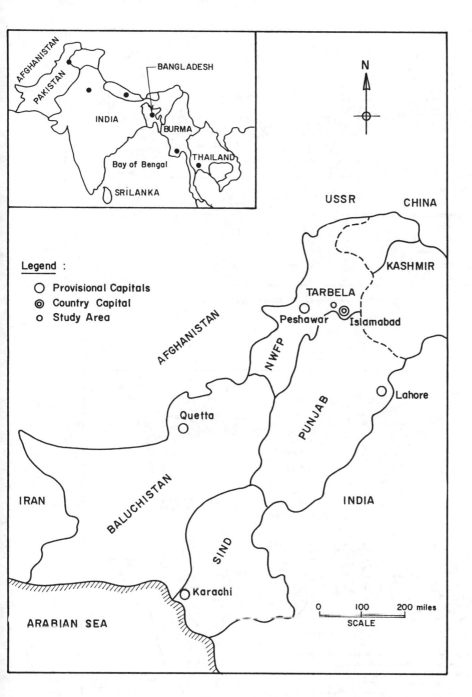

Fig. I   Location Map of Tarbela Dam

221

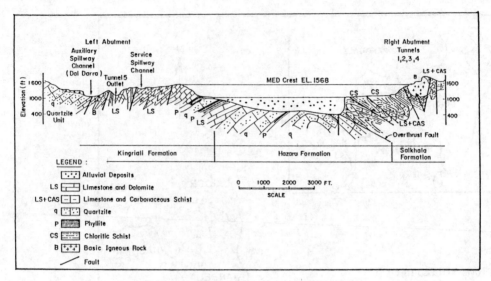

Fig.2(a) A Geological Section Across Indus River Valley

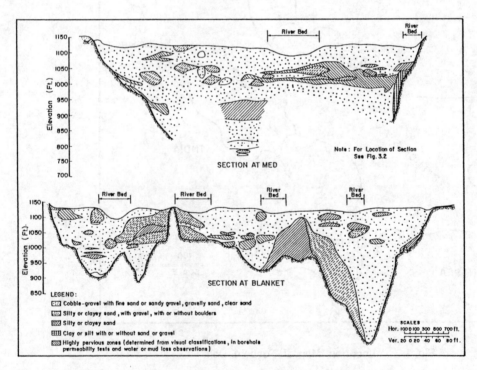

Fig. 2 (b) Geologic Sections Through Alluvium under the MED and Upstream Blanket

222

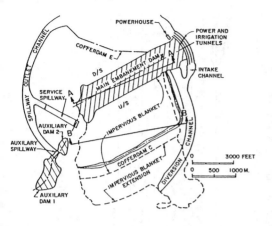

Fig. 3   General Project Plan of Tarbela Project

(AD-2) , are to close saddles at the upstream end of a side valley.

b) Two spillways, Service spillway and Auxiliary spillway, discharging into the side valley.

c) A tunnel through the left abutment to provide controlled irrigation releases downstream.

   3. Right bank :

a) A Group of four tunnels through the right abutment to provide for :

  - River diversion during the last phase of main embankment construction.

  - Regulated power releases.

  - Regulated irrigation releases.

b) A Powerhouse and Switchyard at the downstream side of MED.

1. Main Embankment Dam (MED) : an earth and rockfill embankment across the entire width of the main Indus river valley and the attributable reservoir. Exceptionally huge upstream blanket was employed to control the seepage through the foundation which consists of thick alluvial deposits.

2. Left bank :

a) Two auxiliary earth and rockfill embankments, Auxiliary Dam 1 (AD-1) and Auxiliary Dam 2

2.3 Seepage control of Main Embankment Dam (MED) and Blanket

The Main Embankment Dam (MED) extends some 2750 m (9000 ft) across the main river valley, which has a flat bottom of about 1800 m (6000 ft). The height of the valley section of the dam averages about 137 m (450 ft). Integral with the dam is an impervious blanket extending generally 2350 m (7700 ft) upstream of the

Table 1. Tarbela project particulars

RESERVOIR	
Length	97 km
Maximum Depth	137 m
Area	24,300 ha
Usable Capacity above El. 396m	11,600 mil. m^3
Dead Storage below El. 396m	2,300 mil. m^3
Gross Capacity El. 473m	13,900 mil. m^3
Mean Annual Inflow	79,000 mil. m^3
MAIN EMBANKMENT DAM	
Length at Crest El. 477m	2,743 m
Maximum Height	143 m
Embankment Volume	106 mil. m^3
Blanket Volume	23 mil. m^3

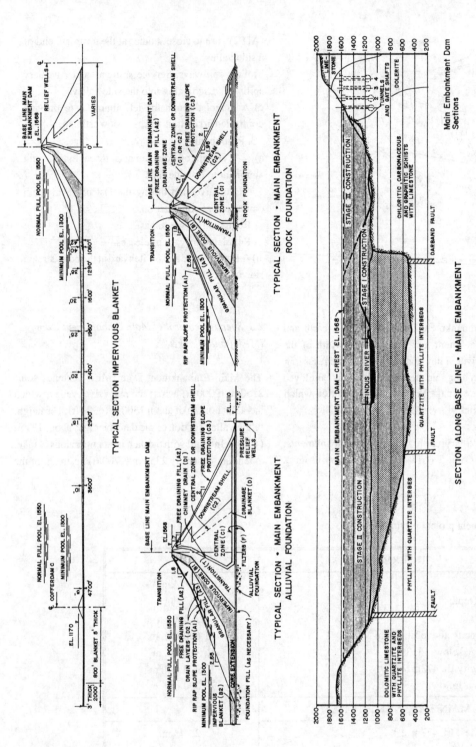

TYPICAL SECTION IMPERVIOUS BLANKET

TYPICAL SECTION - MAIN EMBANKMENT
ROCK FOUNDATION

TYPICAL SECTION - MAIN EMBANKMENT
ALLUVIAL FOUNDATION

SECTION ALONG BASE LINE - MAIN EMBANKMENT

Main Embankment Dam
Sections

Fig. 4   The Section of Main Embankment and Upstream Blanket

224

center line of the dam. The section of MED and impervious blanket is shown in Fig. 4. Control of seepage of water from the reservoir through and under the embankment dam is necessary for three reasons :

- To conserve water in the reservoir by reducing quantity of seepage flow through the embankment and the foundation.

- To preserve the stability of the dam against the seepage force induced by the flow in the embankment and the foundation.

- To prevent migration of soil particles from the embankment, foundation and abutment by seepage force in order to avoid internal erosion.

2.4 The foundation soils

The subsurface condition for the Main Embankment Dam shows that the major part of the dam, approximately 1830 m (6000 ft) from the left bank, is founded on deep pervious alluvium. Remainder of the dam, adjacent to the right bank, about 610 m (2000 ft), is founded on a bedrock bench. The bedrock appeared to be predominantly metamorphosed limestones, quartzites, schists and phyllites, which dip steeply toward the banks at the dam abutments at about 45^0.

The alluvial soils are composed predominantly of boulder-gravels choked with fine to medium sand. Part of the foundation area of the dam and blanket where islands exist was covered with up to 5 m of loose, fine uniform sand, silty sand and silt. Occasional sand and silt strata interspersed throughout the deeper alluvium. Area of high permeability, attributable to open work zones are present in the alluvium.

The Foundation soils were divided into four horizontal layers or zones :

Zone F1 (0 to 5 m) : variable loose surface deposits of silty fine sand and fine to medium sand.

Zone F2 (5 to 24 m) : loose to medium-dense boulder-gravel generally choked with sand; occasional layers and pockets of fine to medium sand; zones of openwork boulder-gravel were believed to be discontinuous.

Zone F3 (24 to 34 m) : loose to medium-dense boulder-gravel-sand mixtures containing some highly pervious openwork zones.

Zone F4 (34 m to bedrock) : medium-dense to dense boulder gravel-sand mixtures, having sufficient sand to choke the material enclosing discontinuous but occasionally thick zones of hard silt and dense sand. Permeabilities of these layers were assumed as shown in Table 2 .

Table 2. Permeability of the alluvial layers

Zone	Depth (m)	K_{mean} (cm/sec)
F1	0 - 5	2×10^{-2}
F2	5 - 24	2×10^{-2}
F3	24 - 34	4×10^{-2}
F4	34 - 91	1×10^{-2}

To control the seepage flow in this very thick and pervious alluvial deposit was one of the major issues in designing Tarbela Dam. It was decided to introduce exceptionally large upstream blanket to reduce the quantity of flow and release the pore pressure beneath the embankment dam.

2.5 Section of MED

A typical section of MED is also shown in Fig. 4 . The major zones of the MED are descrbed in the following:

1. Free-draining cap and shell : The free-draining rockfill or angular boulder-gravel cap and upstream shell zone (A2 and A3), located directly under the riprap on the upstream face, are constructed of good quality granular material. The free-draining zone (A2) is for stabilization of the upstream slope upon reservoir drawdown. Drainage from A3, during drawdown, is carried out in drainage layers (D2) into the free-draining zone.

2. Impervious core : The sloping core of the dam (Zone B1) is believed to make a more natural and economical connection with the impervious blanket than a vertical core would have. The core (Zone B1) was specified to be a well-graded mixture of angular

gravel-sand and silt, and was designed to be shelf-healing along settlement cracks, if they should occur. The core slopes upstream from the crest, thereby reducing the quantities of sound, strong , select material necessary for the extension connecting the core to the blanket (Zone B2). To take advantage of the high shear strength of the free-draining cap and shell material (Zone A2), the slope of the core was steepened above El. 439 m (1440 ft) from 1V : 1.8H to 1.5V : 1H. The inclined core was also expected to be affected less by differential settlement because of the smaller load imposed on the core after its placement. The foundation and the heavy center portion of the dam would be already partially consolidated when the core was placed, and like blanket, the inclined core would follow differential ground settlements more flexibly than a central core, in which the entire height would be affected by such settlements.

3. Transition zone : Immediately downstream of the core is a transition zone (T), which contains enough fines to serve as a secondary core. It is relatively thicker than B1. It was formed of reasonably well-graded angular boulder gravel (ABG) mixtures containing not more than 15 percent fines. This zone provided a dense, essentially incompressible foundation for the core.

4. Downstream shells : Zones C1 and C2, located downstream of the transition zone T1, and comprising the largest portion of the downstream shell, are formed of sound durable rock and boulder-gravel-sand mixtures. The central zone C1 was designed to support zone T, to be essentially incompressible to minimize core settlement, and to be more permeable than zone T to drain any seepage that managed to flow through that zone. The essential requirement for zone C2 next to the downstream face of the dam was that it has adequate strength to maintain the stability of the 1V : 2H downstream slope.

5. Drainage features : A chimney drain (D1) just downstream of the transition zone (T) intercepts seepage through possible cracks in the core and thus protect the downstream shell against saturation. Drainage from the chimney drain flows into the drainage blanket (D) and discharges at the downstream toe. The drainage blanket is protected from contamination by fines from both above and below, by filter layers of the dam overlying the foundation alluvium

to collect underseepage and prevent it from saturating the downstream shell of the dam. The drainage blanket was therefore located above tailwater level, and was protected by layers of filter material top and bottom. The other function of the drainage blanket was to collect the small amount of seepage that was expected to seep through the core, and runoff penetrating the downstream shell.

3 SEEPAGE ANALYSIS

The seepage monitoring data of Main Embankment Dam (MED) and upstream blanket of Tarbela Dam Project were collected to compare the predicted resultsby the saturated-unsaturated flow model calculations by UNSAF2D. The data related to Tarbela Dam have been cited in many references (Khan and Naqvi, 1970; Binger, 1978; Agha, 1981; Low III And Fox, 1982; Izhar-Ul-Haq, 1991). The analysis calculations were done for various reservoir filling and depleting stages, considering steady state as well as unsteady state of seepage. The calculations were also done to simulate the drop in piezometric heads due to the sedimentation accumulated on the upstream blanket of MED.

3.1 F.E. discretization and boundary conditions

Figure 5(a) shows the model considered for the flow domain of the MED and its upstream blanket. We have considered the central part of the MED and its upstream blanket for the study. This zone has maximum thickness of alluvium. The rock depth is considered as impermeable boundary. On the upstream side, the end of upstream blanket has been considered as upstream boundary of the model where the head was fixed to that of the reservoir level in each corresponding case. The downstream boundary has been considered at 564 m from the edge of the MED. In the MED body only upstream granular fill (A3), Impervious Core (B1), and downstream transition zone (T) are considered in the flow domain. All the other zones of MED have sufficiently high permeability to be assumed as free flow zones (i.e. no head loss in the zone).

Table 3 : Material properties used for analysis

PARAMETERS	ZONES		
	Upstream Shell	Core & Blanket	Foundation
Permeability (m/day)			
K_{xx}	0.864	0.00432	43.20
K_{yy}	0.864	0.00432	4.32
Porosity	0.02	0.17	0.2
Negative Pressure Head (m)	-0.50	-1.00	-
Density (t/m^3)			
Total	-	2.35	2.36
Dry	-	2.20	2.24

The F.E. mesh of the discretized flow domain is shown in Fig. 5 (b). Detailed discretization scheme of MED and foundation is presented in Fig. 6. The total number of the nodes is 483, whereas that of the elements is 403. The maximum band width is 19. The boundary conditions applied to the flow domain considered are described as follows : the upstream end of the flow domain is considered as prescribed head boundary,i.e. the total head at this boundary was set same as dam reservoir level. The constant head boundary at El. 338 m (1110 ft) was also assumed, at the downstream , 564 m from the toe. The rock below MED and upstream blanket are considered as impermeable, no flow boundaries. The Chimney Drain of the MED has been considered as seepage face.

3.2 Material parameters

The parameters to assign for various zones of flow domain are taken same as used for the design of the project. The summary of the material properties used is given in Table 3. Fig. 7 shows the characteristic curves which relate moisture content, negative pres- sure heads and relative permeability. Only dam body and upstream blanket were considered in unsaturated conditions before the reservoir impounding. The initial negative head for the shell was taken as 50 cm while for the core and upstream blanket material a value of 1.0 m was assigned.

3.3 Analyzed cases

The analysis of seepage conditions was carried out by considering various stages of reservoir filling and depleting :

1. Steady state analysis at reservoir EL. 369 m (1210 ft) : To simulate the conditions of just before the filling of the reservoir a steady state analysis was carried out at reservoir El. 369 m (1210 ft). This result would provide the initial condition for the following analysis.

2. Steady state analysis at reservoir El. 467 m (1530 ft), 1975-1976 : The first filling of the reservoir, in 1975, was taken up to El. 467 m (1530 ft). The reservoir had stayed at this elevation for about 3 months.The steady state analysis for this elevation was carried out by assuming that 3 month time was enough to attain the steady state condition.

227

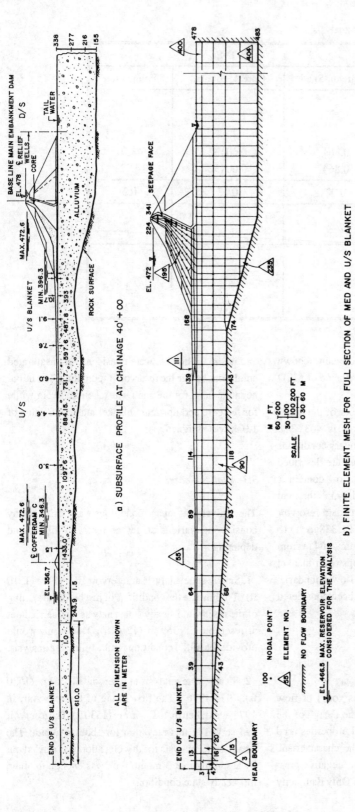

a) SUBSURFACE PROFILE AT CHAINAGE 40' + 00

b) FINITE ELEMENT MESH FOR FULL SECTION OF MED AND U/S BLANKET

Fig. 5   Actual Conditions and Overall F.E. Discretization of Flow Domain

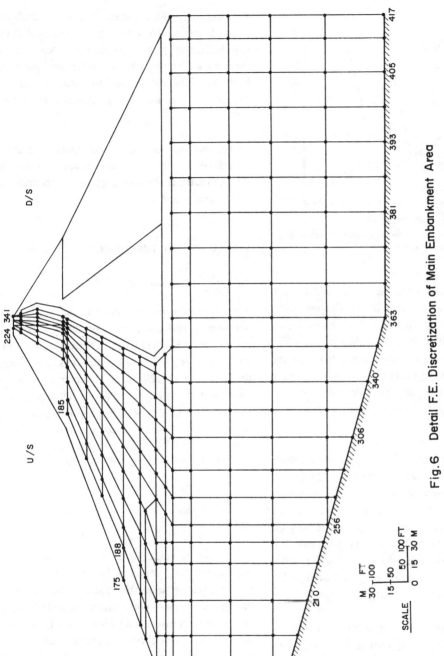

Fig.6   Detail F.E. Discretization of Main Embankment Area

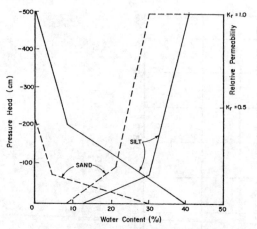

Fig. 7 Characteristic Curve Relating Relative Permeability, Pressure Head and Water Content

3. Steady state analysis at reservoir El. 473 m (1550 ft), 1980 : Tarbela Dam reservoir was designed for the maximum El. 473 m (1550 ft), considering only steady state conditions at this level. Analyses have been carried out under the same condition. After 1977, the reservoir had been brought to its maximum elevation several times , therefore, the calculations can be compared with actual monitoring data of the maximum reservoir level.

4. Unsteady state analysis during reservoir filling, 1975 - 1976 : The reservoir filling of 1975-76 was simulated by considering unsteady state conditions during reservoir filling. In this analysis, it was assumed that reservoir is raised from El. 369 m (1210 ft) to El. 467 m (1530 ft) at the rate of about 1.0m/day. The reservoir had stayed at El. 467 m for about 3 months. This was also simulated by keeping the reservoir at the same elevation for 95 to 180 days.

5. Unsteady state during reservoir operation,1981 : The reservoir filling and depleting curve of 1981 which is considered as a representative curve was employed.

6. Steady state analysis at El.467 m (1530 ft) with sedimentation effect : Over the years, the sediments brought by Indus river water are accumulated in the reservoir. This accumulation of sediments has increased the thickness of upstream blanket. The

sedimentation on the upstream blanket was simulated by using the equivalent permeability. The calculations have been done by assigning the total heads of 1981, observed at Piezometer B-145, to the upstream head boundary. The calculation was carried out by decreasing the permeability of upstream blanket to 25%.

7. Unsteady state during reservoir operation of 1981 with sediment effect : The effect of sedimentation on the piezometric readings during reservoir operation of 1981 was also performed.

4 RESULTS AND DISCUSSIONS

4.1 Steady state analysis at El. 369 m

The initial reservoir condition in the flow regime was obtained by performing a steady state analysis at reservoir level 369 m (1210 m). The calculations were done in one step which took 5 iterations to converge to a head tolerance of 0.008 m. The comparison of the results of this analysis with the relevant piezometric readings shows that the difference between these two was -6.6 m to +4.7 m. The results of this analysis provided the initial condition for all the unsteady seepage analysis of the reservoir operations.

4.2 Steady state analysis at reservoir EL. 467 m, 1975-76

Comparison of nodal heads with the piezometric readings of 1975-76 reservoir El. 467 m was made in upstream blanket area, MED fill and MED foundations. The comparison graphs are shown in Fig. 8(a),(c) and (e). In most cases , comparison points are located near 1:1 line, which shows good agreements. The seepage for this case was calculated as 191 m³ per day per meter length.

4.3 Steady state analysis at reservoir El. 473 m, 1981

The comparison between the calculated results and the

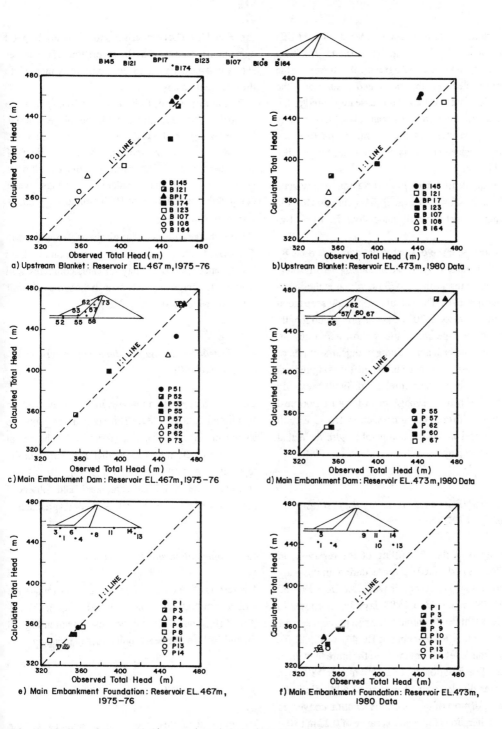

Fig. 8   Comparison of Observed and Calculated Total Heads for Steady State Analysis

231

observations for upstream blanket, MED fill and MED foundation are shown in Fig. 8(b), (d) and (f). These results also show a good agreement. However, the differences between the calculated results and the observations are greater for this case compared to the previous case under the upstream blanket. This may be due to the effect of the sedimentation over the blanket, which significantly changed the permeability of the blanket.

The equipotential lines plotted for this steady state condition is shown in Fig. 9. The function of the upstream blanket is clearly understood from this figure.

The actual seepage design of the dam was done by using flow nets considering steady state conditions at reservoir El. 473 m, which is the maximum design level. The calculations show seepage at the downstream toe as 201 m^3 per day per meter length. The actual seepage observations show about 267 m^3 per day per meter length. The geological sections of the flow domain are not similar in the longitudinal direction of the dam, therefore, it is difficult to exactly compare the actual seepage record with theoretical predictions with one geological section. However, still the theoretical analysis is comparable with the actual monitored seepage.

4.4 Unsteady state analysis during the first reservoir filling of 1975-76.

The analysis of the first filling of the reservoir, in 1975-76, was performed considering an unsteady state condition. The reservoir was risen from El. 369 m to El. 467 m, i.e. 98 m, in 95 days. For purposes of analysis, 1.0 m per day monotonic rise of the reservoir was assumed. After reaching at El. 467 m (1530 ft), the reservoir level was maintained for more than three months. The duration of calculation was 180 days which was divided into 13 time steps with minimum of 5 days. In most of the cases, the solution converges with 4 - 6 iterations for head tolerance of 0.23 m to 0.9 m.

The comparison of various calculated heads with the relevant piezometric readings are shown in Fig. 10(a),

(c), (e) and (g). The comparison shows, generally, good agreement except at P-55. However, trend of the curves shows that the calculated values ultimately merge into the observed values.

The flat response of P-55 for about 70 days may be due to the initial unsaturated conditions of the material. This piezometer showed quick response after the core was saturated as indicated by the readings of subsequent years. The same piezometer showed good agreement with calculated values when compared with readings of 1980 monitoring data, as shown in Fig. 10(h).

The discharge with the reservoir rise is also calculated. The peak discharge value at the downstream to drain is calculated as 196 m^3 /day /m. This compares well with the discharge of steady state at the same reservoir elevation which was 191 m^3/day/m .

4.5 Unsteady state analysis during the reservoir operation of 1981.

The reservoir operation analyses are compared with the monitored data of 1981. The comparison plots are shown in Figs. 10 (b), (d), (f) and (h). The actual reservoir operation was somewhat different from the representative curves as shown in Fig. 10 (b) etc. However, the trend of calculated and observed potential variation, in general, is in good agreement.

4.6 Sedimentation effect

The results of calculations carried out to simulate the sedimentation effects are also plotted in Fig. 10(b), (d), (f) and (h). The effects can be clearly seen in Fig. 10 (b) and (d) , i.e. the head right under the blanket.

5 CONCLUSIONS

The purpose of this study was to demonstrate the capability of the Finite Element Analysis and the usefulness of consideration of saturated-unsaturated zones together in earth dam analysis. The numerical

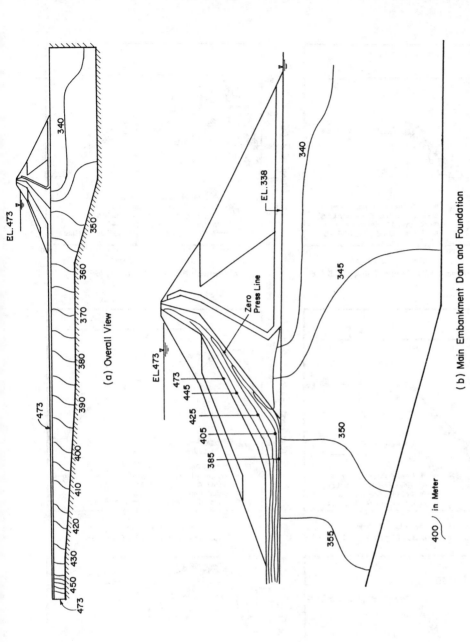

(a) Overall View

(b) Main Embankment Dam and Foundation

400 ⟋ in Meter

Fig. 9   Equipotential Lines of Flow Regime Steady State Analysis at EL.473

233

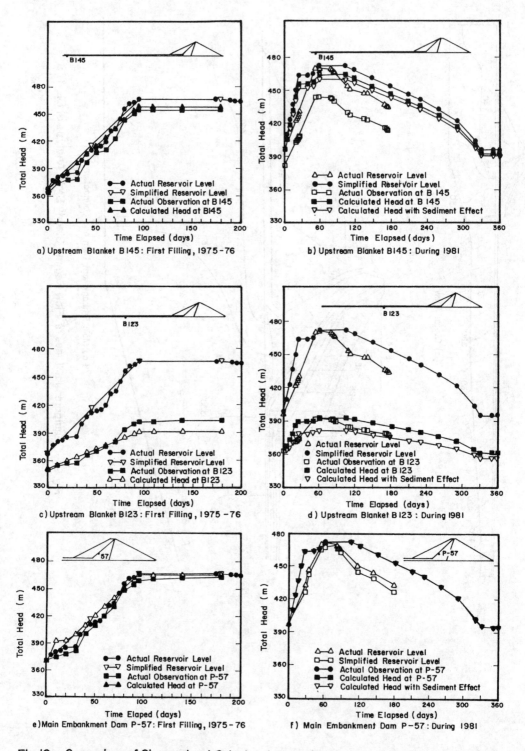

Fig. 10 Comparison of Observed and Calculated Total Head for Unsteady State Analysis

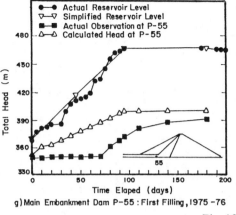

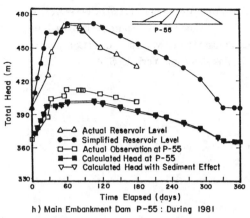

g) Main Embankment Dam P-55 : First Filling, 1975 -76 h) Main Embankment Dam P-55 : During 1981

Fig. 10 (Cont'd)

analysis results can be considered authentic if they compare well with the actual monitoring data.

The comparison of theoretical results with the actual monitoring data revealed, in general, a good agreement. Therefore it is considered possible that the computer code used, UNSAF2D, is capable of predicting the behavior of flow regime very effectively, for both steady and unsteady cases.

From the results, it is appropriate to conclude that Tarbela Dam is performing very well. The exceptionally large upstream blanket was effective to control seepage in the alluvial deposit beneath the Main Embankment Dam. The effects of the sedimentation are also explained by the analysis. The response of piezometers in the foundation both under the blanket and MED are considerably quick which is reflecting very high permeability of alluvial deposit.

ACKNOWLEDGEMENT

The authors would like to show their deepest appreciation to Dam Safety Organization (DSO), a subsidiary of Water and Power Development Authority (WAPDA), in Pakistan for providing the monitoring data of Tarbela Dam. Special thanks and appreciation are due to General Manager of Tarbela Dam, Mr. A.K Khan and Director, Mr. M.A. Bhatti.

REFERENCES

Agha, A. 1981. Dam Engineering in Pakistan. *Proceeding of the Symposium on Problems and Practice of Dam Engineering*, A.I.T, Bangkok.

Binger, K. J. & Khoshgoftaar, M. R. 1979. Tarbela : Plans, Problems and Success. *Inter. Water Power and Dam Const.*

Freeze, R.A. 1971. Influence of the Unsaturated Flow Domain on Seepage through Earth Dams. *Water Resources Research*, vol. 7, No. 4, pp. 929-941

Izhar-Ul-Haq, 1991. Foundation Problems of Tarbela. *International Conference on Large Dams*, Vienna.

Khan, S. N. & Naqvi, S. A. 1970. Foundation Treatment for Underseepage Control at Tarbela Dam Project (Pakistan). *Transaction 10Th Inter. Conference on Large Dams*, Montreal.

Low III, J. & Fox, H. R 1982. Sedimentation in Tarbela Reservoir. *International Conference on Large Dams*, Rio de Janero.

Neumann, S. P. 1973. Saturated-Unsaturated Seepage by Finite Element. *Journal Of the Hydraulics Div., ASCE*, Vol. 99, No. HY 12, pp. 2233-2250.

Nishigaki, M. & Takeshita, Y. 1987. *FEM Program UNSAF2D*, University of Okayama, Japan (in Japanese).

Szalay, K. & Marino, M. 1981. Instrumentation Of Tarbela Dam, Recent Development in Geotechnical Egg. For Hydraulics Projects. *Proceed. Of ASCE International Convention*, New York City.

Water And Power Development Authority (WAPDA) 1985. Project Completion Report of Tarbela Dam Project, Prepared by TAMS, The Project Consultants, Tarbela, Pakistan /New York, U. S. A.

Back analysis of cut slope instability in soft Bangkok clay

Prapote Boonsinsuk
Asian Engineering Consultants Corporation Ltd, Bangkok, Thailand

ABSTRACT: A number of slopes formed by excavation of the soft Bangkok clay had been unstable during the construction phase of a large housing complex. The mode of slope movement was the classical rotational slip surface type normally observed in clay. The slope stability when major movement was imminent was back analyzed based on the slope configuration before the movement. Measured in-situ vane shear strength was used in the total stress analysis to determine the short-term stability at the end of construction. The effective stress analysis was applied to establish the strength parameters of the soft Bangkok clay which would lead to the factor of safety of 1.0. The suitable method of analysis and the appropriate strength parameters for slope stability analysis are recommended.

INTRODUCTION

Slope stability problems in the soft Bangkok clay have been encountered by many infrastructure projects, particularly roads and housing development. Fill embankments are normally required for roads in order to prevent flooding and allow for settlements caused mainly by consolidation of the soft Bangkok clay and ground subsidence due to groundwater withdrawal. For large housing projects, it is common to construct man-made lakes for flood control and architectural purposes, and to use the excavated soil to raise the elevation of the adjacent area. Because of the low shear strength of the soft Bangkok clay, slope instabilities in both embankments and excavations often take place. The unstable slopes are normally a few metres high and the mode of slope movement is usually rotational, unlike landslides which involve destructive movement of soil mass.

In order to design against slope instability, it is always difficult for geotechnical engineers to select appropriate soil parameters and design criteria. Often a full-scale test embankment/excavation has to be constructed at the proposed site to verify the design assumptions. Such an approach demands substantial construction budget and time, which is warranted only for very large development projects. For example, a few full-scale embankments and an excavation had been constructed at the proposed site for the Second Bangkok International Airport (Boonsinsuk, 1974; Essa, 1974), which is located about 25 km from the Bangkok metropolis (Fig. 1). The soft Bangkok clay in this area is well known for its very low shear strength and large settlement as observed in the performance of the Bangna - Bangpakong Highway (Bergado et al, 1990).

In an area close to the proposed airport recently developed for a residential housing complex (Fig. 1), more than ten slopes were unstable during construction while the majority of the slopes formed were stable.

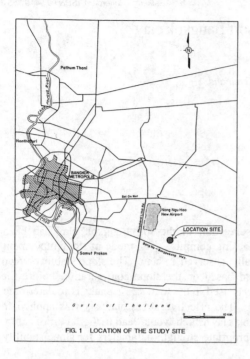

FIG. 1 LOCATION OF THE STUDY SITE

divided depthwise into Soft Bangkok Clay and Stiff Bangkok Clay. The top portion of the Soft Bangkok Clay is a hard crust of a few metres thick caused likely by weathering process.

The Soft Bangkok Clay is about 10 - 15 m thick with its natural water content reaching as high as 130 % in some places. The natural water content is normally higher than its liquid limit, confirming its soft natural state. Its plasticity index varies considerably, indicating different proportions of clay minerals and grain size distribution. In consequence, the shear strength of the Soft Bangkok Clay is very low (about 1.5 ton/m^2 at shallow depths), leading to low bearing capacity and high compressibility.

A cursory site investigation was carried out early in the site formation planning stage. Four boreholes (BH 1 - BH 4) were drilled along the western boundary of the site (Fig. 2). The thickness of the soft clay is approximately 10 - 15 m as indicated by the fact that the natural water content is significantly higher than its corresponding

Most of the unstable slopes were constructed by the cut and fill method to form man-made lakes and raise the existing ground elevations. Because of the large quantity of fill required, the existing ground was excavated as deep as possible with the intention of stabilizing any unstable slope at a later stage. All unstable slopes were eventually stabilized by flattening the slopes, adding toe berms and impounding water in the lakes. This gave an opportunity to study the slope instability characteristics and to determine the appropriate design parameters by back analyzing the unstable slopes. Both total and effective stress analyses were used, of which the results are presented in this paper.

GEOLOGY AND SOIL PROPERTIES

The study site is located on the Chao Phraya Plain along the Chao Phraya river which travels through the central part of Thailand. The recent sediments in this area are mainly marine deposits with a thick top layer of clay overlying a sand layer located at about 20 - 25 m depth. The clay layer is generally

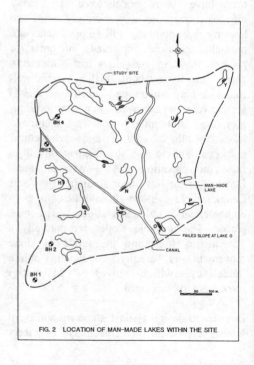

FIG. 2 LOCATION OF MAN-MADE LAKES WITHIN THE SITE

238

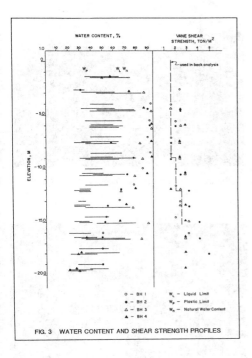

FIG. 3 WATER CONTENT AND SHEAR STRENGTH PROFILES

SLOPE CHARACTERISTICS

The locations of the unstable slopes within the site are indicated in Fig. 2 together with other man-made lakes which showed no signs of slope distress. The majority of the site was relatively flat, prior to development, with two canals crossing the site. Of all the man-made lakes constructed, approximately half of the lakes experienced slope instability as marked in Fig. 2. The other half did not have any slope instability problems even though they were constructed using the same configuration and construction method. It is therefore very difficult to predict when a slope in this area will become unstable.

During the period from mid 1990 to mid 1991, man-made lakes were constructed by excavating about 3.5 m below the existing ground surface with a side slope of about 1 (vertical) to 4 (horizontal). The excavated soil was used to fill the existing ground surface in the vicinity of the man-made lakes. In effect, the height of the side slope of each lake was increased to about 4.5 m or more above the bottom of the lake. The cut/fill operation was generally carried out in the dry season to avoid problems usually caused by rainfall. Once the cut/fill process was completed, the side slope was left bare and the lake was dry, pending additional work. A number of slope movements occurred after some time had elapsed, ranging from a few days to a few months. The movements generally took place without causing much attention prior to rotational sliding. Some relevant construction details of the man-made lakes including the time for slope movement to occur after completion of construction are summarized in Table 1.

In general, the movement was rotational in which the crest of the slope cracked and moved downward about 1 m while the toe of the slope heaved about 1 m (Photograph 1), resembling rotational movement of the soil mass. Slope instabilities usually occurred along a straight stretch of side slope (Photograph 2), implying negligible

liquid limit (Fig. 3). Below the top soft clay layer is a stiff clay layer which is characterized by its natural water content being close to or below its plastic limit. The soft clay is dark grey with some silts, shell bits and traces of organic matters. The stiff clay is light grey and light brown with traces of silt and sand. Some parts of the site were filled with sandy and clayey soils. The groundwater table was measured to be about 1.5 - 2.0 m below the ground surface at the end of the dry season. The undrained shear strength of the clay was measured by field vane testing. The results reveal that the vane shear strength of the soft clay varies slightly with depth and ranges from about 1.5 to 2.5 ton/m^2. In the stiff clay layer, the vane shear strength increases gradually with depth. From the water content and vane shear strength profiles, the soft clay layer appears to be relatively uniform although some local variations exist.

contribution from the three-dimensional effects of any side slopes nearby. The ground surface was dry at the time of movement as indicated by numerous shrinkage cracks of the surface soil. A closer view of the unstable slope crest is illustrated in Photograph 3 with its heaved toe shown in Photograph 4. It would appear that the top of the moved mass simply rotated without seriously being disturbed except in the vicinity of the slip surfaces. The cracks along the crest were dry without any visible free water. Another view of the rotational movement is shown in Photograph 5 in which the upward movement of the slope toe is evident. While the majority of the slope instabilities took place without any water impoundment in the lakes, the movement of the slope at Lake L occurred even though there was water in the lake (Photograph 6). In addition, there seemed to be multiple rotational movements involved as noticed from many shear surfaces visible at the lower portion of the unstable slope.

PHOTOGRAPH 3   CREST OF FAILED SLOPE AT LAKE O

PHOTOGRAPH 4   TOE OF FAILED SLOPE AT LAKE O

PHOTOGRAPH 1   FAILED SLOPE AT LAKE E

PHOTOGRAPH 5   UNSTABLE SLOPE AT LAKE U

PHOTOGRAPH 2   UNSTABLE SLOPE AT LAKE G

PHOTOGRAPH 6   FAILED SLOPE AT LAKE L

From about ten slopes which moved without water impoundment in the lakes as listed in Table 1, it is noticeable that the shortest time leading to slope instability after the construction was completed was about 4 days (Lake U and Lake Y). The longest time to slope movement was about 144 days as exhibited by Lake F. Other times to movement ranged from 6 days to 96 days. All the slopes were excavated to approximately the same elevation of -3.5 m

except Lake U and Lake Y which were excavated slightly deeper (i.e., -4.0 m and -3.8 m respectively). Such a small difference in the bottom elevations of the lakes should not be the main cause for the major difference in the time to slope movement. Other factors such as the maximum height and the average slope inclination appear to contribute insignificantly to the instabilities of the excavated slopes. It is obviously very difficult to predict the performance of the slopes in this area. Nevertheless, it is reasonable to state that the majority of slope movements observed were caused by the redistribution of negative porewater pressures resulting from the release of overburden pressures due to excavation. The time to slope movement after the excavation is completed depends possibly on the time for re-establishing the groundwater level in the vicinity of the slope.

Unlike other slope movements in the study area, the slope movement at Lake L was probably triggered by stockpiling at the crest of the slope. Furthermore, it was possible that the movement involved a series of

TABLE 1 SUMMARY OF EXCAVATION FAILURE CHARACTERISTICS

Lake	Excavation Commenced	Excavation Completed	Excavation Failure	No. of Construc- tion Days	No. of Days to Failure (1)	Maximum Height, m (2)	Length of Failed Slope, m	Bottom Lake Elevation, m	Average Slope (V:H) (3)
E (both sides)	5/10/90	12/12/90	2/3/91	63	80	5.0	55	– 3.5	1 : 4.3
F	8/12/90	22/12/90	15/5/91	14	144	6.7	60	– 3.2	1 : 5.5
G	13/12/90	1/4/91	10/4/91	109	9	5.5	40	– 3.5	1 : 4.0
H	21/7/90	4/10/90	14/10/91	75	10	4.7	50	– 3.5	1 : 2.8
I	3/8/90	16/8/90	28/9/90	13	42	4.7	40	– 3.5	1 : 3.2
L (both sides)	16/6/90	6/12/90	12/3/91	173	96	4.7	40	– 3.5	1 : 5.5
N	23/12/90	7/2/91	10/3/91	46	31	4.5	55	– 3.5	1 : 4.0
O	10/2/91	23/2/91	28/2/91	13	6	5.0	75	– 3.5	1 : 4.4
P	23/2/91	1/3/91	11/3/91	6	10	4.7	55	– 3.5	1 : 4.0
U	6/3/91	18/3/91	22/3/91	12	4	5.0	70	– 4.0	1 : 4.0
Y	12/3/91	22/3/91	26/3/91	10	4	5.2	30	– 3.8	1 : 3.8

Note: (1) After completion of excavation   (2) Excluding stockpile   (3) Averaged from the top of slope to the bottom

circular slip surfaces as evidenced by a number of shear surfaces in the lower portion of the unstable slope (Photograph 6). The slope movement at Lake L was also accompanied by another slope movement adjacent to it and close to a canal (Fig. 2). Although the sequence of movement could not be confirmed, it was likely that the slope movement at Lake L led to the movement of the adjacent slope since the stockpiles were closer to the crest of the slope at Lake L.

BACK ANALYSIS

The appropriate methods of analysis for a cut in clay have been recommended by Bishop and Bjerrum (1960) as shown in Fig. 4. The ø = 0 method (total stress analysis) is considered to be applicable for the short term stability analysis during excavation while the c'- ø' method (effective stress analysis) is suitable for the long- term stability analysis since porewater pressure should be taken into account.

In order to determine the strength parameters for design of remedial works and other earth structures in the study area, the unstable slopes were back-analyzed by both total and effective stress analyses. The back analysis had to be based on certain basic assumptions since there were many unknowns involved that could not be measured or determined at the time of slope movement. The only available information of each unstable slope was its configuration both before and after movement, including the time to slope movement. Detailed site investigation of the slope instabilities was not warranted for such a project. Nevertheless, back analyses of the unstable slopes could be performed by imposing the following common assumptions on every unstable slope :

1.   The original ground surface elevation was at 0 m. Any material existing above the ground surface elevation was assumed to be fill material having a bulk density of 1.5 ton/m^3. The fill material was assumed to offer no resistance to slope movement, i.e,

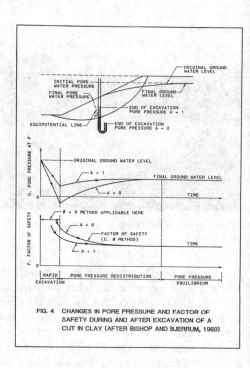

FIG. 4   CHANGES IN PORE PRESSURE AND FACTOR OF SAFETY DURING AND AFTER EXCAVATION OF A CUT IN CLAY (AFTER BISHOP AND BJERRUM, 1960)

tension crack in the fill material was imposed. Such an assumption seemed to be justified since the ground was dry prior to slope movement resulting in many deep cracks. Many cracks along the unstable slope crest were approximately vertical, indicating small contribution of the fill material to resisting any slope movement.

2.   In the effective stress analysis, the thickness of the weathered crust was assumed to be 2 m and the same strength parameters were used for all slopes. This was introduced to limit the number of variables in the back analyses. The contribution of the 2 m thick weathered clay crust was small when compared with the role of the thick soft clay layer.

3.   The final groundwater level (Fig. 4) was assumed to be at -2.0 m elevation behind the slope and daylighted at the bottom of the excavation which was normally at the elevation of -3.5 m. The groundwater level was observed to be about 1.5 to 2.0 m below the ground surface during the site

investigation. The groundwater level was kept constant for all analyses to reduce the number of variables involved.

Based on the above assumptions and the known slope configuration before movement, back analyses of the unstable slopes were carried out by both total and effective stress analyses. The fact that slope instability occurred at least a few days after the construction was completed implied the significant role of porewater pressure in the vicinity of the slope. Thus the effective stress analysis should be the appropriate method of slope stability analysis in this case. The effective stress analysis was therefore used in the back analysis to determine the strength parameters that would yield the factor of safety of 1.0. Because of the variations in actual subsoil conditions, the strength parameters that led to the theoretical factor of safety of about 1 ± 0.05 were accepted for calculation purposes. The total stress analysis was also performed by using the measured vane shear strength profile (Fig. 2) for comparison. Both approaches of slope stability analysis employed the simplified Bishop method (Bishop, 1955) which was applied by using a computer program to search for the slip circular surface with the minimum factor of safety. The location of the theoretical critical slip surface was then compared with the observed surface of the unstable slope. Only the locations of the cracks at the slope crests and the heaved toes of the unstable slopes could however be compared with the theoretical slip surfaces.

A. EFFECTIVE STRESS ANALYSIS

One of the most important tasks in the effective stress analysis is the selection of the appropriate strength parameters. Triaxial tests are usually used to establish the required strength parameters, i.e., the cohesion intercept c' and the angle of internal friction, ø'. For the proposed Second Bangkok International Airport site which is located near the study site (Fig. 1), the following strength parameters have been established

from consolidated-undrained triaxial compression testing (Boonsinsuk, 1974):

Depth, m	c', ton/m^2	ø', deg
0 - 1.8	1.09	12.3
1.8 - 3.4	1.27	17.5
3.4 - 5.1	0.40	16.1
5.1 - 7.35	1.01	10.7
> 7.35	1.44	13.9

From the strength parameters listed above, the following values were used in the back analysis so as to yield the theoretical factor of safety close to 1.0:

Depth, m	c', ton/m^2	ø', deg
0 - 2.0	0.5	15.0
2.0 - 15.0	0.25-0.40	16.0

In order to limit the amount of variables, the depth and the strength parameters of the crust from 0 to -2 m elevation were kept constant. The angle of internal friction ø' of the soft clay was kept constant at 16 degrees while the cohesion intercept was varied from 0.25 to 0.40 ton/m^2 in order to arrive at the factor of safety of close to 1.0 ($\pm$ 0.05).

The results of the effective stress analysis are summarized in Table 2 showing the strength parameters used and the theoretical factor of safety. The locations of the critical slip surfaces with respect to the slope configuration after movement are presented in Figs. 5 - 15. It is obvious that by varying the cohesion intercept c' within a narrow range from 0.25 to 0.40 ton/m^2, the theoretical factor of safety calculated also varied slightly from 0.94 to 1.07 Although the variation appears to be small, its implication on the selection of the appropriate strength parameters and the appropriate factor of safety for design is quite significant. The cohesion intercept c' of the soft Bangkok clay that should be used in the slope stability analysis should apparently be lower than the values derived normally from the consolidated-undrained triaxial compression testing. Even though the lowest cohesion intercept obtained from triaxial testing is used in the analysis, the factor of safety for a permanent slope should not be less than 1.2 for a slope which will cause little property damage in case of slope movement.

Comparing the critical slip surfaces obtained from the effective stress analysis with the unstable slope configuration as shown in Figs. 5 - 15 reveals relatively good correlation. In general, the locations of the tension cracks of the theoretical critical slip surfaces are close to the cracks at the crests of the unstable slopes observed. Similarly, the toes of the theoretical slip surfaces are in good agreement with those of the unstable slopes. The use of the effective stress approach and the parameters chosen are therefore justified.

B. TOTAL STRESS ANALYSIS

Although the effective stress analysis is evidently the appropriate method for slope stability analysis in this study, the total stress method should not be totally ignored, unless proven otherwise. The vane shear strength measured at the site (Fig. 3) was used in the total stress analysis by selecting representative values for each layer as shown. The factors of safety obtained are listed in Table 2 while the critical slip surfaces are presented in Figs. 5 - 15. The factor of safety calculated should represent the factor of safety at the end of construction as suggested in Fig. 4. As a result, the factor of safety calculated from the total stress analysis was generally higher than 1.0, indicating relatively stable slope. In fact, the calculated factor of safety should be higher since the lower bound of the measured vane shear strength was selected for analysis (Fig. 3). Even though the measured vane shear strength could be lowered by a correction factor of about 0.9 as suggested by Bjerrum (1972), most of the theoretical factors of safety calculated from the total stress analysis are still higher than 1.0 . This confirms the validity of the effective stress approach used in analyzing the long-term stability problems faced in this study. If one uses the total stress analysis for the excavated slopes, the factor of safety for design should be at least 1.3 when the average measured vane shear strength is used with Bjerrum's correction factor.

TABLE 2 RESULTS OF BACK ANALYSIS

Lake	Soft Clay		Factor of Safety at Failure (Effective Stress Analysis)	Factor of Safety at Failure (Total Stress Analysis)
	C' (ton/m²)	∅'(deg.)		
E	0.30	16	0.94	1.25
F	0.25	16	1.05	1.10
G	0.30	16	1.06	1.47
H	0.40	16	0.97	1.19
I	0.40	16	0.99	1.20
L	0.25	16	1.07	1.00
N	0.30	16	1.04	1.47
O	0.35	16	1.01	1.34
P	0.35	16	1.05	1.37
U	0.40	16	0.97	1.51
Y	0.40	16	1.05	1.14

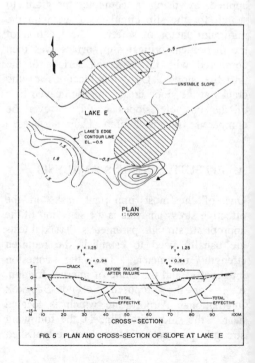

FIG. 5   PLAN AND CROSS-SECTION OF SLOPE AT LAKE E

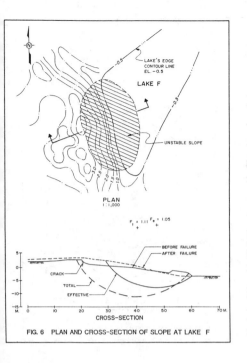

FIG. 6   PLAN AND CROSS-SECTION OF SLOPE AT LAKE F

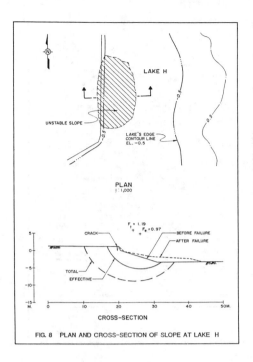

FIG. 8   PLAN AND CROSS-SECTION OF SLOPE AT LAKE H

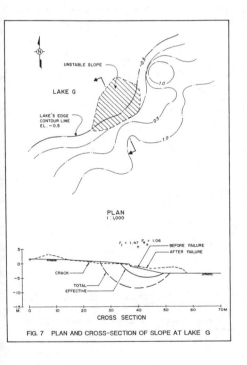

FIG. 7   PLAN AND CROSS-SECTION OF SLOPE AT LAKE G

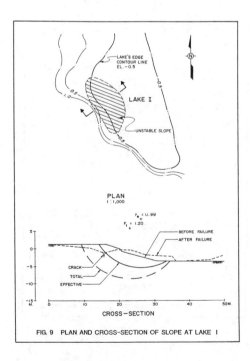

FIG. 9   PLAN AND CROSS-SECTION OF SLOPE AT LAKE I

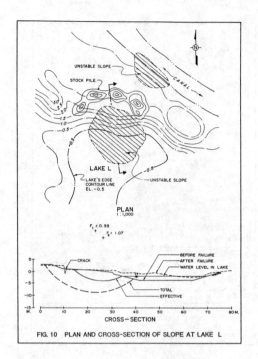

FIG. 10   PLAN AND CROSS-SECTION OF SLOPE AT LAKE L

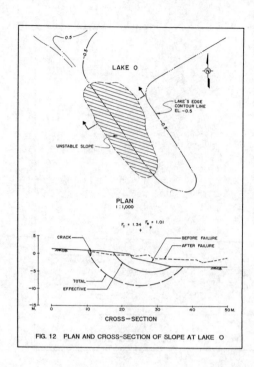

FIG. 12   PLAN AND CROSS-SECTION OF SLOPE AT LAKE O

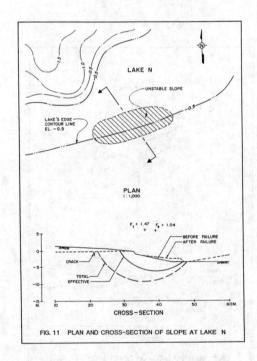

FIG. 11   PLAN AND CROSS-SECTION OF SLOPE AT LAKE N

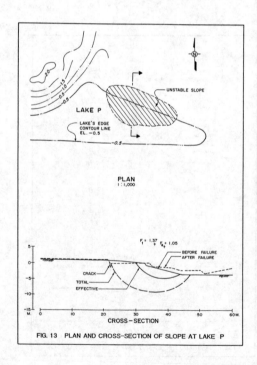

FIG. 13   PLAN AND CROSS-SECTION OF SLOPE AT LAKE P

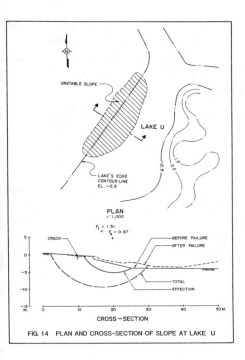

FIG. 14   PLAN AND CROSS-SECTION OF SLOPE AT LAKE U

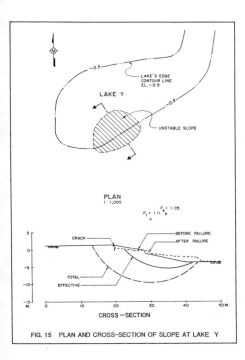

FIG. 15   PLAN AND CROSS-SECTION OF SLOPE AT LAKE Y

In general, the critical slip surfaces obtained from the total stress analysis presented in Figs. 5 - 15 are larger than those determined by the effective stress analysis. In some cases, the former is in better agreement with the observed surface after movement than the latter. The deeper critical slip surface determined from the total stress analysis should be due to the fact that the representative vane shear strength chosen for analysis (Fig. 3) was kept constant for the top 9 m thick layer. In the effective stress analysis, the shear strength of the soft clay increased with depth as the effective stress increased.

CONCLUSIONS

From the results of the back analysis of the unstable cut slopes, the following conclusions can be made:

1.   In designing cut slopes in the soft Bangkok clay, the first step is to ensure that the factor of safety at the end of construction should be at least 1.3. This should be calculated by using the average range of in-situ vane shear strength in conjunction with Bjerrum's correction factor. Other methods of measuring undrained shear strengths that are suitable for slope stability analysis in the soft Bangkok clay remain to be proven.

2.   The next crucial step is the application of the effective stress analysis to determine the long-term stability of the cut slope. The appropriate cohesion intercept c' of the soft Bangkok clay should be in the range of 0.25 and 0.40 ton/m^2 while the internal friction angle should be approximately 16 degrees. The cohesion intercept c' is significantly less than the values normally obtained from the consolidated-undrained triaxial compression testing. The final groundwater level should be similar to the level observed prior to excavation. The minimum factor of safety for design should be 1.2.

3.   Movement of cut slopes should be prevented by construction control. The fact

that the short term stability is not a problem should be taken advantage of. For example, water should be stored in the artificial lake immediately after excavation to prevent long-term slope instability.

4. Remedial measures of the unstable slopes should use the configuration of the unstable slope as a guide. The most economical remedial method is to use the heaved toe as the stabilizing berm and to slightly lower the height of the slope. Immediate water storage in the lake will further enhance the stability of the slope.

5. Although the soft Bangkok clay is relatively uniform, local variations in groundwater level and intermittent weak layers often exist. Detailed site investigation should be carried out for certain structures where slope movements, if occurred, will be costly.

ACKNOWLEDGEMENT

The assistance offered by Mr. Bunjong Wijackanawong in compiling the slope configurations and relevant information is acknowledged.

REFERENCES

Bergado, D.T., Ahmed, S., Sampaco, C.L. and Balasubramaniam, A.S. (1990), Settlements of Bangna - Bangpakong Highway on Soft Bangkok Clay, Journal of Geotechnical Engineering, Vol. 116, No. 1, pp. 136 - 155.

Bishop, A. W. (1955), The Use of the Slip Circle in the Stability Analysis of Slopes, Geotechnique, Vol. V, No. 1, pp. 7 -17.

Bishop, A. W. and Bjerrum, L. (1960), The Relevance of the Triaxial Test to the Solution of Stability Problems, Proc. ASCE Research Conference on Shear Strength of Cohesive Soils, Boulder, Colorado, pp. 437 - 501.

Bjerrum, L. (1972), Embankments on Soft Ground, Proc. of the Specialty Conference on Performance of Earth and Earth Supported Structures, Purdue University, Indiana, Vol. 2, pp. 1-54.

Boonsinsuk, P. (1974), Stability Analysis of a Test Embankment on Nong Ngoo Hao Clay, M. Eng. Thesis, Asian Institute of Technology, Bangkok, Thailand.

Essa, R. (1974), Stability of a Trial Excavation in Nong Ngu Hao Clay, M. Eng. Thesis, Asian Institute of Technology, Bangkok, Thailand.

Prediction versus Performance in Geotechnical Engineering, Balasubramaniam et al. (eds)
© *1994 Balkema, Rotterdam, ISBN 90 5410 355 8*

The performance of hydraulic structures predicted from model tests

Gunther E. Bauer

Faculty of Engineering, Carleton University, Ottawa, Ont., Canada

ABSTRACT: Most cofferdams are constructed with an adequate factor of safety regarding the structural components. Therefore failure can occur only due to piping or bottom heave within the excavation caused by water seeping from the higher outside to the lower inside water table. This paper scrutinizes the different analytical and semi-empirical methods used in the analysis of sheeted cofferdams for hydraulic considerations and these are compared to the results of a series of model tests carried out in a flow tank. The model test investigation in turn is used to analyze a failure of a prototype cofferdam.

1 INTRODUCTION

Groundwater is a familiar problem on construction projects. It is frequently encountered in open excavations. Methods for handling it by pumping from open sumps, from deep wells, or, preferably by the use of a well point system, are now accepted features of construction practice, but they must always be used with care and under expert guidance.

It is now generally accepted that the presence of groundwater is part of normal subsurface conditions to be studied and taken into consideration, together with the geological features of the site, in the design and construction of sheeted excavations.

There are many different ways water can affect the stability of a cofferdam. Among the most important ones are: hydrostatic pressure on the walls, upward seepage forces within the excavation, changes in pore volume, and capillary tension. The actual area of contact between the soil grains and the sheet piles of a cofferdam is relatively small; therefore, virtually the full water pressure is exerted on the structure. Since the effective weight of the soil is reduced by submergence, the lateral earth pressure exerted by the soil grains is reduced proportionally. But the total lateral thrust (soil and water) on the walls of the excavation is increased by two to three times the dry soil pressure for most sands. This fact is well recognized.

Among the important considerations in the design of cofferdams is the hydrodynamic analysis of the seepage condition. The basic principles of flow through soils represent an acceptable approach for the prediction of the quantity, the velocity, and the direction of seepage flow, as well as of the seepage forces. The failure of full-scale sheeted cofferdams led to an investigation consisting of a series of model tests in the laboratory. The factors leading to the hydraulic failure of the prototype were discussed previously (Bauer, 1984).

2 STATIC WATER AND SEEPAGE PRESSURES

The excavation of a cofferdam is, if possible, done in the dry which means that the water level inside the cofferdam has to be kept at or below the bottom of the excavation.

As the excavation proceeds the inside water level is lowered and the hydraulic head, which is the difference between the inside and outside water levels, will be increased proportionally. For soils having a low permeability such as silty clays, silts, or fine silty sands, or where the sheet pile wall terminates in an impervious layer, the seepage velocity will be relatively small and therefore hardly any water will be seeping into the excavation. For such cases the sheet pile wall has to be designed for the full hydrostatic pressure. For design purposes the maximum high water level behind the retaining wall should be considered, keeping in mind, that if no provision for drainage is made, the water level could be raised quickly during a rainstorm or any other source of water supply. The submergence of the soil will cause a reduction in active and passive earth pressure due to the decrease of effective unit weight of the soil, but the total force on the active side of the cofferdam will be increased by two to three times of the dry soil pressure due to the influence of the water.

The theory of seepage flow in porous media is based on a generalization of Darcy's law. For sheeted excavations in pervious soils under a hydraulic head the pressure loss of the seeping water due to the viscous friction of the fluid should not be neglected. The static water pressure will be decreased at the high water table side by the value of the hydraulic gradient and will be increased by a similar amount inside the excavation. In order to compute the direction and rate of flow of water under a long sheet pile wall, it is necessary to determine the intensity and the distribution of the pore water pressures. The water pressure and the direction of flow can very easily be determined from a flow net, which represents the flow of water through an incompressible soil. Finite element modelling has been used in recent years to simulate the flow behaviour of water. In this study the program SEEPAGE developed by Desai as modified by Felio (1987) was used to predict seepage pressures and flow quantities. This program uses 4-node elements and simple Euler integration. The code allows the solution of the following problems:

1. Steady seepage: confined and unconfined
2. Anisotropic, non homogeneous soil systems
3. Transient seepage and rapid drawdown.

The program was used to model the various conditions created in the flow tank of the experimental investigation.

3 BOTTOM HEAVE AND PENETRATION DEPTH

Terzaghi and Peck (1967) have found from model tests that within the excavation the upward directed seepage flow will cause the sand to rise within a distance of D/2 of the sheet pile. The zone of danger of bottom heave is confined to the shaded prism a b c d indicated in Figure 1. The average excess hydrostatic pressure on the base of the prism is $\tau_w h_a$. From the condition of equilibrium at the horizontal section, bc, we have the total excess hydrostatic force acting upwards which is

$$U = 1/2\ D\ \tau_w h_a \qquad (1)$$

where τ_w = unit weight of water

h_a = average excess hydrostatic head on the base of the soil prism.

The submerged or buoyant weight of the prism of soil acting downward is

$$W' = (1/2)\ D^2 \tau' \qquad (2)$$

where τ' = submerged unit weight of soil. Therefore the factor of safety with respect to piping is given by

$$F = W'/U = D\tau'/(h_a \tau_w) \qquad (3)$$

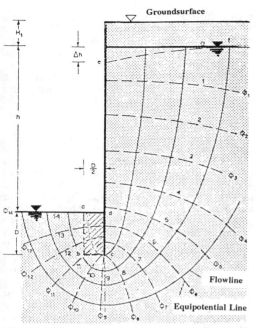

Figure 1. Steady Seepage under Sheetpile Wall

Increasing the penetration depth, D, of the sheet piles results in a smaller excess hydrostatic head at the base of the prism, which results in a smaller excess hydrostatic head at the base of the prism, which yields in a smaller hydraulic gradient, a lower flow velocity, and a larger factor of safety.

The penetration depth necessary to guarantee an adequate factor of safety against boiling was also given by Ordujanz (1954). Bottom heave in the excavation is imminent when the submerged unit weight of the soil is equal to the flow pressure directed vertically upwards. The maximum hydraulic gradient is given by

$$i = h/(h + 2D) \qquad (4)$$

To consider a factor of safety, F, with respect to bottom heave the equation will be

$$\tau'/F = i\tau_w = h/(h + 2D)\tau_w \qquad (5)$$

from which

$$D = (h/2)[(F_w/\tau') - 1] \qquad (6)$$

From the English Civil Engineering Code of Practice, the following minimum values for the sheet pile penetration depth are given as

B > h, D_{min} = 0.4 h   B = halfwidth
B = h, D_{min} = 0.5 h
B = 0.5 h, D_{min} = 0.7 h   (7)
As can be seen the penetration depth, D, does not only increase with an increase of hydrostatic head but also increases with a decrease in B. McNamee (1949) summarized the work done at the Building Research Station in England and presented the result of the model tests in a form of graphs. Figure 2 gives the penetration depth required for excavations in sands of different widths for factors of safety ranging from 1.0 to 2.5.

Marsland (1953) undertook extensive model studies and concluded that for excavations in dense homogeneous sands, failure occurs when the exit gradient equals the critical gradient of flotation, and also that for loose homogeneous sands, failure occurs when the pressure difference between the pile tip and the excavation base is sufficient to lift the column of submerged sand adjacent to the inside of the sheeting. He suggested also values of the penetration required for various factors of safety for narrow excavations in loose and dense sand as given in Figure 3(a). Figure 3(b) gives the penetration depth in dense sand also for different factors of safety, but for two different ratios of T/H. In all cases the water levels were taken to be at the corresponding sand levels, inside and outside the excavation. His conclusions were similar to those of McNamee, that in loose sand local failure by piping or boiling adjacent to the sheet pile would occur and in dense sand an upheaval of the bottom was more dominant.

4 FLOW TANK TESTS

A two-dimensional flow tank, shown schematically in Figure 4, was used in the experimental investigation to explore the hydraulic failure of sheeted cofferdams. Several standpipe piezometers were installed on the side of the tank to monitor

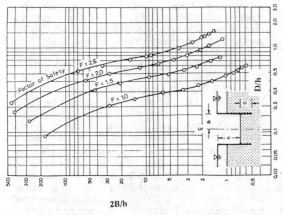

Fig. 2. Penetration Depth (McNamee 1949)

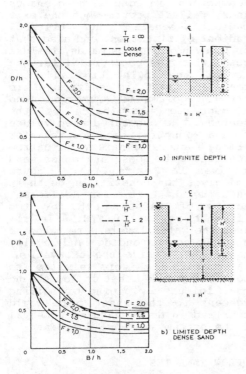

Fig. 3. Penetration Depth (Marsland 1953)

the seepage pressures during a test. The overall dimensions of the tank and the location of the piezometers are given in Figure 4. The variables in these model tests were: sand density, sheet pile penetration depth, width of cofferdam and the distance of the bottom of the excavation to the impervious stratum, e.i., bottom of the tank. The difference of the hydraulic head was varied in order to cause incipient soil instability. For a given geometry of a cofferdam and soil density, the porewater pressures and seepage quantities were recorded during a test. Flow patterns were made visible with dye. In approaching hydraulic failure the construction procedure, as commonly employed in the field, was simulated. The water level was kept the same initially on both sides of the sheet pile wall. As the excavation within the cofferdam proceeded, the water level was lowered to the base of the cofferdam until boiling or bottom heave occurred. At incipient failure, **pore water pressure** measurements, seepage quantities and flow patterns were obtained at this state. A typical result of porewater pressure distribution is shown in Figure 5.

5 FINITE ELEMENT MODELLING

The required properties as input for the program, besides the geometry of the cofferdam, were the soil permeability and porosity. The permeability coefficient was back-calculated from the model tests measuring the seepage quantities. Several mesh configurations for the finite element analysis were employed (i.e., sensitivity analysis) before a mesh of 175 elements was adopted as shown in Figure 6(a). It was necessary to ensure that some nodes coincided with the boundaries and the location of the piezometers. Therefore the mesh had to be changed to accommodate various geometries for the cofferdam. The output from the analysis was in the form of seepage quantity at any specified cross-section and pressure potentials at the nodal points. A typical result for the critical case shown in Figure 6(b) is given in Table 1.

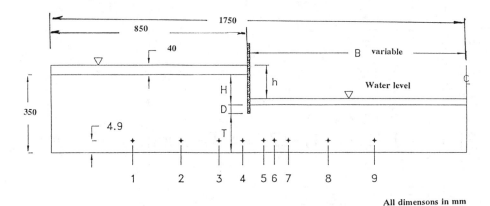

All dimensons in mm

Figure 4. Overall Dimensions of Flow Tank

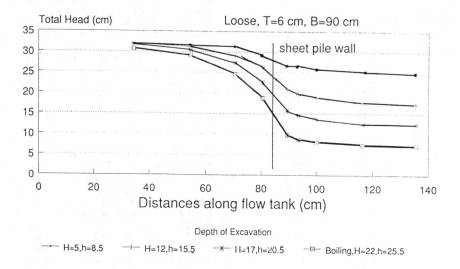

Figure 5. Measured Porewater Pressures

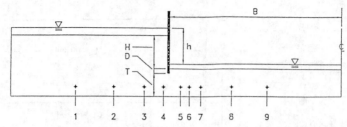

Loose,T=12cm,H=17.5cm,B=90cm,h=17.5cm

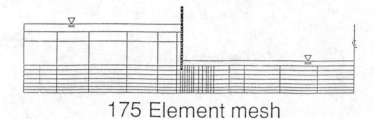

175 Element mesh

Figure 6. Finite Element Mesh

TABLE 1.Test and F.E.M. results

Piezometer Location	Pore Water Pressure (mm)	
	Expt.	F.E.M.
1	310	350
2	295	332
3	253	296
4	207	252
5	167	206
6	156	192
7	148	181
8	140	172
9	136	170

The experimental (i.e. model tests) and the F.E. analysis results are given. It should be kept in mind that for the case shown the excavation was close to hydraulic failure and the soil within the cofferdam started to loosen up resulting in a higher permeability as compared to a steady seepage condition. This higher coefficient was used in the numerical analysis which accounts for the greater values in porewater pressures as compared to the measured values. It was also noticed that the standpipes had a certain time lag before indicating the correct head. This phenomenon also would give lower experimental values.

6 FIELD CASE

In connection with an extension for a sewage treatment plant, a braced open cofferdam, 13 m by 21 m in plan was constructed. A sectional view through the cofferdam is shown in Figure 7 indicating the dimensions of the structure, water levels, and soil properties. During the construction the inside of the excavation was kept dry by pumping from sump pits. Three water pressure indicators were mounted at different elevations on the sheet pile wall. The groundwater level was also determined from borehole and standpipe observations. Flow nets were constructed for different steady seepage conditions incorporating the observed water pressures. The flow nets were considered to be sufficiently accurate in determining the water pressures, direction of seepage and

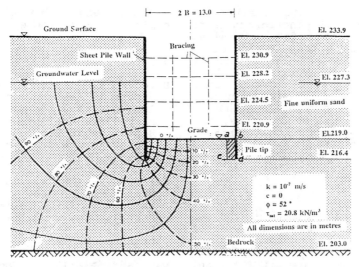

Figure 7. Hydraulic Failure of Braced Cofferdam

seepage quantity. Figure 7 shows the flow net for the most severe condition.

Before the excavation had reached its final grade the bottom became soft, indicating local quicking of the initially very dense sand. Well points were installed but the filters proved to be unsatisfactory. As a consequence the well points became silted up and inefficient. In one corner of the cofferdam the sand became 'quick' and rapid erosion occurred providing a water passage behind the sheeting. This caused ground settlement and extensive damage to the excavation and adjacent structures (Bauer, 1984). Ultimately, it was necessary to partially flood the cofferdam so that the well points could be reinstalled with large-diameter graded sand filters. This field case was used to scrutinize the various methods discussed in this paper and to estimate the stability of this cofferdam.

7 DISCUSSION OF RESULTS

The porewater pressures in a braced cofferdam (Figure 7) were monitored during the construction stage. As the excavation reached its specified grade as shown in Figure 7, a

hydraulic failure occurred. This situation was used to compare the various semi-empirical and analytical methods. The results are given in Table 2.

It should be pointed out that the actual depth of sheetpile penetration below the grade of the excavation (Figure 7) was 2.6 m and the difference in head between the outside and inside water levels was in the order of 11 m when piping occurred close to the sheetpile wall. Therefore, all methods listed in Table 2, except Ordujanz's

TABLE 2. Comparison of results

Reference	Safety Factor F	Seepage* (m³/s)	Penetration depth (m) F=1.0	F=2.0
Terzaghi & Peck (1967)	1.15	4.8(10⁻⁷)	2.4	4.6
Eng. Code	1.2	-	-	-
McNamee(1949)	0.8		3.2	7.0
Marsland(1963)	1.1		2.6	6.4
Ordujanz(1954)	1.8		1.6	3.5
Model Tests	1.0	6.1(10⁻⁵)	2.4	4.9
F.E. Analysis		7.3(10⁻⁶)	2.7	5.3

* Seepage per metre of wall from one side only

analysis, are adequate to estimate, within reason, the penetration depth for a factor of safety of unity, i.e., at the instant of incipient hydraulic failure. The method by Ordujanz takes the full shear strength of the soil into consideration. This is an unsafe assumption since when boiling or piping within the sand occurs, the shear strength, due to the increased **pore water pressure**, is considerably reduced.

8 CONCLUSIONS

This paper has examined various analytical and semi-experimental methods to design, and/or analyze a sheeted cofferdam against hydraulic failure. The method or the combination of methods to be used depend on the economic feasibility of each, the soil condition, the availability of equipment and skilled workers, the workmanship of construction and the geometry of the cofferdam.

The method proposed by Ordujanz (1954) assumes that the full shear strength is developed on the failure planes of the soil body. This assumption does lead to rather unsafe results, because prior to a hydraulic failure, the voids of the sandy soil will increase considerably and the intergranular stresses tend to become zero. This means that at failure the full shearing resistance is no longer available to resist the buoyant forces.

The methods proposed by the other investigations listed in Table 2 yielded fair to good agreements with the actual field case. The finite element analysis proved to be a convenient and powerful technique to investigate various boundary conditions of a cofferdam. This method will even be more advantageous if layered soil systems were encountered. This complex system would be difficult to simulate in a flow tank. The flow tank, however, proved helpful to investigate various physical parameters responsible for a hydraulic failure, and also it was possible to visualize the flow pattern and **pore water** pressure distribution in the standpipes.

9 REFERENCES

Bauer, G.E. (1984). Dewatering, Hydraulic Failure and Subsequent Analysis of a Sheeted Excavation. Proc. Int. Conference on Case Histories in Geot. Eng., St. Louis, Mo., U.S.A., May, p. 1415–1421.

Felio, G.Y. (1987). SEEPAGE: A Microcomputer Program for Seepage through Porous Materials. User's Manual, Civil Eng. Dept., University of California, Los Angeles, U.S.A.

Marsland, A. (1953). Model Experiments to Study the Influence of Seepage on the Stability of a Sheeted Excavation in Sand. Geotechnique III, 6:53, pp. 138–159.

McNamee, J. (1949). Seepage into a Sheeted Excavation in Sand. Geotechnique I, 4:21, pp. 229–241.

Ordujanz, K.S. (1954). Gruendungen fuer Bauwerke, Berlin Ch.5

Terzaghi, K. and Peck, R.B. (1967). Soil Mechanics in Engineering Practice, John Wiley and Sons, Art. 24.

Prediction versus Performance in Geotechnical Engineering, Balasubramaniam et al. (eds)
© *1994 Balkema, Rotterdam, ISBN 90 5410 355 8*

Transmission line anchor design and construction in volcanic clays

I.G.Bruce
Bruce Geotechnical Consultants Inc., Vancouver, B.C., Canada

D.J.Armstrong
B.C.Hydro, Vancouver, B.C., Canada (retired)

ABSTRACT: Increasing development in areas of the tropics provide engineering challenges with respect to both design and construction. A transmission line project in Papua New Guinea provided both when conventional dead-man anchors were installed by local labour forces in low density, sensitive volcanic soils of the Tari Basin. Anchor pull out tests, undertaken to verify that design techniques used in temperate climate soils were applicable to the residual clays, are described along with some of the construction challenges which had to be addressed.

INTRODUCTION

Placer Dome Technical Services, (PDTS) is currently developing stage 3 of the Porgera gold mine for the Porgera Joint Venture (PJV), in the Enga Province of Papua New Guinea, 300 km by road west of Mt. Hagen along the Enga Highway (Figure 1). Power for the project was originally provided by diesel generators on the plant site from fuel delivered daily from Mt. Hagen. Future expansion of the mine required a less expensive alternative power source to be developed. After searching unsuccessfully for an economic hydro electric scheme in a number of areas, an agreement was struck between the PJV and the Hides Joint Venture (HJV) to supply gas from the newly discovered Hides field for the generation of power by the PJV.

The power is now generated by gas turbines at Hides, 73 km south west of the mine site and the power is transmitted via a 132 kV transmission line to Porgera. The transmission line route is shown on Figure 1. Construction of the powerline was completed in June of 1991. Power generation started in November of 1991 and it has now been running successfully for over one year.

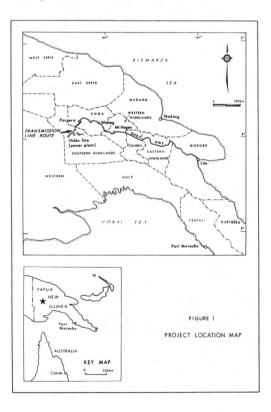

FIGURE 1

PROJECT LOCATION MAP

The transmission line route traverses areas of steep, creeping, unstable terrain. Terrain analysis using colour stereo air photographs in conjunction with geologic maps published by the Geological Survey of Papua New Guinea, (Davies 1983), was used to identify the potential geotechnical problems along several routes thereby allowing the client to make a preliminary selection of the most favourable corridor. The typical terrain units along the route were identified. Detailed mapping and strength assessments using vane shear tests were undertaken at an early stage to augment the terrain analysis and define the limitations on the type of anchoring system which could be used successfully, (Bruce, 1990).

Guyed steel towers were selected for this transmission line as their light weight facilitated helicopter installation along a route of largely inaccessible terrain (Ian Hayward International, 1988). It was originally assumed that the tower guys would be anchored by shallow buried "dead-man" anchors while the towers would be supported on grillage type footings, all hand installed by local labour. However, the presence of soft to very soft residual soils of volcanic origin up to several metres thick, soft to very soft colluvium, stiff but sensitive, low density volcanic clays, and exposed bedrock, all identified during the preliminary geological assessment, appeared to preclude the use of hand excavated dead-man anchors.

Preliminary reconnaissance indicated that the stiff volcanic ash soils appeared to be the most suitable for dead-man anchors. Initial estimates of the soil engineering properties were obtained from Wesley (1973). The volcanic clays underlie the first half of the transmission line route through the Tari Basin. It was expected that most of the dead-man anchors to be used along the route would be installed in the volcanics while deeper drilled anchors would be used in the remaining part of the line.

In order to confirm that dead-man anchors and conventional design formulae would provide acceptable results in the light weight sensitive volcanic clay, in-situ anchor pullout tests were undertaken in the volcanic soils, prior to final design and the commencement of construction.

ANCHOR TEST PROGRAM

A total of four tests were undertaken at two separate sites using two types of dead-man anchors. The first anchor type consisted of 250mm diameter steel pipe with a wall thickness of 10mm. The anchors were 2.0m long. The second anchor type consisted of a timber anchor approximately 2.0m long and approximately 300mm in diameter.

The anchors were bolted to 30mm diameter steel rods which passed through the anchors and through 150mm square and 9.5mm thick steel plate washers. The anchor rods were deformed in their upper ends into closed eyes so that a wire rope could be attached to the anchors for the pull out tests. The anchors were pulled at an angle of approximately 35° to the vertical by jacking upward against the wire rope attached to two diametrically opposed anchors. The jack was placed on a reaction frame consisting of a spread footing and a central column. At test site #1, the spread footing and column was constructed of reinforced concrete 0.3m thick and 1.4m square. A concrete column 0.28m by 0.30m in section and approximately 3m long was constructed as an integral part of the footing and was used to support the hydraulically operated jack. A photograph of the test configuration is shown on Plate 1.

At test site #2, a 10mm thick steel plate reinforced with angle iron was used as a footing and a 230mm O.D. steel pipe, with a wall thickness of 13mm was used as a column. A 25mm thick steel cap was used to support the jack.

Anchor deformations were measured using rulers with 1mm gradations taped to the wire rope connecting the anchors. A string line was stretched across the site and used as a reference to measure deformations. A drafting set square was used to provide a right

Plate 1 - Test Configuration at test site #1. The anchors are
installed and ready to test. Note the reference string line for
measuring anchor deformations.

angle to the cable for measuring
deformations relative to the
string.
    Settlement of the central footing
and heave of the stakes placed
above the buried anchors were also
measured using the rulers taped to
the central column of the footing
or taped to the stakes above the
anchors. The deformations were
monitored using a survey level set
up a short distance from the test
site.

LOAD PREDICTIONS

Prior to testing, the design
formula proposed by Meyerhof (1973)
was used to predict the ultimate
uplift capacity of the dead-man
anchors. The formula proposed by
Meyerhof is;

$$P_{ult} = A(S_u N_c' + \delta H) + W$$

where

P_{ult} = ultimate anchor load
A = area of anchor plate
S_u = undrained shear strength
N_c' = breakout factor
δ = unit weight of the soil
H = anchor depth
W = weight of the anchor.

Meyerhof (1973), indicates that for
clay soils, the breakout
coefficients may be expressed as;

$$N_c' = 0.6 \ H/B$$

for anchors inclined at less than
45° to the vertical and where B is
the equivalent width of the
footing.   Theoretical uplift
loads were calculated for each
anchor at each site. The anchor
widths were assumed to be equal to
the diameters of the dead-man
anchors. Vane shear strengths were
measured at 0.5m intervals between
the top of the anchor and the
ground surface. The vane shear
values were corrected for
plasticity, (Azzouz et.al., 1983),
and an average value chosen to

259

define the undrained shear
strength. A soil bulk density of
12kN/m³, as measured in the field,
was used for design.

The breakout factors were
calculated using the anchor depths
and anchor dimensions for each
anchor configuration. The
resulting breakout factors and
ultimate design loads are
summarized in Table 1. The actual
pull-out values measured during
testing are also shown for
comparison.

Table 1   Calculated and Measured
Pull-out Resistance

	B (m)	H (m)	S_{ult} (kPa)	N_c (kN)	P_{ult} (kN)	$P_{measured}$ (kN)
Test #1	0.25	2.5	64	6.0	220	169[1]
	0.3	2.5	64	5.0	221	161[1]
Test #2	0.25	2.5	29	3.6	65	66
	0.3	2.5	29	3.0	66	66

Notes
Full anchor loads were not
established by these tests due to
failure of the footing by punching
through of the column.

TEST CONCLUSIONS

On the basis of the anchor pull-out
tests undertaken in the Tari Basin
it was concluded that shallow dead-
man anchors can be used in the firm
to stiff sensitive low density
volcanic soils of the Tari Basin
and that anchor design capacities
could be determined using the
corrected vane shear strengths
measured in the field in
conjunction with the design formula
of Meyerhof (1973).

CONSTRUCTION

A total of 562 dead-man anchors
were constructed during the period
between October 3, 1990 and March
23, 1991. On the basis of field
vane testing undertaken at the
centre of each proposed tower site,
an average corrected design shear
strength of 50 kPa was defined for
the soils of the Tari Basin. The
design depth of a typical trench
for a dead man anchor then becomes
2.6m. Safety precautions
prohibited the excavation of a
narrow trench to 2.6 m depth

without the use of shoring.
Consequently, the trenches were all
excavated by hand to depths of
approximately 1.5m, and 19mm thick
plywood shoring was installed
against the excavation walls.
Several incipient failures were
observed during excavation and
shoring was at times installed at a
shallower depth. The plywood
shoring was braced using threaded
Dywidag bar and flat steel plates
welded onto Dywidag couplers. The
trench was then excavated to a
depth of 2.6m, the depth required
to provide the design uplift loads,
and an undercut notch was installed
on the face of the trench closest
to the tower. Construction of a
typical trench using local labour
is shown on Plate 2.

The majority of the anchors
installed were 3m long, 0.3 m
diameter, 17 mm thick galvanized
steel pipe. The anchor was lowered
into the excavated trench by rope

Plate 2 - Local Labour Installing a
dead-man anchor in the Tari Basin.

and a Dywidag thread bar was placed through the pipe and passed through a curved washer and attached using a Dywidag thread bar nut. A second nut was attached on the front end of the anchor and tightened down so that the anchor rod could not be removed by hand.

Inspection was undertaken during excavation of the trenches and the undrained shear strengths of the soil were determined using a hand vane. In areas where the soils were slightly softer than the minimum allowed for a single anchor, a pair of dead-man anchors were installed. Generally the second anchor was placed 5m farther from the tower. If the soil was significantly softer than the minimum strength or if the trench was unstable and could not be excavated safely, the trenches were backfilled and the anchors were changed to deep drilled and grouted anchors. In some instances where drilling was not possible, special deadman anchors were installed. These consisted of four 3m long 150mm diameter steel pipes bolted together and backed by a single steel channel to form a grillage type of anchor.

The soil characteristics of the volcanic clays were found to be extremely variable over short distances. In some places, where shallow dead-man anchors were installed no more than 15m apart, the ground varied from being so soft that double or special dead-man anchors were required to being so hard that the dead-man anchors trenches could not be excavated by hand, and deep drilled and grouted anchors were installed.

All towers were set onto their footings by helicopter between January 24 and April 30 1991. The guys were attached to the anchors and the towers left to be plumbed and tensioned at a later stage. Some minor movements were noted in some anchors as tensioning occurred. Much of the movement is considered due to imperfect shaping of the undercut notches where a neat excavation could not be constructed. Ongoing minor movements are expected.

ACKNOWLEDGEMENTS

The authors would like to thank Placer Dome Technical Services for permission to publish this paper and would like to acknowledge the help of the Porgera Joint Venture Engineering Geology staff in Papua New Guinea and the Ian Hayward International Staff for their assistance.

REFERENCES

Azzouz, A.S, M.M. Baligh, and C.C. Ladd, 1983. Corrected field vane strengths for Embankment Design. Proc. ASCE J. of Geotech. Eng., V109, GT5, May.

Bruce,I., 1990. Geotechnical Foundation Report, Porgera Project, Hides Porgera Power Line Anchor Design. Report for Placer Dome Inc. 30 pp.

Davies, H.L., 1983. Geology of Wabag, Papua New Guinea. Department of Minerals and Energy, Geological Survey of Papua New Guinea. Sheet SB/54-8 84pp.

Ian Hayward International, 1988. Preliminary Design Report, Hides Porgera Powerline. Report for Placer Dome Inc., 17 pp.

Meyerhof, G.G., 1973. The Uplift Capacity of Foundations Under Oblique Loads. Canadian Geotechnical J. Vol. 10., pp.64 -70.

Wesley, L.D., 1973. Some Basic Engineering Properties of Halloysite and Allophane Clays in Java, Indonesia,. Geotechnique Vol. 23, No. 4, pp. 471-494.

Prediction versus Performance in Geotechnical Engineering, Balasubramaniam et al. (eds)
© *1994 Balkema, Rotterdam, ISBN 90 5410 355 8*

Contribution to the design history of the gravity wall

P.Ventura
Dipartimento di Ingegneria Strutturale e Geotecnica, Facoltà di Architettura, Roma, Italy

ABSTRACT: The paper tries to merge acquired values of the old technical criteria with that of the innovation. Purpose is tested by examining design evolution and the safety control of the gravity wall and back-analysis some old and recent observed works.

RESUME: L'étude tend à fondre les valeurs des anciens critères techniques avec ceux de l'innovation. Le fin est prouvé avec l'examen de développement du projet et du contrôle de la sécurité des murs à gravité et avec l'analyse à postérieur de quelques vielles et nouvelles oeuvres observées.

1 INTRODUCTION

The study of the design history is a continuous invitation to search simple correlation and to decant essential criteria.

Artistic, Naturalistic, Experimental, Physic-Mathematic, Technical, Technological and Economical values are nowadays developed with very dense intensity and interaction.

The parallel development of the human work, founded in the further Ethical values, needs the simplicity and the mitigation of the new complex criteria.

The paper examines this aspect for Technical value, between the above-mentioned, employing gravity wall design as leitmotif.

2 COMPARISON BETWEEN DESIGN CRITERIA

The history of the gravity wall design from the beginning is articulated versus the safety control.

Geometry and experimentation have

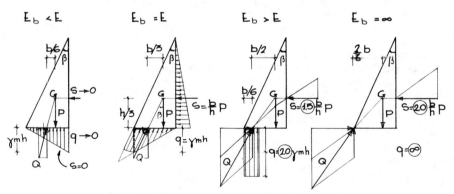

Fig. 1. Elastic design model wall: correlation between the thrust S and bearing stress q, and between the overturning and bearing safety factor

SLIDING	GLOBAL SLOPE	OVERTURNING	BEARING

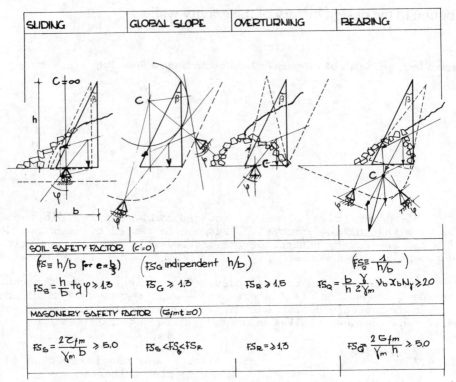

SOIL SAFETY FACTOR $(c'=0)$

$\left(FS \equiv h/b \text{ for } e=\frac{b}{3}\right)$ $\left(FS_G \text{ indipendent } h/b\right)$ $\left(FS_Q = \frac{1}{h/b}\right)$

$FS_S = \frac{h}{b} tg\,\varphi \geq 1.3$ $FS_G \geq 1.3$ $FS_R \geq 1.5$ $FS_Q = \frac{b}{h}\frac{\gamma}{2\gamma_m}\nu_b \chi_b N_\gamma \geq 2.0$

MASONERY SAFETY FACTOR $(\sigma/mt=0)$

$FS_S = \frac{2\tau fm}{\gamma_m b} \geq 5.0$ $FS_G < FS_S < FS_R$ $FS_R = \geq 1.3$ $FS_Q \frac{2\sigma fm}{\gamma_m h} \geq 5.0$

Fig. 2. Limit design model wall: correlation of the safety factor with the various rotation center kinematisme as for soil as for masonry, with no tension strength

governed initially the megalithic walls or the Roman "sostruzionis" masonry for terraced slope.

The history of the applied mechanics is notoriously evolved by: classic static analysis of the Belidor and Poncelet, stability analysis of the Coulomb and Prony and elastic analysis of the Boussinesq and Poulos.

In retrospect, from this evolution the elastic *design criteria* based on a wall with uniform stiffness (thrust model) is indicated in Fig. 1.

Traditional serviceability condition is for $E_b = E$, whose base modulus is equal to backfill modulus or medium is homogeneous, and the eccentricity is at the limit of the inertia kern.

The rapport thrust/weight is therefore proportional to the base/height or to the external slope β, in the usual elastic design criteria.

The reduction of the base modulus $E_b < E$ minimizes rapidly versus zero value S thrust and q admissible bearing stress, increasing $E_b > E$ viceversa allows S and q to augment.

Figure 1, by graphical static analysis, shows that the phenomena is very different for rigid soil base $(E_b = \infty)$.

The theoretical infinite value of the bearing stress q corresponds infact to the double value of the S thrust.

The safety factor at the limit overturning failure, deduced by rigid-elastic analysis, is therefore 2.0 for infinite bearing capacity.

The maximum q vector at the external wedge of the wall, for $E_b = E$ in Fig. 1, is inclined to β and is greater than vertical q component as proved by Rankine on the gravity dam outline design.

Figure 2 extends the analysis at the *stability design* criteria reduced to the 8 traditional mechanism of failure, distinguishing the soil from the masonry wall.

The traditional design is formed on the deterministic safety factor F.S., or better on the stability index (Cestelli Guidi, 1992) to each one scalar equilibrium equation.

The F.S value can be deduced by comparison between elastic and wedges methods of stability analyses.

The kinematic analysis is appoints that the instantaneous rotation center is infinite for sliding failure, over the base of the wall for the global slope failure, at the base for overturning and below the base for bearing failure.

Consequently the safety factor results to proportional h/b for sliding failure (in concomitance to the soil reaction at the limit of the inertia kern: $e = b/3$ in Fig. 1), independent for overturning and global stability and inversely proportional to h/b for bearing capacity.

Recalling that the shape factor $v_b = 1,0$ for strip load, inclination factor $X_b = (1 - tg\delta)^4$ and bearing factor $N\gamma = 0,9 (Nq - 1) tg\varphi$, and the above mentioned proportional terms are always a function of the friction angle.

The mechanism of the masonry wall failure is similar to the soil, and the described inversion is between F.S. and σ_{fm}/h for bearing capacity and τ_{fm}/b for sliding internal at the structure.

The σ_m admissible in masonry wall is deduced by the old gravity dam (Arredi, 1988) still in exercise since VI century ($\sigma_m \leq 0,8$ MN/m²).

The safety factor values are taken to Italian code respectively for the geotechnical and the masonry code, based on homogeneous and no tension ($\sigma_{fmt} = 0$) model for masonry, similar to a non-cohesive ($c' = 0$) soil.

It's noted that the traction strength is, on the contrary admissi -ble for masonry in the presence of valid brick or stone texture, specially by whole transversal element or "diatoni". These mitigate the elevated values of the safety factor.

Figure 2 also shows in particular the typical failure investigation aspect: the backfill is over the crumbled stones in overturning mechanism and below the fill in sliding, and is always synchronized with the rotation center of revolution.

The *modern design* criteria is concentred on the stress paths versus dominium failure, as in figure 3, and on safety reliability.

Shear and flexural stiffness and strength versus axial load of the wall is correlated.

The M - N failure dominium in particular is simplified by Heyman (1969) model linearized so far as $N = P$. The limit dominium is the intercept at the origin with slope $b/2$, corresponding to the maximum eccentricity of the axial load (v. Fig. 1) and in analogy with slope of the Coulombian dominium T - N.

The safety factor of the stress paths versus shear and flexural strength is shown in figure 3 for limit design.

At the N = P load of the wall, the

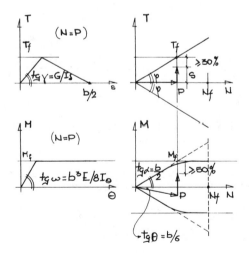

Fig. 3. The stiffness and failure's dominium with the stress paths for evalues safety factor

elastic parameters of the constitutive curves are evidenced in figure 3 by stiffness G/I_γ and $b^3 E/8 I_\theta$.

The absence of the local permissible stresses criteria in stability analysis is correlated.

The described figure 3 shows in fact the strict correlation between the various mechanism and the interdependence between constitutive parameters and the safety factor.

The *predesign* base/height ratio of the retaining wall for soil without cohesion (c' = 0) from the limit sliding analysis results (Fig. 2) in:

$$\frac{b}{h} = \frac{\gamma}{\gamma_m} \frac{K_a}{tg\varphi} FS_s$$

in particular for e = b/3 (Fig. 1) results $FS_s = tg\varphi/tg\beta$ equivalent to indefinite slope stability (C = ∞), or:

$$\frac{b}{h} = tg\beta = \frac{1}{FS_s} tg\varphi \qquad (1)$$

from the comparison between elastic and limit design.

Moreover, from the elastic design, in the absence of accidental action, results to:

$$\frac{b}{h} = tg\beta = \sqrt{\frac{\gamma}{\gamma_m} K_a} \qquad (2)$$

for $\gamma_m = \gamma$

$$\frac{b}{h} = \sqrt{K_a} = tg\,(45° - \frac{\varphi}{2}) = tg\,(\frac{90° - \varphi}{2})$$

and for E = cost, in figure 1, being K_a the Rankine active stress ratio, and γ/γ_m soil/masonry total unit weight.

The sliding safety factor consequently (1) and (2) results:

$$FS_s = \frac{tg\varphi}{tg\,(\frac{90° - \varphi}{2})} = 1,3 \qquad (3)$$

for $\gamma = \gamma_m$ and c = 0, φ = 35° corresponding to elastic soil with limited settlements, and b/h ≅ 0,5.

The overturning analysis comparison between elastic (Fig. 1) and limit (Fig. 2) design, near the rotation failure, appoints:

$$FS_R = 1,5 \text{ for } e_{max} = b/2 \qquad (4)$$

corresponding to the bearing capacity safety factors

$$FS_Q = \frac{q_{overt}}{q_{el}} = \frac{2 \gamma_m h}{\gamma_m h} = 2,0 \qquad (5)$$

and the limit thrust is $S_{lim} = 2,0 S_{el}$ corresponding q = ∞.

Then the safety factor, to each one scalar equation, for masonry structures is more elevated than for soil. The factor in fact gives:

$$FS_s > 5,0$$

$$FS_Q > 5,0 \qquad (6)$$

$$FS_R = \frac{b/6}{e} > 1,3$$

caused by the: lack of the stresses overlap, overmentioned $\sigma_{fmt} = 0$ hypothesis and the accidental action incidence.

Remembering that for pseudo-static predesign

$$\frac{b}{b_e} = \frac{tg\varphi - a/g}{tg\varphi} \frac{K_a}{K_{ae}} \text{ for (c' = 0)} \quad (7)$$

being b_e the base in earthquake a/g condition; gives $b_e > b$ for elevated a/g intensity and $K_{ae} > K_a$

The mechanic contribution to the design is very direct, and the simplicity in the technical work is an important necessity in this hurried epoch.

The comparison between elastic and limit design, with traditional friction model, focalizes the clear and practical safety analysis, "partialized" for every mechanism of the failure.

Predimensioning and security in design are only the starting goals.

Correct divulgability specially in Third World and Human development

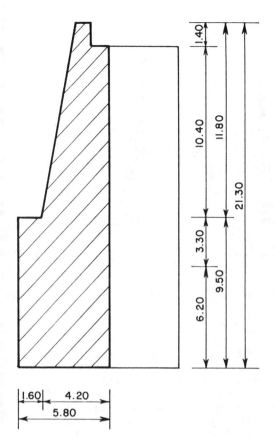

Fig. 4. Reproduction from the Record Office, of the wall's execution along Tiber river in the Rome historical center (Medici, 1887)

delivered from the excessive use of the technology and scientific criteria in the current technical application (Sgreccia, 1990) is a much more important question.

The numerical risk analysis, the assurance value and the life counter value in money, are very delicate ethic problems.

The partial safety factor in semi-probabilistic criteria, increasing to 50% the action and penalizing the 50% the strength of the material, can be inserted in the above mentioned ancient "partialization" of the stresses.

3 CONTRIBUTION TO THE CASE HISTORY

The simple old rules are

useful in particular for back analyzing the old wall.

Figure 4 indicates the old counter-wall along the Tiber river, realized during Capital development and Ministry edifices.

The relief is reproduced from the work's Journal deposited in the Record Office (Medici, 1887).

The foundation of the wall is executed by pneumatic coffer-dam until at the depth 2÷4 m below mean sea level. This level is testified by Ripetta hydrometer, next the filled old luvial port, and actually is 33 cm lower due to settlements induced by backfill and by subsidence for recent general Tiber regression.

The top of the foundation or the base of the wall as been fixed at minimum Tiber level approximatively 6.0 m above sea level in the City center.

Characteristic dimension of the masonry wall is h = 10.4 m and b = 4.2 m, whose b/h = 0.4 in harmony with the previously examined design consideration.

The absence of cohesion (C'=0) has been assumed for all the alluvional submergible backfill of the wall.

Figure 5 then indicates the strengthening realized in Orvieto Cliff in Confaloniera locality.

Tuffaceous platform has been regularized at the perimeter with many bastion and wall in particular to realize the ascent at the city.

The old tuff masonry wall, indicated at the center of figure 5, is 11 m high and 1.4 m wide at the base, backfilled for 6.7 m and embedded in the tuff for 4.3 m, or b/h ≅ 0.2 in the zone of the crumbling.

Collapses were caused by mine excavated below the foundation of the wall to extract the pozzolana. The stability has been improved by reinforced concrete wall, modelled with cliff conformation, and by injecting refilled grottoes across partially prestressed rock anchorage at the base.

The backfill (Fig. 5) has been monitored to test the retaining execution, and deduce the active stress ratio K_a = 0,2 from back-analysis of in situ measured elastic

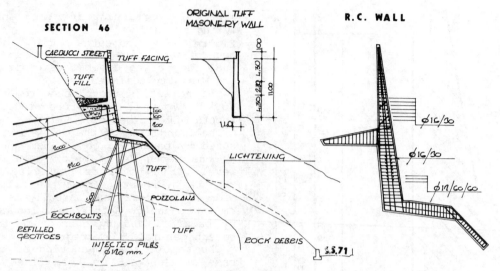

SECTION 46

CARDUCCI STREET — TUFF FACING

TUFF FILL

ROCKBOLTS

REFILLED GROTTOES

TUFF

INJECTED PILES ⌀110 mm.

TUFF

POZZOLANA

TUFF

ROCK DEBRIS

ORIGINAL TUFF MASONERY WALL

LICHTENING

R.C. WALL

⌀16/30

⌀16/30

⌀17/60/60

Fig. 5. Strengthening of the crumbled tuffaceous wall and cliff, by reinforced concrete wall in Orvieto

displacements of the r.c. wall (Ventura, 1987).

The pozzolana height fill without cohesion results from (2):

$$b/h \cong \sqrt{K_a} \cong 0.4 \qquad (8)$$

Consequently $c' > 0$ in the backfill of the centuries-old tuffaceous masonry retaining wall, according to pozzolana chemistry and to the contrary of the wall in figure 4.

In general this back-analysis of the cohesion or seismic (7) effect allows to value the critical zone of the masonry wall, whose strengthening is necessary if the stiffness are not admissible.

4 CONCLUSIONS

Typical ratio of the base/height of the wall is $b/h = 0,5$ according (2) to backfill without cohesion $c' = 0$ and $\varphi' = 35°$ (for gravity dam design $tg\varphi = 0,67 \div 0,75$), and equivalence between the elastic and limit pre-design penalized by the discussed safety factor. Comparison then shows the validity of the 3 deterministic safety factor assumed in the code for soil (3), (4) and (5).

The greater influence of the no tension ($c' = 0$ and $\sigma_{fmt} = 0$) hypothesis for soil and masonry, impends probabilistic safety criteria, especially in seismic predesign (7).

The limit design has been examined for evidence that b/h is proportional (for $e = b/3$), independent or inversely proportional to the safety factor simultaneously with the evolution of the instantaneous rotation center in various mechanism failure.

The back-analysis of the old wall dimensions finally allows to value the cohesion absence extension.

REFERENCES

Arredi, F. 1988. Costruzioni idrauliche. Vol. 3, Cap. VI, UTET, Torino

Cestelli Guidi, C. 1992. Geotecnica e tecnica delle fondazioni. Vol. I Cap. 9 e Vol. II Cap. 16, Hoepli Milano

Coulomb, C.A. 1773. Essai sur une application des règles de maximis et minimis à quelques problèmes de statique relatifs à la Architecture. Ac. Royale de Science, Parigi

Heyman, J. 1969. The safety of the

masonry arches. Int. Journal Arch.
Vol. 11: 383-385, Cambridge

Medici, L. 1887. Giornale dei lavori
dei muraglioni del Tevere. Archi-
vio di Stato, Roma

Sgreccia, E. 1990. Scienziati e tec-
nologi per l'etica dello sviluppo.
Rivista di "Medicina e Morale",
Università Cattolica di Medicina,
Roma

Ventura, P. 1987. Analisi pseudosta-
tica di una antica opera di soste-
gno. Ass. It. Ing. Sismica, Prob-
lemi storici, tecnici e normativi
per la conse vazione dei centri
urbani in zona sismica. 29 sett.,
Roma

Prediction versus Performance in Geotechnical Engineering, Balasubramaniam et al. (eds)
© *1994 Balkema, Rotterdam, ISBN 90 5410 355 8*

Limit states design applied to bridge abutment walls

Gunther E. Bauer
Faculty of Engineering, Carleton University, Ottawa, Ont., Canada

ABSTRACT: This paper discusses the concept of applying the ultimate limit states (mainly instability against sliding, bearing capacity and overturning) and the serviceability limit states (mainly total and differential movements) analyses to the desin of bridge abutments founded on spread footings. This concept of applying reduced strength parameters and factored loads in geotechnical engineering is incorporated in design codes in Canada.
A typical example for a bridge abutment founded on spread footings is given in this paper, and the predicted performance is compared to the actual behaviour of the structure.

1 INTRODUCTION

The limit states design principle has been widely accepted in structural design for over several decades, whereas this concept has found only recently application in foundation engineering . Prior to this, the conventional design methods led to confusion and contradiction between structural and foundation engineers involved in the design of such load carrying elements as footings, piles and retaining walls using codes which specify factored loads for superstructures and unfactored loads for substructures. A further difficulty in extending the limit states philosophy to foundation design has been the lack of uniform practices, and, therefore, the required process of calibrating load and resistance factors is rendered difficult by the lack of precise standards governing present day criteria. The new Ontario Highway Bridge Design Code and more recently the Canadian Standard Association (Standard CSA-S6-1988) have attempted to estabilish uniform analysis and design procedures, and to provide adequate safety factors for the limit states principle.

At this point it is probably expedient to define the various terms associated with limit states design in geotechnical engineering. Limit states are states beyond which the structure will no longer satisfy the design performance requirements. Ultimate Limit States (ULS) are those associated with instability, collapse, loss of equilibrium or incipient collapse of any structure including supports and foundations. The Serviceability Limit States (SLS) is defined as states beyond which excessive deformations, movements or deflections affect the appearance or effective use of structures. This paper discusses the design concept of applying the Ultimate Limit States (ULS) and the Serviceability Limit States (SLS) principles to the design of foundation elements of highway bridges. This new design concept is illustrated by an example of a bridge abutment retaining wall founded on spread footings.

2 ULTIMATE LIMIT STATES

In the design of abutment structures the geotechnical engineer who commonly works with total safety factors, where factors of safety are generally defined as the ratio of resistance of the soil or structure to applied loads in order to safeguard against unacceptable risks. The Canadian Foundation Engineering Manual (1985) recommends the total safety factors given in Table 1. The higher values apply to normal load and service conditions, whereas the lower values should be used under extreme loading conditions. The lower values can also be used for temporary works or where performance observations or back analyses from actual failures of similar structures are available. Meyerhof (1984) has shown that these total factors of safety can be related to the nominal lifetime probabilities of stability against failure in the order of 10^{-3} for retaining structures, such as bridge abutments, and 10^{-4} for foundation elements, such as footings and piles. Abutment retaining structures of highway bridges should therefore have the following minimum total factors of safety:

1. F > 1.3 against a slope failure, that is, the soil shears through the backfill material as well as through the underlying soil.

2. F > 2.0 against a bearing capacity failure of the footing.

3. F > 1.5 against sliding along the base, if the passive resistance at the toe is neglected, and

4. F > 1.3 against overturning about the toe of the abutment wall

In order to obtain the above safety factors, full or unfactored soil strength parameters should be used in the analyses. Also the soil

Table 1. Total factors of safety

Failure	Item	F
Shearing	Earthworks	1.3 - 1.5
	Ret.wall	1.5 - 2.0
	Foundat.'s	2.0 - 3.0
Seepage	Uplift	1.5 - 2.0
	Piping	2.0 - 3.0

loads, such as selfweight and earth pressures are not factored, that is, not multiplied by a factor greater than unity. The concept of ultimate limit states design has been used in Denmark for over three decades (Brinch Hansen, 1956 and the Danish Code of Practice for Foundation Engineering, 1978) and it also forms the basis of the new Ontario Highway Bridge Design Code (1988). Some of the principles are included in the new edition of the National Building Code of Canada of 1988 and in the Eurocode No. 7 to be published in the near future.

In order for the load and resistance factors used in the limit states analysis to be compatible to the conventional total safety factors (Table 1) a careful calibration is necessary. The new adopted factors fall into two categories, those applied to loads and those applied to material strengths. The factors of the first category are generally greater than unity and are applied to dead loads, live and environmental loads, and water pressures to obtain factored loads and to forces as in structural analysis of ultimate limit states. The factors in the second category are less than unity and are applied to the shear strength parameters, cohesion and friction, to obtain the reduced strength parameters of the soil to estimate the factored soil resistance, factored earth pressure or factored bearing capacity. The values of the minimum load factors to be used in the ULS analysis are given in Table 2 for earth retaining structures, such as highway bridge abutment walls. The corresponding reduction factors applied to the soil shear strength and to some loads and water pressures are listed in Table 3, if their effect is beneficial to increase the abutment stability. A reduction factor, for example, should be applied to the passive resistance in front of the abutment wall, or to soil weight on top of the heel in order to reduce the calculated stability of the abutment.

The main considerations for stability of bridge abutments on spread footings are for overall stability, sliding, overturning,

Table 2. Min.load factors - U.L.S.

Item	Load Factor	Comment
Dead loads	1.25	
Live loads	1.50	
Water	1.25	instability
Resultant	1.25	Overturning bearing cap.
Act.earth	1.25	Overturning sliding

Table 3. Reduction factor - U.L.S

Item	Reduction Factor	Comment
Soil strength		
cohesion	0.65	earth press. footings
cohesion	0.50	piles
friction	0.80	stability, earth press. footings, piles
Resist.loads	0.85	> stability < sliding < overturn'g
Water	0.85	Uplift boiling
Pass. press.	0.80	sand
Pass. press.	0.65	clay

bearing capacity failure and intolerable vertical and horizontal movements. Trial calculations using the load and reduction factors specified in Tables 2 and 3 yielded total factors of safety compatible to those listed in Table 1. It has also been found that a load factor of 1.25 used in the ULS analysis and applied to the resultant force acting on the bridge abutment, will give total factors of safety 2 and 3 against overturning and a bearing capacity failure, respectively. These total safety factors are generally used in the conventional analysis to minimize the risk of failures of bridge abutments.

3 SERVICEABILITY LIMIT STATES

The allowable movements of a highway bridge depends to a large extent on the intended function of such a structure. In a study by Moulton et al. (1980) the performance of over 200 highway bridges in the United States of America and Canada were investigated, and they offered the following main conclusions:

1. All bridge foundations move.

2. Large horizontal and vertical movements were often tolerated.

3. Differential or rotational movements were more damaging than uniform movements, and their effect on the performance of the structure was often rated not tolerable.
4. Bridge structures were more sensitive to large horizontal movements than to large vertical movements.

Based on this survey, Bozozuk (1978) proposed the following classification for the movement of bridges given in Table 4.

Most of the movements considered not tolerable were due to differential or rotational movement of the structure. It is suggested to use a value of 1/500 for relative rotation as a guide in the preliminary design of small to medium highway bridges. For computing initial settlements of bridge footings on clay and total settlements on granular soils, both dead and live loads shall be considered. In computing consolidation settlements in clay, only the permanent loads have to be included. For bridge footings located directly on compacted granular pads, the permeability is such that consolidation usually takes place quite rapidly and is completed by the end of construction. No rational method is yet available for the prediction of

Table 4. Classification of movement

Description	Movement (mm)	
	Horiz.	Vertical
Tolerable	< 25	< 50
Harmful	25<m< 50	50<m<100
Intolerable	> 50	> 100

273

these settlements, however, a number of empirical and semi-empirical methods have been developed. Since correlation between predicted and observed results is not generally good, extreme care should be exercised in applying these methods.

The following section goes through an abbreviated design example of a highway bridge abutment using the ULS and SLS principles. The calculated values for movements and settlements are compared to a field study.

4. DESIGN OF BRIDGE ABUTMENTS

The trial section for the bridge abutment is shown in Figure 1. All dimensions and elevations are given in metres. The applied loads together with the abutment dimensions are shown in Figure 2. The footing for the abutment was located on a granular compacted gravel pad, shown as granular "A" in Figure 1. The geotechnical properties of this material were determined as follows:

unit weight γ = 18.2 kN/m³ (100% Standard Proctor)
internal friction angle: ϕ = 47° (from triaxial compression tests)
modulus of deformation: E_s = 50 * 10^3 kPa

The granular backfill (sand) had the following properties:

unit weight: γ_s = 18.0 kN/m³
internal friction angle: ϕ_s = 38°

The density of the concrete was taken as 24.0 kN/m³. According to the ULS design principles discussed before, the following load and reduction factors were selected from Tables 2 and 3:

Load or Force	Factor
Concrete weight	1.20
Soil Weight	1.25
Active earth pressure	1.25
Friction	0.80

Overturning and Sliding

The calculations to check the structure (Figure 2) against overturning and sliding are summarized in Table 5 below.

Table 5. Stability calculations

Material	Force (kN)	Arm (m)	Moment (kNm)
Concr(1)	46.5	1.80	83.7
Concr(2)	76.5	1.15	88.0
Concr(3)	98.5	1.25	123.1
Soil (4)	142.8	2.15	307.0
Bridge	540.0	1.10	594.0
Earth	(131.5	-2.03	-267.0
Σ	904.3	Σ	928.3

The resultant force will cut the the footing at a distance x' from the toe:

$$x' = \frac{\Sigma \text{ Moments}}{\Sigma \text{ V.forces}} = \frac{928.3 \text{ kNm}}{904.3 \text{ kN}} = 1.03 \text{ m}$$

Therefore, the eccentricity, e, is

e = B/2 - x' = 2.50/2 - 1.03 = 0.22m

Since the eccentricity e, is < B/6, or 0.42m, the resultant falls within the middle third of the footing base and the abutment wall is safe against overturning.

The factored sliding resistance between the footing base and the soil was found to be considerably larger than the pushing force (active earth pressure). Therefore, the abutment wall is also safe against sliding.

Bearing Capacity

The safe bearing capacity of the footing should be based on the factored (reduced) shear strength parameters of the granular pad. From Table 3 the reduction factor for friction is 0.80. Therefore, for an internal friction angle of ϕ_f = 47°,

$$\tan \phi_f = 0.80 \tan 47° = 0.858$$

and

$$\phi_f = 40.6°$$

The general bearing capacity is

274

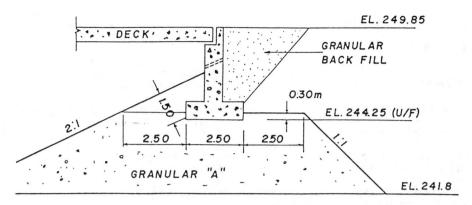

FIGURE 1(a): Length Section Through Bridge Abutment

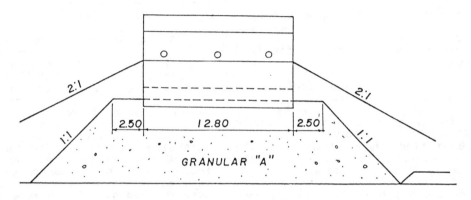

NOTE: ALL DIMENSIONS AND ELEVATIONS IN METRES.

FIGURE 1(b): Cross-Section of Bridge at Abutment Wall

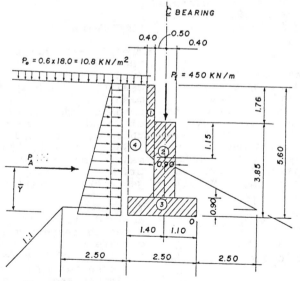

Figure 2: Forces On and Dimensions of Abutment Wall

275

given by the following equation :

$$q_f = 0.5 \gamma B N_{\gamma q}$$

where
q_f = factored bearing capacity

γ = unit weight of soil

B = width of footing

$N_{\gamma q}$ = bearing capacity factor, depending on and depth

For inclined and eccentric loading, above equation must be corrected as follows:

$$q_f \text{ (corrected)} = q_f * R_i * R_e$$

where
R_i = correction factor for load inclination

R_e = correction factor for eccentricity of load

For the case on hand, $N_{\gamma q}$ = 62, R_i = 0.83 and R_e = 0.70 (Bauer,1982) Substituting these values in the above equation, the factored bearing capacity is calculated as

$$q_f \text{(corrected)} = (0.50) (18.2 \text{ kN/m}^2) (2.50m) (62) (0.83) (0.70)$$
$$= 820 \text{ kN/m}^2$$

The applied footing pressure is in the order of 440 kN/m², therefore, the bridge abutment footing is safe against a bearing failure.

Settlement

In the settlement calculation the compression of the compacted fill, under selfweight and the underlying soil was not considered, since the fill material was placed several months before the footing was constructed. **The settlement of the footing can be estimated from the following equation** (Canadian **Foundation Engineering Manual, 1985):**

$$S = \frac{B * q_{net}}{E_s} * f_c$$

where
S = settlement of the foundation

q_{net} = net footing pressure

E_s = modulus of soil compression, and

f_c = a factor depending on footing size, geometry and depth of compressible layer

For the abutment footing and soil conditions given, q_{net} = 272.5 kN/m², f_c = 0.61 (from Canadian Foundation Engineering Manual, Figure 5.3), E_s = 50 x 10³ kN/m₂ and B = 2.50 m (see Figure 2).

Therefore,

$$S = \frac{(2.50m) \ 272.5 \text{ kN/m}^2}{50 \text{ x } 10^3 \text{ kN/m}^2} * 0.61$$

$$S = 8.4 \text{ x } 10^{-3}m \text{ or } 8.4 \text{ mm} < 25 \text{ mm}$$

The tilting of the abutment was estimated from the differential settlement of the toe and the heel of the footing and was found to be in the order of 10 mm.

A bridge abutment wall with dimensions as shown in Figures 1 and 2, and soil properties similar to those given in the above example, **was instrumented and monitored for** over two years. The measured post-construction compression of the granular pad was 6 mm and the average horizontal movement was 12 mm away from the backfill.

CONCLUSIONS

The ultimate limit states (ULS) and the serviceability limit states (SLS) principles applied to the design of highway bridges is now mandatory in the province of Ontario. The National Building Code of Canada has incorporated some of these criteria in its new edition of 1988. This paper has discussed its major implications for a highway bridge abutment from the geotechnical point of view.

The load and reduction factors applied to the loads and soil strength respectively, of the ULS, compare well with the conventional factors of safety used in geotechnical engineering. Allowable movements and settlements of bridge foundations at the SLS depend mainly

on the function of the structure,
the underlying soil and the soil-
structure interaction. It is
suggested that performance
observations on different types of
bridge structures be continued to
obtain further quantitative
information in order that reliable
estimates can be made on the actual
safety of such structures.

REFERENCES

Bauer, G.E., 1982. Design Guide for
    Highway Bridges-Geotechnical
    considerations. Special Report to
    the Ministry of Transportation and
    Communications of Ontario,
    Downsview, Ontario, p. 127.

Bozozuk, M., 1978. Bridge
    foundations move. Transportation
    Research Record, 678, pp. 17-21.

Brinch Hansen, J., 1956. Limit
    design and partial safety factors
    in soil mechanics. Danish
    Geotechnical Institute, Bulletin
    1, 4 p.

Canadian Foundation Engineering
    Manual 1985. Canadian
    Geotechnical Society, Rexdale,
    Ontario, Canada

Meyerhof, G.G., 1984. Safety
    factors and limit states analysis
    in geotechnical engineering.
    Canadian Geotechnical Journal,
    Vol. 21, pp. 1-7.

Moulton, L.K., GangaRao, V.S.,
    Tadros, M.K. and Halvoren, G.T.
    1980. Tolerable Movement Criteria
    for Highway Bridges-Analysis
    Studies, ASCE Convention, Florida.

National Building Code of Canada,
    1988. National Research Council
    of Canada, Ottawa, Canada.

Ontario Highway Bridge Design Code,
    1988. Ministry of Transportation
    and Communication of Ontario,
    Downsview, Ontario, Canada.

Preliminary groundwater modelling of Mae Moh basin for lignite development

Yusuke Honjo, Pham Huy Giao & Noppadol Phien-wej
Asian Institute of Technology, Bangkok, Thailand

Somchai Plangpongpun
Mine Engineering Department, EGAT, Bangkuarai, Nonthaburi, Thailand

ABSTRACT : The Mae Moh basin is an intermontane basin of about 135 km^2 located 28 km east of Lampang city in Northern Thailand (Fig. 1), where a large scale lignite open pit mine is developed by EGAT (The Electricity Generating Authority of Thailand). Recently, an aquifer with very high artesian pressure was found right under the open pit excavation area which need to be treated to ensure the stability of the excavation when the depth increases in the future.

This study is a first attempt to model the relatively small but complicated aquifer system which occurs in the basin. The groundwater flow system in Mae Moh Basin Area is conceptualized as a leaky and fractured multi-aquifer system with some typical hydrogeological characteristics of a small localized basin; namely, upward hydraulic gradients, the presence of a weathered limestone layer at the top of basement with high permeability, fractured zones associated with basement faults and along the basin margins could be significant location of groundwater flow etc. In the study, several possible models of steady analysis are presented to explain some observed evidences so far (up to Fall, 1991). It seems that the quasi-3D model is suitably applicable to hydro-geological characterization of Mae Moh Basin.

1. INTRODUCTION

The Mae Moh Mine is the largest lignite mine in Thailand. Its operation started in 1955 and since then the production capacity has been increased gradually. At present, the daily-maximum lignite production is 30,000 tones which makes it possible to generate 1,425,000 KW at Mae Moh Lignite Power Plant. According to EGAT power development plan, the ultimate generation capacity of 4,725,000 KW will be reached in the year 2001; and the lignite production will be increased to 30 million tons annually. Related to this expansion, one of the critical problems is the influence of groundwater on excavation and slope stability when deepening the excavation from the actual depth of 150 m to a larger depth of about 500 m below Ground Surface .

So far there was not a comprehensive study on the excavation and the slope stability, taking into account the influence of high pressure of groundwater at large depths. This is because , first of all, the investigation and excavation works have been done only in the upper layers, hence the data related to the larger depths are not sufficient enough.

In order to study the possible groundwater conditions based on the limited information in the initial stage of groundwater modelling, numerical analysis can be a useful tool to simulate different groundwater flow scenarios, especially for a small scale basin with complicated structural geology like Mae Moh Basin. " This view is based on the presumption that an observed phenomenon is related to only one feature of flow system , whereas it might be brought about by different causes in different situations. For instance, a decrease in hydraulic head with depth may be caused either by head loses due to the vertical downward component of water motion or by the water being-perched in the permeable layers of a geological

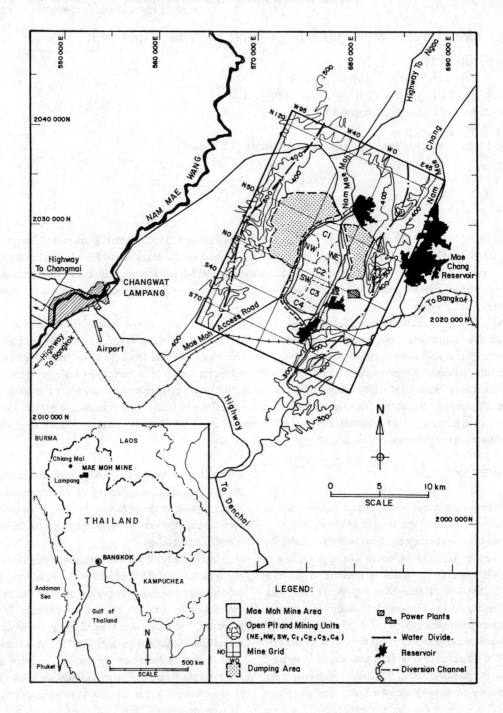

Fig. I : Mae Moh Basin : Location and Extent

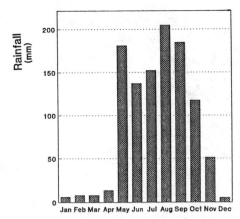

Fig. 2 : Average Monthly Rainfall

formation consisting of a series of more or less pervious beds" (Toth, 1963). This study has the aim of determining the groundwater pressure through groundwater flow modelling by means of FEM (the Finite Element Method) to help understand the important features of the Mae Moh groundwater flow system and, thus, it would become possible for the feasibility study of dewatering or depressurizing systems to reduce adverse groundwater pressures for the safe excavation.

The Quasi-3D model has been proposed and modified by many authors (e.g., Neumann & Witherspoon, 1969; Herrera & Rodatel, 1973; Fujinawa, 1977; Neumann, Preller & Narasimhan, 1982; Huyakorn & Pinder, 1983 etc.), it is an efficient tool to handle a leaky multiaquifer system. For this study on groundwater modelling of Mae Moh Basin, a computer code named QUASI has been developed basing on Smith and Grifiths's FE Subroutines Library (Smith & Griffiths, 1988). The numerical method employed is the combined FE and Convolution Integral Approach (Hossain, 1989).

2. MAE MOH LIGNITE BASIN

2.1 General

1. Mine Landscape and Planning - The Mae Moh Mine is located on a undulated terrain of a graben. The elevation of the valley is about 200 to 300 meters above the mean sea level. It is a big open pit mine, which is divided into several mining pits (Figs. 1 and 6). At present, the lignite production is mined around the North East (NE) pit area. The North West (NW) and the Central (C1) pits will be mined in 1993 and 1996 respectively to supply lignite for additional generating units. From the year 2001 onwards, all the generating units of the Power Plants will need the lignite production from these three pits. When any of the three pits is mined out, the next pit will be mined instead.

2. Mine Grid - In Mae Moh Mine, it is commonly used a local coordinate system introduced by EGAT and called Mine Grid (Fig. 1 and Fig. 6), which is a network of regularly spaced lines, oriented 72^0 NE. Therefore, most graphic representations in this study are drawn according to this coordinate system.

3. Climate - The Mae Moh area has a tropical climate dominated by 2 seasonal monsoon winds. There are three distinct seasons as follows (Tandicul, 1991) : The hot and dry season from mid February to mid May; The rainy season from mid May to mid October; The cool and dry season from mid October to mid February. During the period 1951-1980, it has been observed that the minimum average monthly temperature is 21.3°C, and the maximum 29.7°C .

4. Rainfall and runoff - Monthly rainfall is shown in Fig. 2. A yearly average precipitation is 1,150.6 mm , while the average rainy day is 111 days per year. The runoff in the mine area is mainly induced by the rainfall. The surface runoff, especially from the eastern mountain range, is dammed up by ash and waste damp, is locally pounded thus creating high heads upstream of the low wall (Fig. 3), and flows in an uncontrolled way over surface, partly underground, towards the mine. The coefficient of runoff for the Mae Moh area has been estimated to be 0.75 (Tandicul, 1991).

5. Drainage - The main stream draining Mae Moh basin area is Nam Mae Moh River . The river flows through the mine at an elevation of 300m above the mean sea level . Nam Mae Moh flows from the Northern part of the basin to the Southern part where it meets Nam Mae Chang river (Fig. 1).

6. Surface water - The natural water table fluctuates from about 10m below ground surface during dry season to close to the surface in the wet season.

7. Water Divide - The water divides on the eastern

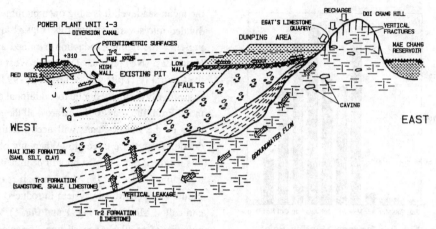

Fig. 3 Conceptual hydrogeological model of Mae Moh basin
(McMahon and Forth, 1988)

western sides as shown in Fig. 1. The northern water divide is outside the study area and the southern one is considered absent in the analysis because of the presumption that the basin is open in the south.

2.2 Geology of the Mae Moh Basin

The Mae Moh Basin is one of the several structural tertiary basins in the mountainous region of Northern Thailand and was formed by post-midle to late Miocene tectonic activities with a series of North - South trending faults (EGAT Geological Report, 1985).

The Mae Moh Basin is mainly bounded by marine Triassic (Lampang Group), it consists of limestone, shale, siltstone and sandstone. These rocks are completely folded and faulted trending from N15°W to N40°E. The most recent geological interpretation of the Mae Moh basin is given in Fig. 4a. There are three major geological units as follows :

1. Pretertiary - In the southern part of the Mae Moh Basin, a sheet of basalt of at least 100 m thick overlies Tertiary sediments, Pleistocene gravel beds, and marine Triassic rocks. Unconsolidated sediments of fluviatile origin cover parts of the basin . They consist of gravel deposits at the bottom and alluvium deposits at the top. The thickness varies from less than 1 meter to 10 meters. The Quarternary deposits overlay the

Tertiary sediments unconformably.

2. Tertiary - Tertiary formations overlay Basement of Triassic rocks. Tertiary succession of Mae Moh Basin can be separated into three formations (Fig. 4b). Each formation consists of sediments that strongly differ in lithology, sedimentary structure, degree of consolidation as follows :

- Huai Luang Formation (HL) : This upper formation consists of red to brownish red sediments with some grey layers interbedded in some parts. Almost all are claystone, siltstone and mudstone with some lenses of sandstone and conglomerate in the central part of the basin. The thickness of this formation varies from less than 5 meters to 250 meters. It is thickest in the central part of the Main Basin , thinning rapidly towards the eastern and western margins, where it is entirely absent or only a few meters thick.

- Na Khaem Formation (NK) : This middle sequence of strata consists of grey to greenish grey, highly calcareous rocks and zones of coal. The thickness varies from 250 meters to 400 meters. The Na Khaem formation consists of five zones of coal which can be separated according to lithological and economic criteria. The two major economic zones of coal are named K and Q (Fig. 4b).

- Huai King Formation (HK) : This is at the bottom of Tertiary succession, in contact with Triassic basement rocks. It consists of mudstone, siltstone, sandstone, conglomeratic sandstone, conglomerate and

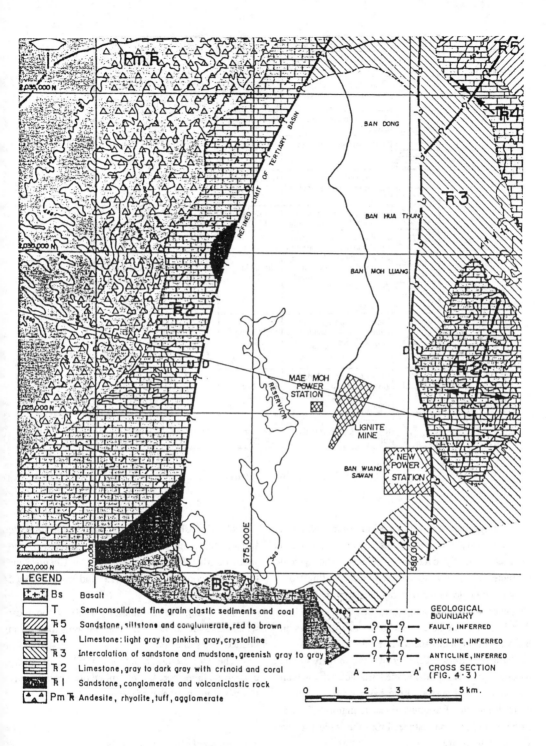

LEGEND

▨ Bs	Basalt	
☐ T	Semiconsolidated fine grain clastic sediments and coal	
▨ Ŧ5	Sandstone, siltstone and conglomerate, red to brown	
▨ Ŧ4	Limestone: light gray to pinkish gray, crystalline	
▨ Ŧ3	Intercalation of sandstone and mudstone, greenish gray to gray	
▨ Ŧ2	Limestone, gray to dark gray with crinoid and coral	
▨ Ŧ1	Sandstone, conglomerate and volcaniclastic rock	
▨ PmŦ	Andesite, rhyolite, tuff, agglomerate	

GEOLOGICAL BOUNDARY

─ ? ─ᵁₒ─ ? ─ FAULT, INFERRED

─ ? ─↕─ ? ─→ SYNCLINE, INFERRED

─ ? ─↕─ ? ─ ANTICLINE, INFERRED

A ──────── A' CROSS SECTION (FIG. 4·3)

0 1 2 3 4 5 km.

Fig. 4a Geological map of Mae Moh basin
(EGAT, 1985)

283

SIMBOL	DESCRIPTION	FORMATION	GROUP	AGE
	River gravel, sand, clay			Recent
	Colluvium, gravel, semi-consolidated sandy clay	Quarternary Terrace	MAE TAENG (PRE-TERTIARY)	Pleistocene
	Red clay and semiconsoli-dated clay	Huai Luang (HL)	MAE MOH (TERTIARY)	Pliocene
	Grey claystone and lignite	Na Khaem (NK)		Miocene
	Yellow and red clay, clayed sand and grey clay-stone	Huai King (HK)		
	Limestone, light grey to pinkish white, crysta- lline	Doi Long (TR4)	LAM PANG (TRIASSIC)	MIDDLE TRIASSIC
	Sandstone, medium grain interbedded with shale and silstone, greenish and grey, conglomerate limestone, recrystaline, well bedded, grey to dark grey	Hong Hoi (TR3)		
	Limestone, pure and crys-talline, massive bedded, grey to dark grey	Doi Chang (TR2)		
	Limestone, banded, dark grey. Basal conglomrate, red sandstone and grey shale.	Phra That (TR1)		LOWER TRIASSIC
	Volcanic rocks, andesite, rhyolite agglomerate and tuff.	Volcanic (P-TR)	VOLCANIC	Permo-Triassic

Fig. 4b : Stratigraphic Sequence

some claystone. It is variegated in color; red, grey, green, yellow, blue and purple, commonly contain calcarete, slightly calcareous cement. The typical character is a fining upward sequence grading from conglomerate to mudstone or claystone at the top. Thickness of the Huai King formation varies from less than 15 meters on the eastern border to 150 meters on the western border of the Main Basin.

3. Triassic - Below Tertiary succession there is the basement consisting of marine Triassic rocks, which also expose and surround the basin on the western, eastern margins. Basement consists mainly of lime-stone, shale, and sandstone. From the oldest to the youngest, there are :

- The Phra That Formation (TR1) overlies the volcanics of Permo-Triassic age. It consists of basal conglomerates and associated limestone .

- The Doi Chang Formation (TR2) is quite wide spread, occurring to the northwest, southwest, and south of Mae Moh, as well as at Doi Chang. This formation consists of massive and recrystallized limestones.

- The Hong Hoi Formation (TR3) is the most widespread formation of shales, sandstones and bedded limestones.

- The Doi Long Formation (TR4) is of finely crystalline limestone. It is of limited distribution, only present at the northeast of mine.

2.3 Structural Geology of the Basin

Mae Moh Basin is divided into three sub basins by basement heights (Fig. 5a) :

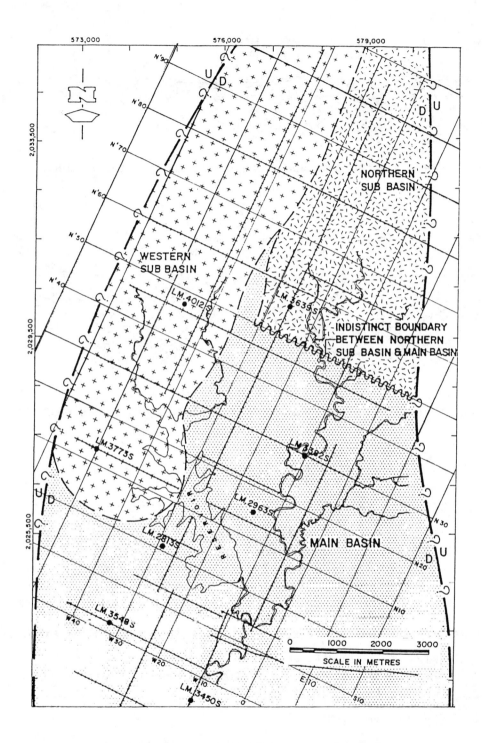

Fig. 5a Mae Moh sub-basins boundaries
(EGAT, 1985)

285

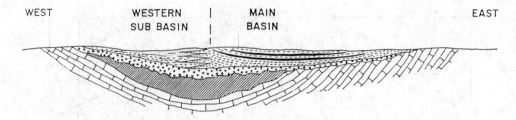

EARLY STAGE OF MAE MOH BASIN DEVELOPMENT (BEFORE FAULTING)

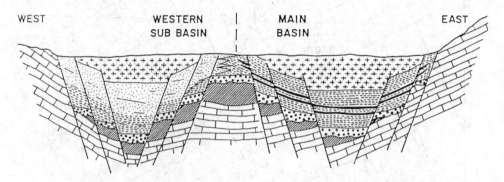

MIDDLE STAGE OF MAE MOH BASIN DEVELOPMENT (FIRST GENERATION FAULTING)

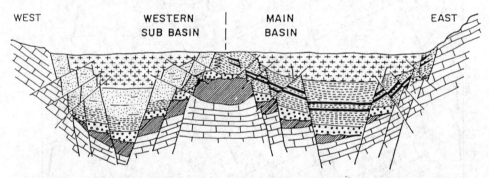

LATE STAGE OF MAE MOH BASIN DEVELOPMENT (SECOND GENERATION FAULTING)

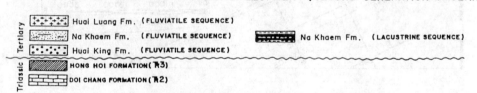

Fig. 5b Schematic structural development of Mae Moh basin
(EGAT, 1985)

286

1. Northern Sub Basin : Located in the northern part of basin. The basement topography is higher than the other sub basins and shows a flat shelf area with depth of about 350-400 meters from surface elevation.

2. Western Sub Basin : Located on the western and northwestern part of the basin. It has a shuttle-like shape with the deepest part of about 750 meters in the central area.

3. Main Basin (or Main Sub Basin) : This is the largest sub-basin and is located to the south and east of Northern and Western Sub Basins, respectively. The southern boundary has not been defined because it continues beneath a younger basalt capping . However, it probably does not extend further than south of S 80 Mine Grid Line. The Main Basin has a shuttle-like shape with deepest part at the central area.

Structural development of the basin is illustrated in Fig. 5b. The sedimentation in the basin was of wide extent and the basin had gradually been faulted and subsided. According to the paleontological study, sedimentation in the basin should have begun in the Miocene age and faulting developed in post-middle to late Miocene age. Mae Moh basin is characterized by a series of normal faults. Some of these formed early in the basin history as subsidence and later deposition of younger sediments occurred. Most faults are epigenetic (post deposition) and displace strata in the Huai King , Na Khaem formations and the disconformably overlying Huai Luang formation. Nearly all the faults are normal, indicating a tensional tectonic regime. Therè are reverse faults with vertical displacement of 30 to 40 m, grouped together on the marginal parts of the basin.

2.4 Hydrogeology

The first alarm on groundwater problem at Mae Moh Mine happened in the latter half of 1988 when a piezometer was drilled in basement (Doi Chang Formation) and confirmed the presence of an artesian confined aquifer with a potentiometric head of 333 m above Mean Sea Level or 13 m above ground surface . Since then a Groundwater Drilling and Testing Project has been initiated by EGAT in order to investigate hydrogeology of the Mine Area and the

whole basin (Fig. 6). At the time of carrying out this study, the Phase 1 has finished. Two production wells and 18 observation piezometers at Site 1 or Groundwater Site 1 (GTS 1) have completed (Fig. 7a). Based on the results of Phase 1, an approximate hydrogeological section and groundwater head distribution has been proposed (Fig. 7b) by Vogwill (1990). The essential data obtained in Phase 1, concerning head measurements and estimated hydraulic parameters, are presented in Table 1. The results have shown that the geology of Mae Moh Basin is more complex than previously thought. However, it should be mentioned that the results of Phase 1 are still insufficient and sometimes ambiguous.

At GTS 1 the Triassic and Permo-Triassic rocks below the HK formation were penetrated. The distribution of these rocks is quite complex and it is difficult to distinguish between the different Triassic formations. The Triassic formations are thought to be aquifers only where permeable secondary structures such as weathered rocks, fractures, and faults are developed. Recent geological studies (EVANS, 1990) have indicated that a brecciated zone can occur locally on top of Doi Chang Formation (TR2). The age of this breccia is unknown but it appears to be pre-Tertiary and post-Triassic. This breccia can be a very important aquifer and therefore represents a depressurization target during mining , depending on its permeability and lateral extent.

In Fig. 3 was shown a conceptual hydrogeological model proposed by Forth (1988), which is developed later by Vogwill (1990). The regional groundwater flow pattern presents three distinct zones :

- Recharge Zone is under and adjacent to the hills on the eastern flank of the basin. Downward vertical flow with steep hydraulic gradients.

- Zone of lateral flow is, in general, under the dumping area. Heads change little with depth. Groundwater flow is mainly lateral.

- Discharge zone occurs under the central portion of the basin and mining area. Groundwater flows upwards with a steep hydraulic gradient. Heads increase with depth. Most boreholes drilled into this zone will flow at ground surface.

Based on the current level of understanding of the hydrogeology of the Mae Moh Basin , there appears

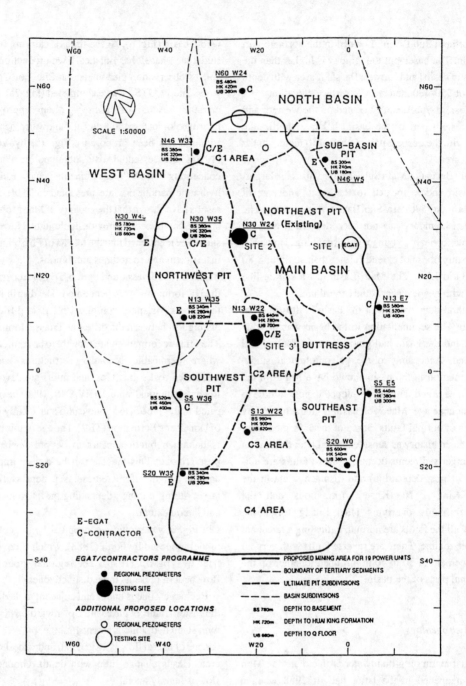

Fig. 6 Groundwater drilling and testing program
(Vogwill, 1990)

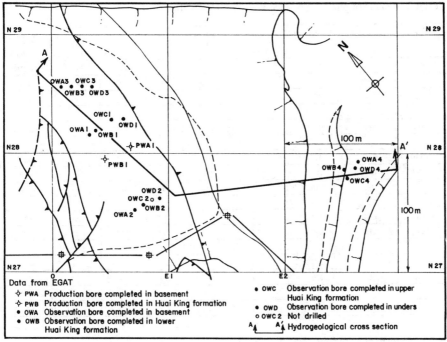

(a) Groundwater Testing Site 1 – GTS 1

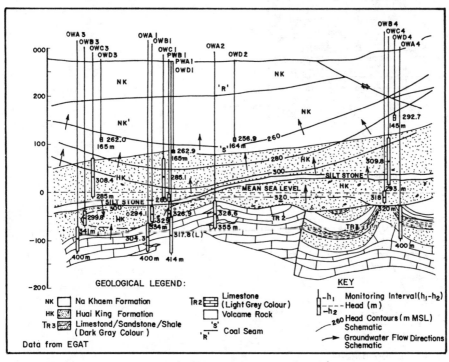

(b) Head Distribution along Section A-A'

Fig. 7   Groundwater Drilling and Testing Program - Phase 1
(Vogwill, 1990)

Table 1 Results of phase 1 obtained at GTS-1 (Vogwill, 1990)

Bore	Formation	Static Head (MSL)	Head when Flowing (MSL)	Drawdown (m)	Flow (m³/d)	Q/s (m³/d/m)	Estimated T (m²/d)	Estimated K (m/d)
OWA1	BMT	288.6	273.8	14.8	2.6	0.18	<1	<0.01
OWB1	LHK	285.2	272.7	12.5	0.4	0.03	<1	<0.2
OWC1	UHK	286.3	272.8	13.5	1.5	0.11	<1	<0.1
OWA2	BMT	328.6	274.5	54.1	203.9	3.77	4-6	0.07-0.1
OWA3/2	BMT	327.8	273.7	54.1	55.3	1.02	1.1-1.3	0.03
OWB3	LHK	295.1	273.8	21.3	0.2	0.009	<1	<0.2
OWC3	UHK	310.8	273.5	37.3	0.1	0.003	<1	<0.08
OWB4	LHK	324.1	298.4	25.7	0.7	0.03	<1	<0.2
OWC4/2	UHK	313.3	298.5	14.8	2.6	0.18	<1	<0.07
PWA1	BMT	317.8	272.8	45.0	28.0	0.62	<1	<0.01
PWB1	HK	326.9	272.7	54.2	13.0	0.24	<1	<0.03
P128	BMT	333.5	320.0	13.5	27.6	2.0	2-4	0.06-0.1

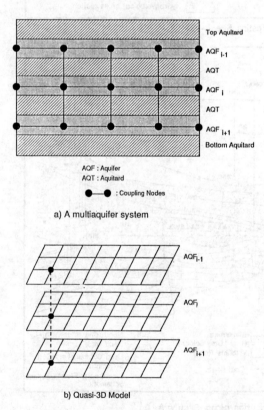

AQF : Aquifer
AQT : Aquitard

●——● : Coupling Nodes

a) A multiaquifer system

b) Quasi-3D Model

Fig. 8   Quasi-3D model of a leaky multiaquifer system

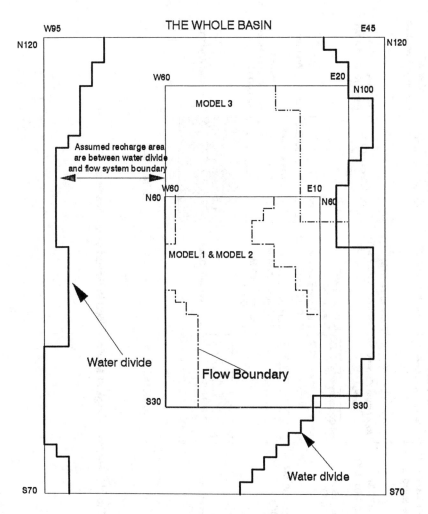

Fig. 9 Flow and FEM Analysis Domains Used in
Groundwater Modelling of Mae Moh Basin

to be two main aquifer systems below the mine area:

1. Triassic Bedrock Aquifer Systems - Most of the Triassic rocks have little or no primary intergranular permeability. The occurrence of Triassic bedrock aquifers, therefore, is a result of secondary structures in the bedrock such as weathering, karst development, joints, fractures, and faults. Permeable zones could occur throughout the Triassic succession, but up to now the Doi Chang Formation (TR2) has been targeted as the main potential aquifer due to the

possible occurrence of karst development in its limestone. It is the authors' opinion that the permeable zones are related firstly to faulted zones, and they are developed not only in Doi Chang Formation (TR2) but also in Hong Hoi Foundation (TR3) and other Triassic Formations. The fractured or faulted zones were most sensible to weathering when Triassic rocks were exposed. Thus, some locally developed brecciated zones could be

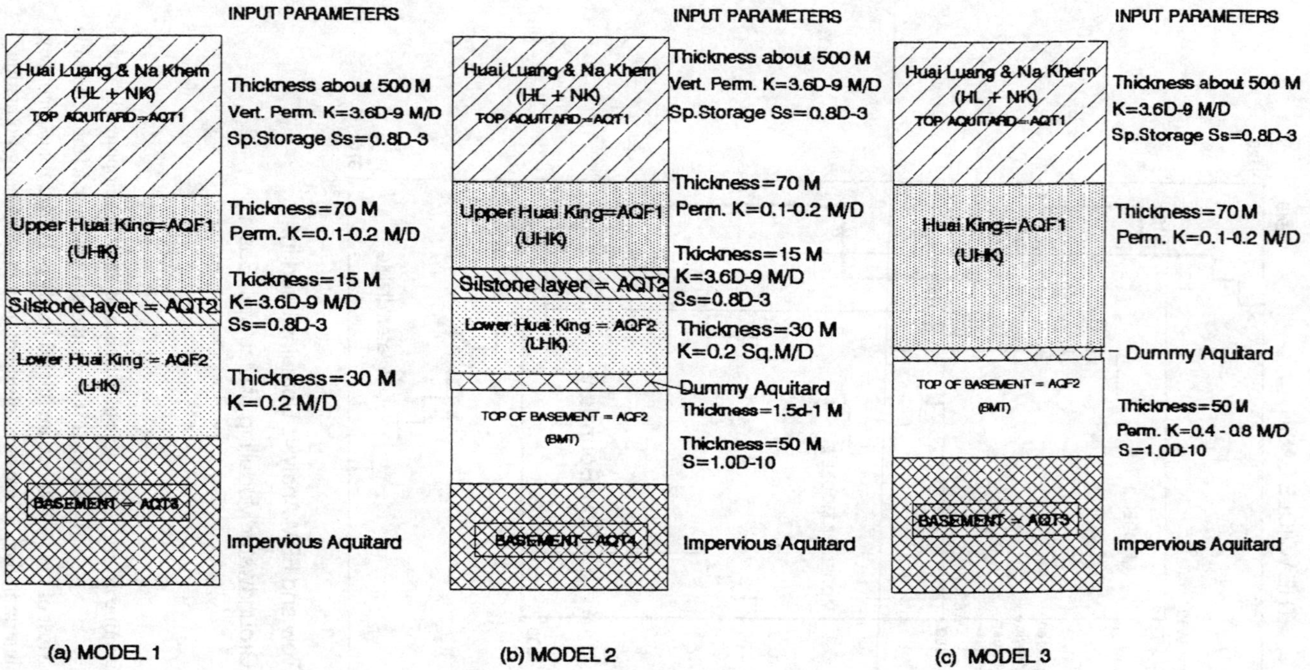

(a) MODEL 1

INPUT PARAMETERS

Huai Luang & Na Khem
(HL + NK)
TOP AQUITARD=AQT1

Thickness about 500 M
Vert. Perm. K=3.6D-9 M/D
Sp.Storage Ss=0.8D-3

Upper Huai King=AQF1
(UHK)

Thickness=70 M
Perm. K=0.1-0.2 M/D

Siltstone layer = AQT2

Thickness=15 M
K=3.6D-9 M/D
Ss=0.8D-3

Lower Huai King = AQF2
(LHK)

Thickness=30 M
K=0.2 M/D

BASEMENT = AQT3

Impervious Aquitard

(b) MODEL 2

INPUT PARAMETERS

Huai Luang & Na Khem
(HL + NK)
TOP AQUITARD=AQT1

Thickness about 500 M
Vert. Perm. K=3.6D-9 M/D
Sp.Storage Ss=0.8D-3

Upper Huai King=AQF1
(UHK)

Thickness=70 M
Perm. K=0.1-0.2 M/D

Siltstone layer = AQT2

Thickness=15 M
K=3.6D-9 M/D
Ss=0.8D-3

Lower Huai King = AQF2
(LHK)

Thickness=30 M
K=0.2 Sq.M/D

Dummy Aquitard
Thickness=1.5d-1 M

TOP OF BASEMENT = AQF2
(BMT)

Thickness=50 M
S=1.0D-10

BASEMENT = AQT4

Impervious Aquitard

(c) MODEL 3

INPUT PARAMETERS

Huai Luang & Na Khem
(HL + NK)
TOP AQUITARD=AQT1

Thickness about 500 M
K=3.6D-9 M/D
Sp.Storage Ss=0.8D-3

Huai King=AQF1
(UHK)

Thickness=70 M
Perm. K=0.1-0.2 M/D

Dummy Aquitard

TOP OF BASEMENT = AQF2
(BMT)

Thickness=50 M
Perm. K=0.4 - 0.8 M/D
S=1.0D-10

BASEMENT = AQT3

Impervious Aquitard

Fig. 10 Hydrogeological Models Used in Groudwater Analysis

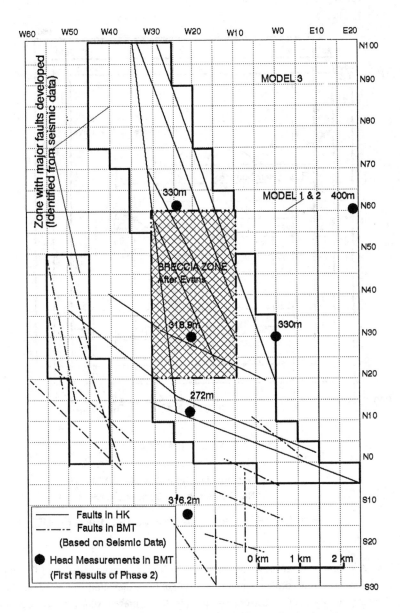

Fig. 11 Synthesized Geological, Hydrogeological and
Geophysical Data

formed on top of Basement. To understand the structure of this complicated aquifer system to achieve efficient depressurization, one of the most important tasks, therefore, is to locate these fractured zones.

2. Basal Tertiary Aquifer System - The Huai King

formation represents a potential aquifer system although it becomes quite thin under the western portion of the mining area. Based on results obtained, which reveal the existence of a siltstone layer, HK can be separated into Upper Huai King (UHK) and Lower

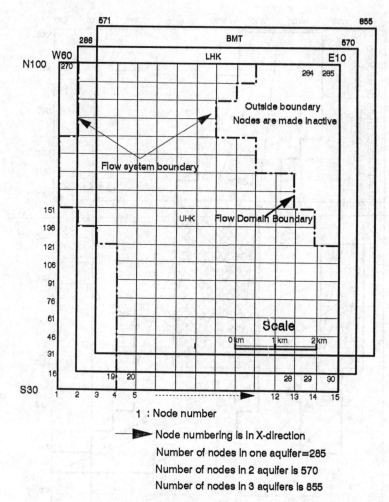

1 : Node number

➤ Node numbering is in X-direction
Number of nodes in one aquifer=285
Number of nodes in 2 aquifer is 570
Number of nodes in 3 aquifers is 855

Fig. 12a   FEM Mesh Used for MODEL 1 and MODEL 2

Huai King (LHK). Justification of this division still needs results from more boreholes, drilled over a larger area in Phase 2 (Fig. 6). If HK is really separated into two effective aquifers, Mae Moh basin aquifer system can be conceptualized as a 3 layered aquifer system, otherwise it is a 2 layered aquifer system. Both situations will be simulated by Quasi-3D models as shown in the following part.

3. GROUNDWATER MODEL ANALYSIS

The objective of groundwater modelling of Mae Moh Basin is to investigate groundwater conditions in

Tertiary sediments and Top of Basement, which can affect the excavation and slope stability of the mine pit. At the present stage, the hydrogeological data are still very few as revealed in the previous section. Therefore, groundwater modelling can only be of preliminary character.

The main purpose of this study is to investigate the important hydrogeological features of Mae Moh groundwater system and to help explain some practical facts, some of which, at the first glance, seem to be confusing; for example, how are boundary conditions of Mae Moh aquifer system; which is the main flow direction; the contribution of recharge ; the role of faults and anisotropy; the interaction between the

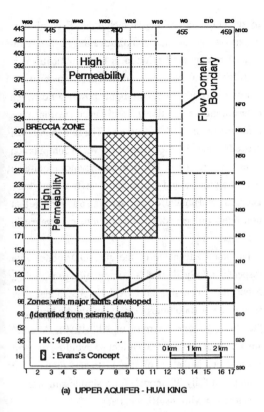

(a) UPPER AQUIFER - HUAI KING

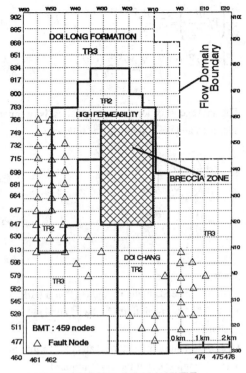

(b) LOWER AQUIFER - TOP OF BASEMENT

Fig. 12b FEM Mesh Used for MODEL 3

aquifers , and, especially, the local abrupt change in head values obtained in some boreholes.

3.1 Quasi-3D model

Geology and hydrogeology of Mae Moh Basin show that groundwater flow here is possibly carried out in a leaky and fractured multiaquifer system. Thus, a quasi-3D model may be most suitably applicable, taking into account actual available hydrogeological data. General features of the Quasi-3D model are :

The quasi-3D model is a groundwater flow model developed for the analysis of a multiaquifer system consisting of several aquifers intercalated by leaky aquitards (Fig. 8). The flow equations used in Quasi-3D are as follows :

The 2-D unsteady horizontal flow equation in the i^{Th} aquifers is :

$$\frac{\partial}{\partial x}\left(T_{xx}\frac{\partial h}{\partial x}\right)+\frac{\partial}{\partial y}\left(T_{yy}\frac{\partial h}{\partial y}\right)+q+V_i-V_{i-1}=S_i\frac{\partial h}{\partial t}\quad(1)$$

where

T_{xx}, T_{yy} : the aquifer transmissivities along X and Y-directions

s_i : the aquifer storage

V_i : flow quantity from the top aquitard to the aquifer

V_{i+1} : flow quantity from the aquifer to the bottom aquitard

The vertical flow in the i_{Th} and $i+1^{Th}$ aquitard are :

295

$$\frac{\partial}{\partial z}\left(T_i'\frac{\partial h_i'}{\partial z}\right) = S_i'\frac{\partial h_i'}{\partial t} \qquad (2a)$$

$$\frac{\partial}{\partial z}\left(T_{i+1}'\frac{\partial h_{i+1}'}{\partial z}\right) = S_{i+1}'\frac{\partial h_{i+1}'}{\partial t} \qquad (2b)$$

The leakage flux terms at boundaries between the aquitards and the aquifers are as follows :

$$V_i = \frac{\partial}{\partial z}\left(T_i'\frac{\partial h_i'}{\partial z}\right)_{z_i=0} \qquad (3a)$$

$$V_{i+1} = \frac{\partial}{\partial z}\left(T_{i+1}'\frac{\partial h_{i+1}'}{\partial z}\right)_{z_{i+1}=b_{i+1}'} \qquad (3b)$$

where

h_i' & h'^{i+1} : heads in the i^{Th} and $i{+}1^{Th}$ aquitard
T_i' & T'^{i+1} : transmissivity in the i^{Th} and $i{+}1^{Th}$ aquitard
S_i' & S'^{i+1} : storavity in the i^{Th} and $i{+}1^{Th}$ aquitard
b_{i+1}' : thickness of the $i{+}1^{Th}$ aquitard

The most fundamental assumption of a quasi-3D model is that it considers only the horizontal flow in aquifers and the vertical flow in the interconnecting aquitards. In other words, it disregards the vertical flow in the aquifers and the horizontal flow in the aquitards. If the permeability of the aquifers is at least two orders of magnitude larger than that of the aquitards, it is known that the calculated results based on this assumption produce an error less than 5% (Neumann and Witherspoon, 1969; 1982). Since this assumption is true in most practical situations the fundamental assumption mentioned above is satisfactory to obtain practical solutions.

The key problem in treating the solution of a quasi-3D model is the technique of coupling of those governing equations of flow in aquifers and in aquitards (Eqs. 1 & 2). There are two ways of coupling, i.e. fully numerical and semianalytical. In the fully numerical technique, Eq. 2 is numerically solved by 1-D FEM procedure; while in the semianalytical one, Eq. 2 is analytically solved and superimposed in time domain

by the convolution integral technique. Each of them has its own advantages and disadvantages. The semi-analytical approach solution is mathematically neat and efficient in saving computation time. However, the iteration process of computing the power series when determining convolution integral can create truncation error and requires a lot of storage. The fully numerical procedure may be 7 to 10 times slower in computation time (Herrera et Al., 1980), but it may not require an iteration process and can account easily for arbitrary geometry and nonuniformity. It has also advantage that non-linear property of aquitards can be relative easily incorporated .

The main advantage of the quasi-3D model with respect to a fully 3D one is computation time saving, hence the processing cost. A fully 3D model may require 100 - 1000 times more computation time than the quasi-3D one (Herrera et Al., 1980). Another advantage of the Quasi-3D might be avoiding overkill error, which is the use of a sophisticated model (for example, a fully 3D model) for a simple situation or the situation with insufficient data.

3.2 Some Important Observations

Some of the important observations taken into account in setting the assumptions for the numerical simulation can be summarized as follows :

1. Hydrogeologically, Huai Luang (HL) and Na Khaem (NK) formations form a very effective top aquitard. HK used to be a continuous confined aquifer, but new geological data indicated that HK can be divided into Upper and Lower Huai King (UHK & LHK) separated by a thin continuous siltstone layer of about 15 meters thick. LHK is speculated to be a major aquifer. Another major aquifer is the Triassic Bedrock Aquifer System, whose permeability depends on the secondary porosity; the latter, in the present study, is represented by its uppermost part of Basement called Top of Basement (BMT) .

2. In the current modelling, the intake area is considered both on the eastern and the western sides for calculating recharge volume. The presence of

fractured limestone on the eastern side (Figs. 3 and 4a) makes it a more important recharge than the western side, where are developed the volcanics of low permeability, especially in its northwestern part (Fig. 4a).

3. The regional flow pattern comprises of 3 distinct zones (Forth, 1988; Vogwill, 1990). By analyzing the head measurements in the boreholes in GTS1, roughly, there was a difference of about 10 meters in head value between LHK and UHK as well as between BMT and LHK (Table 1 and Fig. 7b). This difference is used to introduce fixed head boundaries in the analyzed models.

4. There are two possible groundwater flow directions : one is westwards from the eastern hills, moving across geological structures and Triassic rocks and the other is from north to south . Combination of these two can give a predominat flow direction from northeastern to southwestern (NE-SW).

5. There are interactions between aquifers by upward leakage through aquitards in the central part of the basin, especially area where the breccia zone is suspected to occur .

6. The well observations that are related to the current modelling, i.e some first results of head measurements in Phase 2, are shown in Fig.13 through Fig. 15. The wells or regional piezometers which are being and will be drilled and observed are shown in Fig. 6, in general, their locations are easily identified by means of Mine Grid coordinates.

3.3 Model Parameters

1. Geometry of aquifer systems - The extent of the basin is delimited by the EGAT geologists from N54 to S30 mine grid line. The modelling area for Model 1 and Model 2 is therefore from N60 to S30 mine grid lines at the north and the south, from W60 to E10 mine grid lines at the west and the east, respectively . From seismic data (Vogwill, 1991) HK is more extended than the mine area mentioned above, Model 3 is intended to simulate a larger aquifer system than Model 1 & Model 2. A general view of different flow and FEM analysis domains is illustrated in Fig. 9.

2. Hydraulic Property - The first results obtained at GTS1 confirm the generally low permeability of Triassic rocks. The input parameters employed in current modelling are shown in Fig. 10, for Models 1, 2 and 3, respectively. Some of the input hydraulic parameters, for example the vertical permeability of aquitards or storabilty of Basement, are assumed based on the typical values of soil and rock permeabilities published (CANMET, 1977).

3. Recharge rate - The recharge area , A_{rch}, was determined as 135.25 km² corresponding to 541 elements of 0.5 x 0.5 km each (schematically shown in Fig. 9). The recharge volume is determined as follows :

$$V_{rch} = C \times A_{rch} \times R_f$$

Where
C (%) : The empirical percentage of infiltration (15% in this analysis)
V_{rch} (m³) : Recharge volume
A_{rch} (m²) : Assumed recharge area
R_f : Annual rainfall

Hence, Recharging flux corresponding to one element, say, of 0.25 km² was calculated as 118 m³/day.

4. Boundary conditions - these conditions are specified in every FEM model as shown later, in Figs. 13 through 15.

3.4 Model Settings and Results

3.4.1. Model 1 - This model simulates a two-aquifer system consisting of UHK and LHK separated by a continuous siltstone layer 15 meter thick, on top is a thick aquitard consisting of Huai Luang (HL) and Na Khaem (NK). The Triassic Basement is assumed to be an impermeable substratum (Fig. 10a). From seismic data (VOGWILL, 1991), a lot of faults are identified both in Huai King and Top of Basement . Two zones with major faults developed are attempted to be delimited (Fig. 11) and incorporated in groundwater analysis. In Model 1, for UHK, fractured zone is assumed anisotropic with transmissivity along the fault direction (nearly North- South direction) two times larger than that in East-West direction or denoted as

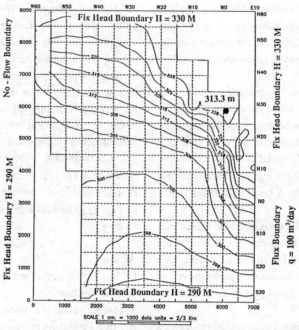

Fig. 13a Head distribution obtained by FEM analysis
for Model 1 - Upper Huai King (UHK)

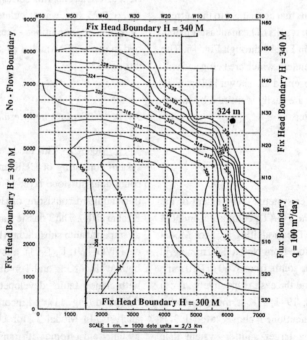

Fig. 13b Head Distribution obtained by FEM analysis
for Model 1 - Lower Huai King (LHK)

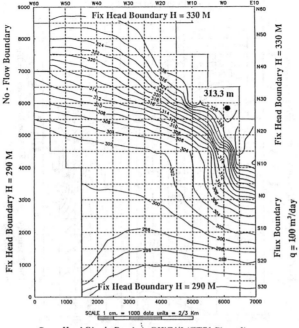

● Head Obs. in Borehole OWC4/2 (GTS1-Phase 1)

Fig. 14a Head distribution obtained by FEM analysis
for Model 2 - Upper Huai King (UHK)

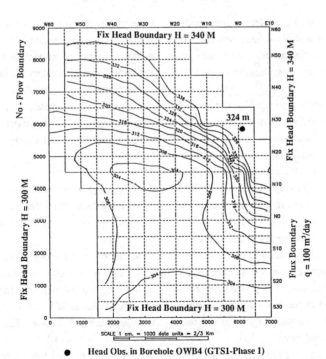

● Head Obs. in Borehole OWB4 (GTS1-Phase 1)

Fig. 14b Head distribution obtained by FEM analysis
for Model 2 - Lower Huai King (LHK)

Fix Head Boundary H = 350 M

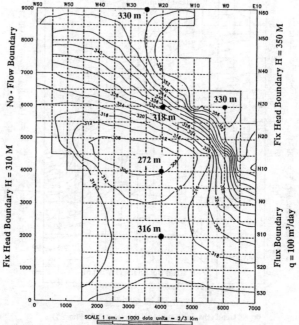

Fig. 14c Head distribution obtained by FEM analysis
for Model 2 - Top of Basement (BMT)

$T_y = 2T_x$, while for LHK, the Finite element mesh nodes where the faults pass through are assigned internal flux nodes, i.e the nodes from which some flux comes into the aquifer system. The intensity was considered 100 m³/day.

The finite element mesh is shown in Fig. 12a. A mesh of rectangular elements 0.5 x 0.5 km has been used to cover the central area of the basin, whose area is 7 x 9 km² corresponding to a set of 285 nodes (15 x 19) for each aquifer and total of 570 nodes for both aquifers. In order to accommodate the irregular shape of the flow area, some nodes are made inactive (i.e. excluded from the calculation).

In Model 1, the northern and southern boundaries are assumed to be fix head boundaries to check the new geological interpretation of the EGAT geologists, which states that main groundwater flow into the mine area in the central part of the basin is from north to south (Vogwill, 1991). The western and eastern ones are set in a more complex way (Fig. 13).

This model is attempted to be realistic by simulating the following characteristics : (1) There are two possible flow directions : one from north to south and the

other is from east to west; (2) The recharge contribution is mainly from the eastern hill side; (3) The presence of Permo-triassic rocks on the northwestern part makes the northwestern boundary no-flow one.

The model results are shown in Fig. 13 a & b. A notable feature of head distribution is steepened hydraulic gradient on the eastern and northern sides; while in the central and southern parts head variation is gentle.

3.4.2 Model 2 - This model is developed from Model 1 by adding the third aquifer, Top of Basement (BMT). The hydrogeological model of Model 2 is illustrated in Fig. 10b.

The finite element mesh - is essentially the same as in the case of Model 1, with only addition of a third aquifer, i.e. Top of Basement (BMT). Number of nodes in each aquifer is 285 and total number of nodes in three aquifers is 855 nodes. The flow domain is kept as the same as Model 1.

In this model, like in Model 1, fractured zone in UHK is assumed anisotropic with $T_y = 2T_x$, while in LHK and BMT, the mesh nodes where the faults (identified

300

from seismic data) are passing, are assigned internal flux nodes. The results of calculation are shown in Fig. 14a, b & c. General picture of heads in all the three aquifers show steep hydraulic gradient on the eastern and northern sides, the central southern parts expose the same smooth head variation in all three aquifers. The pattern of equipotential lines show a predominant resulted flow direction from northeast to southwest.

3.4.3 Model 3 - According to seismic results HK is rather extended and varies in thickness (Vogwill, 1991). The Model 3 includes a wider area than that in Model 1 & 2 (Fig. 9). This model takes into account the two major aquifers, namely the whole Huai King (HK) and Top of Basement (BMT). The main feature of MODEL 3 is that it considers new ideas concerning deep geology of the basin. EVANS (1990) suggested some interesting ideas on deep geology as follows :

1. The current hydrogeological model considers that Doi Chang limestone (TR2) extends below Hong Hoi formation (TR3). Doi Chang (TR2) is assigned the role of important aquifer. However, in Evans's opinion, the Doi Chang limestone is not present everywhere. Hence, in the absence of porous limestone TR2, Hong Hoi formation (TR3) can transmit fluids through its fractures.

2. Evans mentioned of the existence of a limestone breccia zone with significant porosity (Figs. 11 and 12b). He suggested that this limestone breccia be regarded as a part of HK and not of basement. This model is clearly different from the common conceptual model, considering breccia zone or Top of Basement as a well extended aquifer.

It is interesting to note, that Evans's breccia nearly coincided with the zone of major faults in HK, identified from seismic data (Figs. 11 and 12b), the latter is, however, more extended in the north-south direction. The faults in Basement that occurs were present in Hong Hoi Formation. Thus, the possibility of fracture porosity of Hong Hoi formation (TR3), as assumed by Evans (1990), can be realistic.

By reflecting the results of Models 1 & 2 some important modifications are developed in Model 3 as follows :

- The value of aquifer transmissivity is increased , about 4 times larger more than in the previous models.

- The anisotropy in E-W direction plays a significant role in controlling the flow system of Mae Moh Basin. Thus, it was used $T_x = 4T_y$.

- Hydraulic connection between aquifers through aquitards is assumed more pronounced where the fractured zone is developed, and very weak elsewhere, i.e. outside fractured zone the vertical permeability of the aquitards is insignificant.

The finite element mesh is shown in Fig. 12b. An area of 8 x 13 km^2 is discretized into a set of 459 nodes for each aquifer and 918 nodes ,in total, for two aquifers.

The results obtained are plotted in Fig. 15 a & b. The head distribution is much more fitted to the hydrogeological observations and it gives some head values surprisingly near to those measured values, which have just been obtained in Phase 2 and are shown in Fig. 11.

4. CONCLUSIONS

The following conclusions are drawn from the preliminary groundwater modelling of Mae Moh Lignite Basin:

1. The feature of the groundwater flow system in Mae Moh Basin Area is characterized by that of small basins in such aspects as: existence of strong upward hydraulic gradients; existence of limestone with karstic development; existence of a weathered rock layers at the base of the tertiary rock; the faults occur in the basement and along the basin margins, they can be locations of extensive groundwater flow. The high head observed at the piezometer P128 located in GTS1, is related to a highly permeable zone. Such permeable zones are very important in Mae Moh basin aquifer system , they are considered to be formed by the weathering of the Triassic limestones or fault effects.

2. Two major aquifers, i.e HK and Top of Basement (BMT) together with an effective aquitard formed by HL and NK on top and a fine crystalline Triassic Limestone rock substratum are the components of the hydrological model of Mae Moh basin .

3. There are two flow directions, e.g. N-S and E-W and the basin is open in the south . The authors' opinion is that the E-W flow is more important, because only by calibrating the eastern boundary conditions a reasonable head distribution could be obtained. The

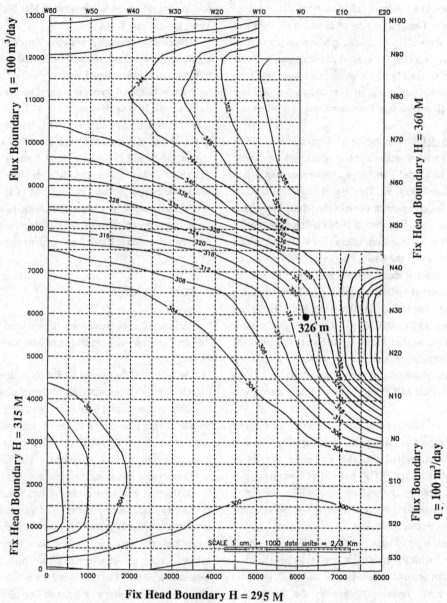

Fix Head Boundary H = 335 M

• : Head Obs. in Borehole PB1 (GTS1-Phase 1)

Fig. 15a Head Distribution Obtained by FEM Analysis
for MODEL 3 - Huai King (HK)

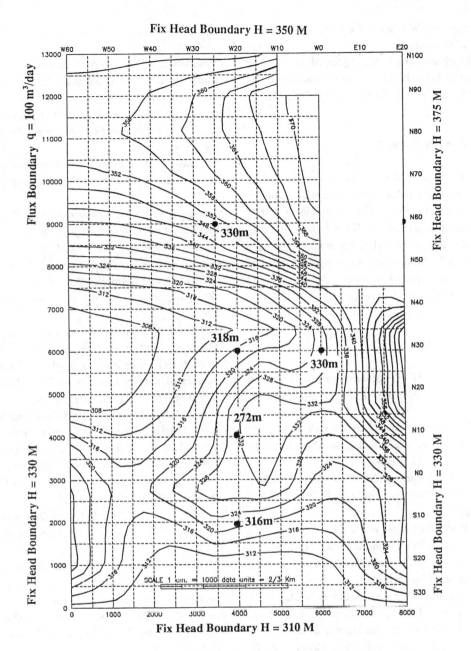

Fig. 15b Head Distribution Obtained by FEM Analysis
for MODEL 3 - Top of Basement (BMT)

anisotropy in E-W direction has a strong influence on the head distribution.

4. For such an aquifer system, a quasi-3D model which consists of FE Galerkin's procedure and the Convolution Integral technique can be successfully used.

ACKNOWLEDGEMENTS

The authors would like to express their sincerest appreciation to The Electricity Generation Authority of Thailand, and namely to Mae Moh Mine Operation Department in general and Groundwater Project in particular for provided data concerning Mae Moh Lignite Basin. Sincere thanks also go to Mr. Richard Vogwill, Mr. Ophas Jariyabhumi, Mr. M.C. (Max) Brown for their useful discussions.

REFERENCES

CANMET 1977. *Pit Slope Manual. Chapter 4, Groundwater*. Ministry of Supply and Services Canada.

EGAT 1985. *Mae Moh Coal Deposit, Geological Report*, Thailand-Australia Lignite Mines Development Projects. Chapter 4.

Evans P. R. & S. Jitapunkul 1989. *Geology of The Mae Moh Basin, Northern Thailand*. EGAT Technical Report.

Evans P. R. 1990. *Notes on the Geology of The Mae Moh Basin*. EGAT Technical Report, Feb., 1990.

Fraser C. I. & Pitt H. A. 1979. Artesian Dewatering Operations at Morwell Open Cut. *Proc. 1st Intern. Mine Drainage Symp*. Denver Colorado, May 1979.

Fujinawa K. 1977. Finite Element Analysis of Groundwater Flow in Multiaquifer Systems, II. A Quasi Three-Dimensional Flow Model. *J. Of Hydrology*, 33 (1977) 349-362

Herrera I. & Rodatel L. 1973. Integrodifferential Equations for System of Leaky Aquifers and Applications, 1, The nature of Approximate Theory. *Water Resources Research*, Vol. 9, pp. 995-1005.

Herrera I., J. P. Henart & Yates R. 1977. A Critical Discussion of Numerical Models for Multiaquifer Systems. *Advanced Water Resources*, Vol. 13, pp. 159-163.

HOSSAIN Z. 1990. *Quasi-3D Modelling of Multi-Aquifer with reference to Ground Subsidence*. AIT Thesis no. GT-89-14, Bangkok, Thai Land.

Huyakorn P. S. & Pinder G. F 1983. *Computational Methods in Subsurface Flow*. Academic Press, Inc., New York.

McMahon B. K & J. Forth 1989. *Review of Geotechnical Evaluation*. EGAT Technical Report, Jan. 1989.

Neumann S. P. & P. A. Witherspoon 1969. Theory of Flow in Two Confined Aquifer System. *Water Resources Research*, Vol. 5, No. 4, pp. 803-816.

Neumann S. P. & P. A. Witherspoon 1969. Applicability of current theories of flow of leaky aquifers. *Water Resources Research*, Vol. 5, No. 4, pp. 817-829.

Neumann S. P., C. Preller & I. N. Narasimhan 1982. Adaptive Explicit - Implicit Quasi Three-Dimensional Finite Element Model of Flow and Subsidence in Multiaquifer System. *Water Resources Research*, Vol. 18, No. 5, pp. 1551-1561.

Smith I. M. & D. V. Griffiths 1988. *Programming the Finite Element Method*. John Wiley & Sons.

Sullivan T. D. & B.C. Burman 1985. Geological and Geotechnical Aspects of Small Basins and Their Effects on Mining. *IMM Conference, Asian Mining 85*, Manila 1985.

Tandicul W. 1991. *Criteria for the Design of Mae Moh Drainage System*. EGAT Technical Report, 1991.

Toth J. 1963. A Theoretical Analysis of Groundwater Flow in Small Drainage Basins. *Journal of Geophysical Research* 68 (10): 4795-4812.

Vogwill R. 1989. *Mining Geology and Open Pit Depressurization*. EGAT Technical Report, Aug. 1989.

Vogwill R. 1990. *Hydrogeological Review*. EGAT Technical Report, Oct. 1990.

Vogwill R. 1991. *Hydrogeological Review*. EGAT Technical Report, Jun. 1991.

The performance of treated mining pond slime

W.H.Ting & R.Nithiaraj
Zaidun-Leeng Sdn. Bhd., Kuala Lumpur, Malaysia

S.J.Ng
Tahir Wong Sdn. Bhd., Kuala Lumpur, Malaysia

SYNOPSIS : The tin mining process in Malaysia has left behind ponds, loose sandy soils, and slime deposits in the pond or on land. The slime is a fine waste material generally comprising soft to very soft silty clay usually containing some fine sand. In order to render the disturbed land suitable for building purposes systems of treatment based on established geotechnical methods have been developed and are briefly described. The method of subsurface investigation appropriate to the deposits are discussed. The geotechnical characteristics of the slime have been identified for the purpose of the design of the treatment. The performance of fill and embankments on the treated slime has been monitored and the results obtained have been analysed.

1.0 INTRODUCTION

The diverse parts in the reclamation process of ponds and soft ground areas in former tin mining land are integrated into an Engineering System.

To account for Site Condition, the materials in which the ponds are formed and the pond geometry are considered; so are the nature and quantity of the infill slime. Also included for consideration is the nature of fill available as earthworks and as backfill material.

In arriving at the Engineering Solution, the usage of the treated ground has to be first established before the design criteria can be arrived at.

The core elements in the proposed Engineering Solution for the reclamation are preloading with acceleration of settlement. Where the soil bearing capacity is adequate surcharging may be applied. In treating the slime by preloading, the problem of disposal is obviated with evident environmental benefits.

The treatment procedure has been implemented for building as well as highway projects. The case history cited in this paper is from a highway project around the Ipoh area.

2.0 PLANNING OF TREATMENT

2.1 DESIGN CRITERIA

2.1.1 Stability and Settlement

The performance of the reclaimed ground with respect to stability and settlement are basic to the treatment process. Stability has to be first taken care off. Acceptable settlement levels vary according to usage.

In general in building areas, treatment to the state of normal consolidation under prevailing loads is recommended. It is generally not viable to treat the ground to such a standard as to support the super-structure loads by spread foundations, and pile supports are recommended.

For road embankments, treatment by stages is required and the procedure is detailed in Section 6.

2.1.1.1 Building Infrastructure Areas

The prepared ground over slime deposits has to be treated in order to avoid damage to facilities built on them. The preloading technique in general is applied. In certain areas where fill load over slime deposits is limited, partial excavation of the slime is carried out in order to accommodate an increase in fill load. A certain amount of surcharging is also carried out if deep excavation is to be avoided.

2.1.1.1.1 Building Platforms

It is not feasible to treat the slime so that it can support spread foundations for buildings. Pile supports are still required. However the improved strength of the soil arising from preloading has beneficial effects especially when slender piles are used.

The slime is to be improved to the extent that it can support ground floor loads with nominally reinforced floor and also apron and parkway loads.

2.1.1.1.2 Channels and Pipes

The standard of treatment for Building Platforms described herein above will suffice for channels and pipes to be constructed without additional foundation support.

2.1.1.1.3 Car Parks and Recreation Areas

In the same token, the proposed Building Platform treatment will be adequate for Car Parks and Recreation areas. A certain amount of selected suitable material may need to be laid if reclamation fill thickness is inadequate.

2.1.1.2 Highway Embankments

The treatment required for the construction of highway embankments is carried out in two stages. The foundation fill is first constructed followed by the embankment load. The treatment under the foundation fill stage is similar to that for building infrastructure areas. It will therefore be sufficient in this paper to deal with the highway problem of ground treatment with foundation fill followed by embankment loads.

2.2 POND DIMENSIONS

2.2.1 Width

The width will affect the decision to fill the pond completely, partially or to provide containment bunds.

2.2.2 Depth of Water and Depth to Pond Bottom

The depth of water in the pond and the depth of the pond bottom from surrounding ground have to be determined as they affect the dewatering and subsequent filling operations. The two depths are different when the water level in the pond is not at outside ground level.

2.3 POND SURROUND

The materials surrounding and forming a pond may have been disturbed by the mining process or they may be deposited waste material.

2.3.1 Side Slopes

The quality of side slopes have to be established as this would affect dewatering procedure and also firmness of platforms for depositing the fill.

2.3.2 Bottom

2.3.2.1 Firm Bottom

Ponds are considered to have firm bottoms when they are underlain by stiff clayey material with strength greater than about 50 kPa or when the bottom strata is sandy.

2.3.2.2 Infill Slime

The soft bottom ponds are those infilled with slime. The thickness of the slime layer is an important consideration. Shallow layers less than about 2.5m may be completely removed. If they are greater than 2.5m thick various types of treatments are then required.

3.0 SUBSURFACE EXPLORATION

A Subsurface Exploration program will have to take into account the known nature of mining deposits. It will also, as in all soft ground problems, have to provide the detail parameters required for the design of the geotechnical treatment. Some of the features specific to mining deposits are discussed below.

3.1 METHODOLOGY

3.1.1 Slime

3.1.1.1 Slurry/Density Probes

A density probe has been designed to determine the thickness of the slurry at the top of slime deposits with densities between 11 kN/m3 and 13.4 kN/m3.

3.1.1.2 Strength/Vane Shear

The field shear strength is determined by the field vane shear instrument. The strength to be measured commences at about 4kPa after the slurry is removed.

3.1.1.3 Consolidation Tests/ Undisturbed Samples

The undisturbed samples to be extracted will have strengths commencing from about 4kPa and thin wall piston samplers are used to extract the samples.

3.1.2 Extent of Poor Material

3.1.2.1 Depth to Firm Bottom

For shallow deposits up to about 6m in depth, which is often the case in the Ipoh project, it has been found that the Light Dynamic Penetrometer (e.g Mackintosh Probe) is an efficient and economical tool to ascertain the extent of poor material. The Penetrometer tests sometimes have to be supplemented with examination of soil samples obtained by hand auguring or from test pits dug by excavators, in order to distinguish between sandy and clayey material. In the case of clayey material, one blow/300mm of the Light Probe Penetrometer can, as an estimate, be equated to 1.0 kPa in undrained shear strength.

4.0 CHARACTERISTIC OF MINING AREA DEPOSITS

4.1 DEPOSITION

In a mining area it is likely that the overburden has been disturbed by the mining process. In its natural condition it may consist of Young and Old Alluvium, and sometimes residual soil of sedimentary deposits associated with limestone.

The underlying rock is in most cases limestone. In the area West of Ipoh the depths of limestone varies from about 6m to 21m. Due to the weathering process the limestone has developed karst features on the surface. The infill material in solution channels and material immediately overlying the limestone can comprise stiff clay, granular alluvium and weathered limestone fragments.

Ponds are formed within the overburden. Some of them are filled with slime but some may have been used just for storage of water.

The slimes are in general deposited into water-filled ponds from tailing pipes. Old ponds with slime may sometimes have been backfilled up to surrounding ground level. This means that the slime to be treated may be encountered in ponds as well as in slime reclaimed ground. The nature of the slime deposit will be discussed in the next section.

4.2 SLIME

The characterisation is carried out for the subsurface investigation results obtained from the Ipoh area prior to implementation of the highway project. (8A-1 in the figures refers to the stretch of road closer to Ipoh while 8A-2 is further South).

4.2.1 Mineralogy

The typical mineralogical composition of slime in the Ipoh project is shown in Table 1. The dominant clay mineral is kaolinite and the next important mineral is illite. Other minerals reported to be present are chlorite, quartz, goethite, gibbsite, paragonite and mica. The absence of montmorillonite may be specific to the area under examination as its presence has been reported for slime samples obtained from other parts of the state.

4.2.2 Particle Size and Consistency

The typical particle size distribution is shown in Fig. 1 with the summary of results shown at the bottom of the figure. It can be seen that the material in general comprise a silty clay with the presence of some fine sand. The silt and clay contents may vary widely, but the fine sand content variation is limited.

The consistency is represented in Fig. 2 by the Clay fraction/PI plot, whose slope gives the activity. Typical results for

307

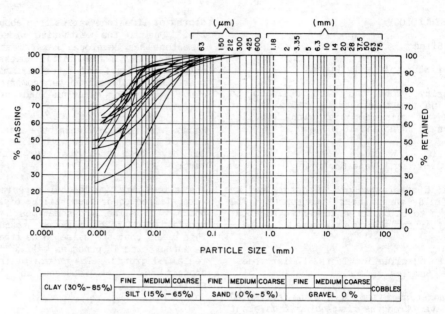

		(μm)		(mm)		

Fig. I   SLIME GRADING

CLAY (30%-85%)	FINE	MEDIUM	COARSE	FINE	MEDIUM	COARSE	FINE	MEDIUM	COARSE	COBBLES
	SILT (15% - 65%)			SAND (0 % - 5 %)			GRAVEL 0 %			

LOCATION	BOREHOLE	SAMPLE DEPTH (m)	CLAY MINERAL TYPE (%)		
			KAOLINITE	ILLITE	MONTMORILLONITE
8A-1	BH 006	1.0	54	-	-
	BH 009	2.0	43	56	-
	BH 009	3.0	40	60	-
	BH 010	6.0	55	45	-
	BH 020	1.25	95	5	-
	BH 020	2.9	95	5	-
8A-2	BA 019	4.5	75	25	-
	BA 031	1.0	81	19	-
	BA 031	2.5	83	17	-
	BA 031	4.0	80	20	-
	BA 039	1.0	75	18	-
	BA 043	1.0	58	42	-
	BA 050	7.0	88	12	-
	BA 055	2.5	75	16	-
	BA 069	6.0	77	23	-
	BA 073	5.0	89	11	-
	BA 110	7.0	54	46	-

TABLE 1   SUMMARY OF CLAY MINERALOGY

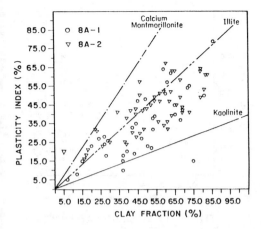

Fig. 2   PLOT OF PLASTICITY INDEX vs PERCENT CLAY FRACTION

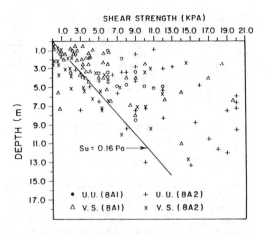

Fig. 3   STRENGTH PROFILE PLOT

Kaolinite, Illite and Montmorillonite from reference sources are plotted on the same figure. There is a small tendency for site 8A-2 samples to be centred on the Illite line when compared with the 8A-1 samples.

4.2.3 Strength

The undisturbed strength profile is as in Fig. 3. It can be seen that the normal consolidation relationship $s = 0.16p_0$ forms the lower bound to the results. The coefficient 0.16 is lower than the usual reported value of 0.2 for silty clays. A possible explanation is the occurrence of mica-related minerals in the deposits. The higher strength values above the lower bound line may be attributed to the presence of granular material or crusting at upper levels and strength values below the line may be due to sample disturbance. For comparison purposes the results of a consolidated drained test is plotted in Fig. 4 in terms of the stress ratios.

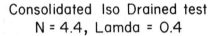

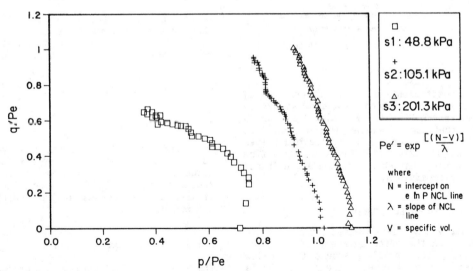

FIG. 4

The sensitivity of the slime values range from 1.8 to 9.8 with a mean value of 4.0. Judging from the test results obtained the slime appears to behave like artificially sedimented silty clays. The supposition appears to be valid because of its young age and method of deposition.

4.2.4 Consolidation Characteristics

4.2.4.1 Compressibility

The compressibility is expressed in terms of the compression ratio $C_c/1+e_0$ and is plotted against logarithm of natural moisture content in Fig. 5. It can be seen that the compressibility values range from about 0.07 to 0.38 with the median value around 0.2. It has been found that the above compression ratio plot shows a broad tendency in the increase of the ratio with natural moisture content. A comparable plot of the ratio against depth showed only scattered results.

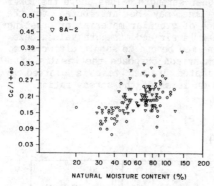

Fig. 5 $C_c/1+e_0$ Vs NATURAL MOISTURE CONTENT (%)

4.2.4.2 Permeability From Cv

The coefficient of consolidation (c_v) values are plotted against silt/clay fraction in Fig. 6. In general it can be seen that the values range from about 1 to 4 m2/yr although a few samples had values below 1 m2/yr. The results from laboratory test show that the slime is not highly impervious.

4.3 GROUND WATER LEVEL

Ground water level varies from site to site. In the affected areas of the Ipoh project, the level is around 0.7m below existing ground level while in the Kuala Lumpur project area it is about 3m below existing ground level.

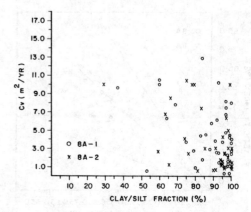

Fig. 6   Cv vs SILT/CLAY FRACTION

5.0 DESIGN OF TREATMENT

The system of the treatment applied is illustrated by the flow chart presented in Fig. 7. The full treatment accorded is that shown in the bottom of the chart.

The treatment process is divided into two stages. In the first stage, the Foundation Fill is placed and a rest period is prescribed until the required performance is obtained. In the second stage the Foundation Fill is loaded by the building infrastructure or the road embankment as the case may be. A surcharge equivalent to the infrastructure load may be applied. Another rest period then follows. Fig. 8 shows in sections the stages of the full treatment.

In the following sections various factors that have to be considered in the detail design are first described followed by a brief statement of the design methodology.

5.1   STABILITY CONSIDERATIONS

5.1.1 Foundation Fill

The Foundation Fill functions as a preload as well as a berm to support super-structure loads. It has a further function as an initial working platform. It will have to possess adequate thickness and width.

5.1.1.1 As Preload

The Foundation Fill serves as a preload to improve the underlying slime when the

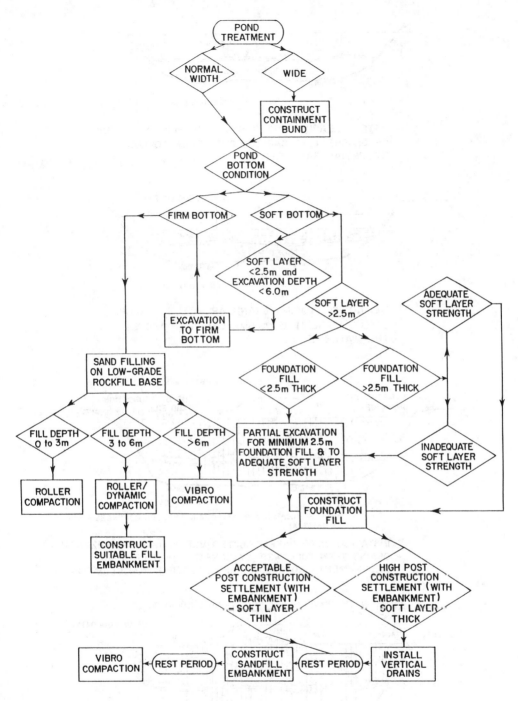

Fig. 7   DETAIL DESIGN FLOW CHART

311

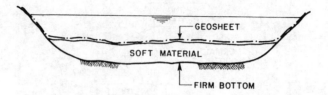

A) AFTER CLEARING AND REMOVAL OF SLURRY, LAYING
OF GEOSHEET IS CARRIED OUT, THEN FOLLOWED
BY INITIAL SAND FILL

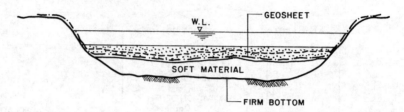

B) FILLING BY PLACING GRANULAR FILL IN UNIFORM
LAYERS UNTIL IT IS A SMALL HEIGHT ABOVE
THE WATER LEVEL

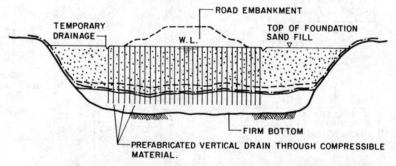

C) INSTALLATION OF PREFABRICATED VERTICAL AND HORIZONTAL
DRAINS THEN FOLLOWED BY CONSTRUCTION OF ROAD
EMBANKMENT AND ALLOWED FOR FURTHER CONSOLIDATION

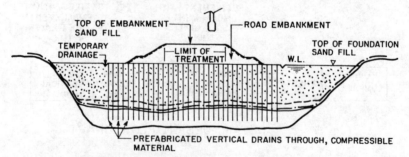

D) COMPACTION USING VIBROFLOT

FIG. 8   CONSTRUCTION STAGES

slime is unable to support the building infrastructure or embankment loads even when acting in conjunction with the Foundation Fill.

5.1.1.2 As Berm

In the case of the highways, the Foundation Fill is to be of adequate width and thickness to serve as a stabilising berm overlying the soft ground to support the embankment.

5.1.1.3 As Working Platform

Together with the underlying geotextile separation sheet an initial granular fill of about 600mm thickness, formed if necessary by excavation, is laid on top of a layer of slime of not less than 7 kPa shear strength to serve as a working platform.

5.1.2 Loading on Foundation Fill

5.1.2.1 Building Load

The nature of loading that may be accommodated has been described in Section 2.1.1.1. Surcharging is usually applied on the foundation fill out of consideration for settlement rather than as a preload for improvement of stability.

5.1.2.2 Embankment Load

Because of the inherent limitations of the treatment process for the given ground conditions, highway embankments in the treated areas are restricted to a height of not more than 2.5m.

5.2 SETTLEMENT

5.2.1 Vertical Drain

When the thickness of slime is such that the treatment process cannot be completed within the practical period of not more than six months, vertical drains are introduced to accelerate the process.

5.2.2 Surcharge

Surcharge may be applied to cater for settlements if the insitu slime in conjunction with the foundation fill can support the surcharge load and the need

for settlement or settlement rate reductions necessitate such an application.

5.3 METHODOLOGY

The various aspects of the design methodology can now be synthesised and presented as available options of different types of treatment. The treatment types are introduced in order of technical complexity. The core procedure of preloading with accelerated consolidation will involve the greatest complexity.

Table 2 gives the summary of the number of ponds that have been subjected to various types of treatment in the Ipoh highway project.

For simplicity of project implementation, the categorization into the types mentioned in Table 2 is in accordance with construction rather than as design procedures.

TREATMENT TYPE	NUMBER
I (FIRM BOTTOM)	7
II (COMPLETE REPLACEMENT)	24
III (PARTIAL REMOVAL & PRELOAD)	8
IV (CONTAIN & PRELOAD)	1
SOFT GROUND	14

TABLE 2 BREAKDOWN OF TREATMENT TYPES

5.3.1 Avoidance

As earlier mentioned the obvious method of avoidance should not be ignored. Such an option was not available in the highway project.

5.3.2 Direct Fill For Firm Bottom

If the pond bottom is firm, the filling can proceed in a direct manner without the need for dewatering or other procedures if free-draining fill material is used.

5.3.3 Replacement

Under the appropriate conditions enumerated in the Design Flow Chart (Fig. 7), the slime may be partially or completely replaced by the Foundation Fill.

5.3.4 Deep-Seated Slime

Where the slime is deep-seated complete replacement will not be feasible, although partial replacement may still be implemented to supplement the treatment process.

5.3.4.1 Contain And Preload By Foundation Fill

The deep-seated slime may be contained in its entirety by the Foundation Fill and the containment bund were applied, prior to further treatment.

5.3.4.2 Partial Excavation

In order to accommodate a Foundation Fill of adequate thickness, or in order to create a viable initial working platform partial excavation may be carried out. The remaining slime is then treated by the procedure described in the previous Section.

5.3.5 Fill

5.3.5.1 Material

As part of the treatment process where the initial Foundation Fill may have to be placed without compaction and because of the general high ground water level, free-draining material has to be used as fill. The embankment may be constructed of suitable or granular fill depending on the compaction procedure to be adopted.

5.3.5.2 Compaction

The treatment procedure proposed does not require compaction of the first layer of the Foundation Fill. Where it has been feasible to dewater prior to placement of fill, roller compaction may be carried out. For thick foundation fills, when no roller compaction has been carried out, the embankment fill needs to be constructed

of granular material and deep compaction procedures such as dynamic consolidation and vibro-compaction are recommended.

6.0 PERFORMANCE OF TREATED SLIME

The performance of the treated slime has been monitored by various measurements that were made. The following sections discuss the performance results obtained in relation to predictions from theories using laboratory test derived parameters.

6.1 SLIME UNDER FOUNDATION FILL AND EMBANKMENT LOAD

The slime is essentially confined by the sides of the pond and monitoring of performance has to take into account this factor.

6.1.1 Measured Strength Profile

The measured strength profile for the slime in one of the ponds under foundation fill load is compared with that predicted from the theoretical line $s = 0.16p_0$ in Fig. 9.

It can be seen that there is a distinct gain in strength after the rest period. The observation is important from the design point of view as the next stage loading is dependent on the gain in strength. It also affirms that the material is capable of developing a gain in strength.

6.1.2 Consolidation Characteristics

6.1.2.1 Settlement/Time of Consolidation

The measured total settlement as well as the rate of settlement are compared with that predicted from laboratory test results. The comparisons are carried out using the parameters $C_c/1+e_0$ (compression index) and c_v (coefficient of consolidation), respectively. The results are given in Table 3 which presents the values from laboratory tests as well as those back-analysed from field observations. The assumed design values prior to the implementation of the project are also presented. At the design stage it was assumed that c_h is equal to c_v.

It can be seen from the table that for the

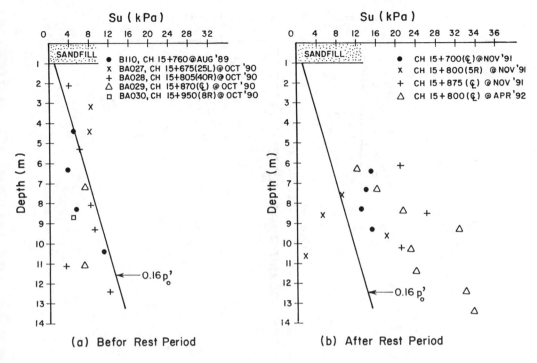

Fig. 9   GAIN IN STRENGTH

	LOCATION	LABORATORY TEST	DESIGN ASSUMPTION	FIELD OBSERVED	VERTICAL DRAINS
C_c ------ $1+e_0$	POND 1 (8A1)	0.113	0.308	0.151	YES
	POND 3A (8A1)	0.160	0.308	0.118	YES
	POND 9 (8A1)	0.287	0.308	0.241	YES
	POND 24 (8A1)	0.305	0.308	0.275	NO
	POND 29 (8A2)	0.200	0.350	0.200	YES
C_v $(m^2/year)$	POND 1 (8A1)	2.0	4.0	4.4	YES
	POND 3A (8A1)	1.0	4.0	2.9	YES
	POND 9 (8A1)	2.5	4.0	2.94	YES
	POND 24 (8A1)	3.8	4.0	3.61	NO
	POND 29 (8A2)	1.5	1.0	3.00	YES

TABLE 3 TABULATION OF CONSOLIDATION PARAMETERS

compression index $(C_c/1+e_0)$, the field observed values follow closely those that are derived from laboratory tests.

For the coefficient of consolidation (c_v), the field observed values when vertical drains are installed, are generally twice that of the laboratory test values. This may reflect the difference between c_h and c_v. In the single case of purely vertical compression, the field observed and laboratory test results are close to each other. It may be of interest to note that the design values were assumed as the average laboratory-obtained values. Therefore, there still remains a possibility that

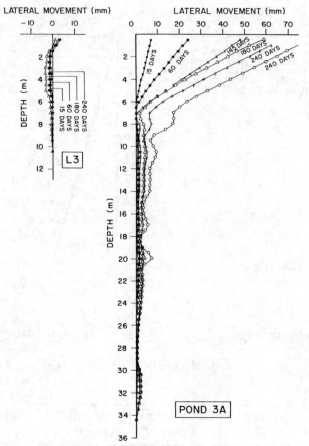

FIG. IO HORIZONTAL MOVEMENT

the differences between field observed and laboratory test values are due to non-homogeneity of the slime deposits rather than differences in horizontal and vertical permeabilities.

6.1.3 Horizontal Movements

Typical horizontal movements as measured by inclinometers in the course of placement of the embankment fills are shown in Fig. 10. The larger displacement in Pond 3A when compared with soft ground L3 may be due to the fact that L3 is confined by the surrounding ground while the measurement point in the former is adjacent to a containment bund.

7.0 CONCLUSION

The recommended planned treatment of ponds and other soft ground areas in ex-mined land, adopted as a total system, have been successfully implemented in several projects.

The treatment design parameters derived from the soil investigation program and subsequent characterisation lead to realistic design values.

The comparisons between predicted and performance values of gain in strength, settlements and rate of settlement as shown in Section 6, by and large demonstrated that the procedure adopted in investigation, design and construction have been a success.

316

5 Dynamic behaviour of soil and earthquake

Prediction versus Performance in Geotechnical Engineering, Balasubramaniam et al. (eds)
© *1994 Balkema, Rotterdam, ISBN 90 5410 355 8*

Case studies on liquefaction-induced ground displacement during past earthquakes in Japan

M. Hamada
Tokai University, Kumamoto, Japan

ABSTRACT: Case studies were conducted on liquefaction-induced ground displacements, and on their caused damage to buried pipes and foundation pipes. The characteristics of the ground displacements were discussed and a regression formula was proposed in order to predict ground displacement depending on the thickness of the liquefiable layer and the gradient of the ground surface as well as the liquefiable layer. Furthermore, the causal relationship between the liquefaction-induced ground displacement and the damage to in-ground structures was examined.

1 INTRODUCTION

The experience of the 1964 Niigata earthquake developed awareness of the following types of damage due to liquefaction: 1) Settlement, tilting, and toppling of structures due to reduction in ground bearing capacity; 2) Floating of buried structures such as manholes and tanks due to their buoyancy in the liquefied soil; 3) Tilting or collapse of retaining walls and quay walls as a result of increased earth pressure. After the Niigata earthquake, research into these types of liquefaction-induced damage has progressed. Various kinds of countermeasures to prevent such damage have been developed, and they are being applied in practice.

By using a set of two aerial photographs taken before and after the 1983 Nihonkai-Chubu earthquake the author and his co-workers showed for the first time that the liquefied soil may induce large permanent ground displacements of several meters depending on the geographical conditions.(Hamada 1986, Hamada 1992, O'Rourke 1992) These large ground displacements severely damaged buried pipes and foundation piles in Noshiro City during the Nihonkai-Chubu earthquake as well as in Niigata City during the Niigata earthquake. Therefore, the liquefaction-induced ground displacements have been strongly recognized as an urgent and serious subject which needs to be resolved for the earthquake resistant design of buried structures.

This paper introduces some results of case studies on liquefaction-induced ground displacements and their related damage during past earthquakes in Japan, and discusses the reliability of regression formula and the accuracy of analytical methods which were proposed for the prediction of the liquefaction-induced ground displacements.

2 LIQUEFACTION-INDUCED GROUND DISPLACEMENT

The 1964 Niigata earthquake with a magnitude of 7.5 caused extensive soil liquefaction in Niigata City and the surrounding areas. Many buildings, bridges, quay walls, and lifeline systems such as electricity, gas, water, and telecommunication suffered severe damage.

The vectors in Figure 1 represent the horizontal ground displacements, which are caused by the soil liquefaction at Ohgata area in Niigata City. The numbers in the parentheses are the vertical displacements. The shadowed area in the figure is a natural levee, while the other area is mostly an old riverbed. The Tsusen River was about 10 m wide at the time of the earthquake, but it had been the old stream of the Agano River with a large width. The horizontal displacements were caused from the natural levee with a little higher elevation toward the old riverbed with lower elevation. However, it must be noted that the mean gradient of the ground surface is very small, at less than 1.0 %.

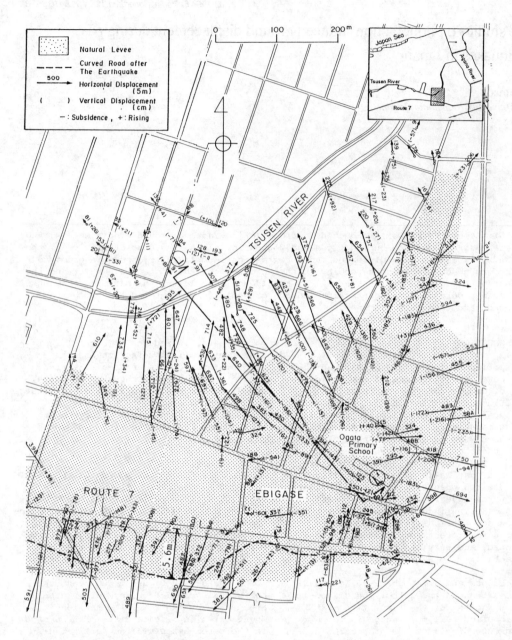

Fig.1 Permanent ground displacement in Ebigase and Ohgata areas

Ground displacements with a remarkable magnitude were observed at Ohgata Primary School and its vicinity. From the playground of the school, displacements occurred in northwest, north and northeast directions. Displacements in northwest direction were particularly dominant, and continued for about 300 m towards the Tsusen River. The maximum displace-ment in this area was over 8 m. Beyond the Tsusen River, the displacement ended.

According to measurements of vertical displacements, shown as numbers in parentheses in Figure 1, most part of the area subsided due to liquefaction. The primary school and its vicinity, where the horizontal ground displacement began, subsided greatly with a maximum of 2.0 m. On the

Photo 1 Risen portion of the Tsusen River

Photo 2 Ground fissure in the playground of the school

contrary, the ground rose at many points in the neighborhood of the Tsusen River where the horizontal ground displacements ended. Many witnesses remarked that near Tsusen River a considerable amount of sand and water spouted from the ground, and the riverbed rose above water level in some places, enabling people to cross the river on foot. Photo 1 shows the Tsusen River after the earthquake where the riverbed rose.

Figure 2 shows the location of sand boils and ground fissures, which were identified in aerial photographs taken after the earthquake. Numerous ground fissures were observed in the area of the primary school, where the ground was under tensile strain and where the surface subsided greatly. Photo 2 shows one example of ground fissures of the primary school. Sand boils were seen in the area of the Tsusen River where the horizontal ground displacements ended and where the surface rose.

Figure 3 shows soil profiles and the liquefied soil layers, which were estimated by the Factor of Liquefaction Resistance F_L(Iwasaki, 1978), along section L-L'. The estimated liquefied layer increases in thickness from 4 m to 6 m between the Ohgata Primary School and the Tsusen River. The boundary between the estimated lique-

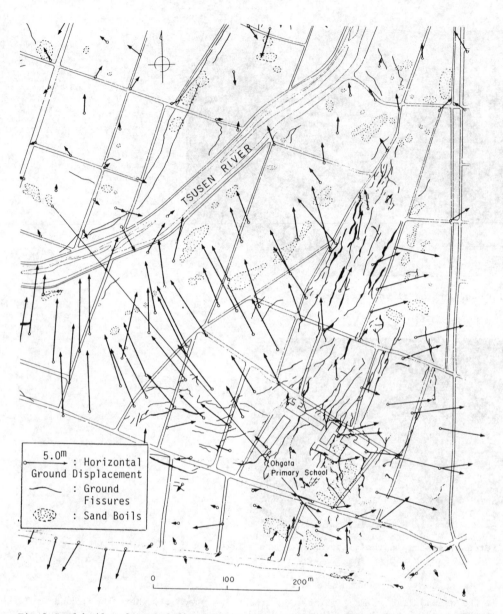

Fig.2 Sand boil and ground fissures in the vicinity of the Ohgata Primary School

fied layer and the lower non-liquefied layer is inclined with the direction of the ground displacements. The ground surface is also inclined toward the river, but with a gradient of about 1.0 %.

3 DAMAGE TO IN-GROUND STRUCTURES BY LARGE GROUND DISPLACEMENT

The Niigata Family Court House, a four-story reinforced concrete building, was located in the Hakusan area on the left bank of the Shinano River. It was constructed on concrete pile foundations, each with a diameter of 35 cm and a length of 6 to 9 m. After the earthquake, the building inclined about 1 degree due to differential settlement of the ground, and it was conjectured that the foundation piles were damaged. However, after minor repairs were made on the inclined floors,

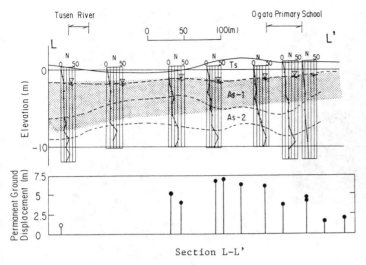

Fig.3 Soil conditions by estimated liquefied layer

the building was used without additional modification for 25 years. When the building was reconstructed, two foundation piles were excavated.

Figure 4 illustrates the damage to No.2 pile and the SPTvalues of the ground, and Photo 3 shows the damage. The pile was damaged at two points. At the upper point the concrete was crushed and the steel reinforcing bars were severely bent as shown in Photo 3(a). At the lower point, there are several horizontal cracks caused by a large bending moment (Photo 3(b)).

The SPT-values down to -8 m, are mostly less than 10, but they increase suddenly below -8 m. It can be assumed that the soil layer below the ground water level (-1.7 m)* and above about -8.5 m liquefied during the earthquake, while the lower layer did not liquefy. It is noteworthy that the two points of damage in the pile roughly coincide with the boundaries between the estimated liquefied and non-liquefied layers.

The permanent ground displacements in the horizontal direction, measured in the vicinity of the building, are shown in Figure 5. The solid vectors shows the displacements on the ground surface around the building as measured from aerial photographs taken before and after the

* It should be noted that the ground water level was measured 25 years after the earthquake.

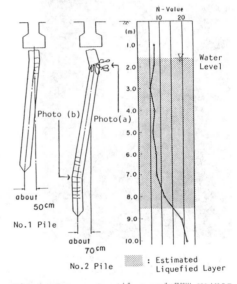

Fig.4 Damage to piles and SPT-values of the ground

earthquake. The ground moved by 1 to 2 m, mostly in the northeast direction.

It was reported that the pile was also deformed mostly in the northeast direction with a relative displacement of about 70 cm between the tip and the top of the pile. Therefore, it can be concluded that the foundation piles failed due to the liquefaction-induced ground displacement.

In Niigata City, numerous buried pipes

(a) Upper part of No. 1 Pile

(b) Lower part of No. 2 Pile
(Bending cracks)

Photo 3 Damage to the foundation piles
of Niigata Family Court House

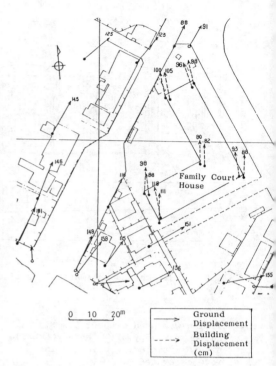

Fig.5 Permanent ground displacement in
the vicinty of Niigata Famaily Court House

Photo 4 Protrusion of a buried gas pipe
above the surface

were severely damaged by liquefaction
and especially by liquefaction-induced
ground displacements. Photo 4 shows one
example of damage to a castiron gas pipe,
which was forced out of the ground. The
gas pipe had a diameter of 15 cm, and
it was connected with mechanical joints.
It is well known that buried pipes with
a specific gravity less than that of the

liquefied soil, such as sewage pipes,
can float. However, such abrupt surfacing
as seen in Photo 4 is less probable as
a result of buoyancy.

Figure 6 shows the permanent ground
displacements in the vicinity of the damaged gas pipe and the ground strain calculated from the measured displacements. A
compressive strain with a magnitude of
about 0.2 %, occurred mostly in the direction of the pipe axis. It can be conclude-

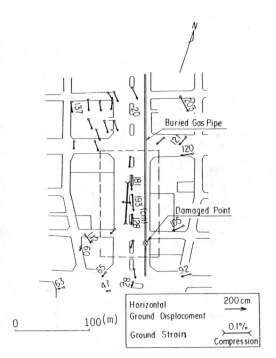

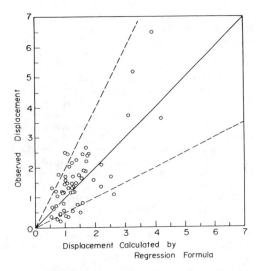

Fig.7 Comparison of the permanent ground displacement estimated by regression formula with the observed ground displacement

Fig.6 Horizontal ground displacement in the vicinity of damaged gas pipe

d that the pipe was buckled and forced above the surface by the compressive strain when the ground lost much of its stiffness against the lateral movement of the pipe due to liquefaction.

4 PREDICTION OF LIQUEFACTION-INDUCED GROUND DISPLACEMENT BY EMPIRICAL FORMULAE AND NUMERICAL ANALYSES

An accurate prediction of magnitude of liquefaction-induced ground displacements is essential for a rational earthquake resistant design of foundation piles and buried structures. For this purpose, some regression formulae and numerical methods have been proposed.

A regression analysis was conducted by Hamada et al. to investigate if there is any relationship between the magnitude of permanent ground displacements, and the thickness of the liquefied layer and the gradient of the ground surface. On the basis of this study, the following empirical formula was proposed:

$$D = 0.75 \sqrt[2]{H} \cdot \sqrt[3]{\theta} \qquad (1)$$

D: Magnitude of permanent ground dis-

placement in horizontal direction(m)
H: Thickness of liquefied layer (m)
θ: The greater gradient of ground surface or lower boundary of liquefied layer (%)

Figure 7 shows the correlation between the ground displacements predicted by the empirical formula and the observed ones. The observed ground displacements are octually scattered between values half and twice the values given by the formula. The proposed formula indicates that the correlation with the gradient of the ground surface is poorer than with the thickness of the liquefied layer, because the cubic root of the gradient is proportional to the displacements.

Several numerical methods have been proposed to predict the magnitude and the distribution of the liquefaction-induced ground displacements. The most important function which is required of these numerical methods is to accurately analyze the behavior of the ground after the liquefaction. The key aspect for the numerical analysis is whether the ground after the liquefaction behaves as a liquid or as a solid. The numerical methods where the ground after the liquefaction is regarded as a solid were proposed by Yasuda (Yasuda, 1991) and Yoshida(Yoshida, 1989). In these methods estimation of residual strength and remaining shear modulus after the liquefaction shall be most influential to the analytical results. According to the case studies of the Niigata and Nihonkai-Chubu earthquakes, the maximum ground displacement reached over 12 m and 5 m, respectively. This means that the shear strain of the liquefied soil with thicknesses of 5-10 m is over 100 %. Generally, it is difficult to properly evaluate the residual strength and the remaining shear modulus under such a large strain.

The methods where the ground after the liquefaction is regarded as a liquid, were proposed by Towhata(Towhata, 1989). In this method, an accurate evaluation of fluid properties of liquefied ground is essential for the accurate prediction of the ground displacements. Some experimental studies have been done by Hamada et al. (Hamada, 1992) in order to investigate the fluid properties.

As for the development of numerical methods which accurately explain the observed ground displacement, the research has been just started, since the phenomena of liquefaction-induced ground displacements was recognized by the earthquake engineering people at the first time during the Nihonkai-Chubu earthquake. For the establishment of rational earthquake resistant design where the effect of the liquefaction-induced ground displacements is fully taken into   consideration, the development of proper numerical methods is strongly required.

5 SUMMARY

This paper introduces some typical examples of liquefaction-induced large ground displacement and their caused damage, and outlines the state of the art about empirical formulae and numerical methods for the prediction of the ground displacements. The contents of this paper can be summarized as follows;

1) Liquefied ground may displace in the horizontal direction with a magnitude of several meters, sometimes over 10 m, depending on the geographical condition.

2) Liquefaction may cause severe damage to in-ground structures such as buried pipes and foundation piles by lateral movements of ground, besides the subsidence of ground and the lifting by buoyancy.

3) The ground displacements predicted by a proposed empirical formula scattered between half and twice of the observed values.

REFERENCES

Hamada, M., Yoshida, S., Isoyama, R., and Emoto, K. 1986. Study on liquefaction-induced ground displacements. A Report by Research Committee, ADEP. Tokyo:20-35.

Hamada, M. and O'Rourke, T.D.(Edition). 1992. Case studies of liquefaction and lifeline performance during past earthquakes, Vol.1, Japanese Case Studies, Technical Report NCEER-92-0001. NCEER.

O'Rourke, T.D. and Hamada, M. 1992. Case studies of liquefaction and lifeline performance during past earthquakes, Vol.2, U.S. Case Studies, Technical Report NCEER-92-0001. NCEER, NY.

Iwasaki, T., Tatsuoka, F., Tokida, K. and Yasuda, S. 1978. A practical method for assessing soil liquefaction potential based on case studies at various sites in Japan. Proc. 5th Japan Earthquake Symposium: 641-648(in Japanese).

Yasuda, S., Nagase, H., Kiku, H. and Uchida, Y. 1991. A simplified procedure for the analysis of the permanent ground displacement. Proc. 3rd Japan-U.S. Workshop on Earthquake Resistant Design of Lifeline Facilities and Countermeasures for Soil Liquefaction, NCEER: 225-236.

Yoshida, N. 1989. Finite displacement analysis on liquefaction-induced large permanent ground displacements. Proc. 2nd U.S.-Japan Workshop on Liquefaction Large Ground Deformation and Their Effects on Lifelines, NCEER: 207-217.

Towhata, I., Yamada, K., Kubo, H. and Kikuta, M. 1989. Analytical solution of permanent displacement of ground caused by liquefaction. Proc. 2nd U.S.-Japan Workshop on liquefaction, Large Ground Deformation and Their Effects on lifelines, NCEER: 131-144.

Hamada, M. Otomo, K. Sato, H. and Iwatate, T. 1992. Experimental study on effects of liquefaction-induced ground displacement on in-ground structures. Proc. 4th Japan-U.S. Workshop on Earthquake Resistant Design of Lifeline Facilities and Countermeasures against Soil Liquefaction, NCEER: 481-492.

Prediction versus Performance in Geotechnical Engineering, Balasubramaniam et al. (eds)
© 1994 Balkema, Rotterdam, ISBN 90 5410 355 8

Evaluation of effective factors causing ground failures during 1990 Manjil-Iran earthquake

S. M. Fatemi Aghda & J. Ghayoumian
Graduate School of Science and Technology, Kumamoto University, Japan

A. Suzuki & Y. Kitazono
Faculty of Engineering, Kumamoto University, Japan

A. A. Nogole Sadat
Applied Geology, Geological Survey of Iran, Tehran, Iran

ABSTRACT: In this work, a study of the Gilan region(north part of Iran) is performed to determine what geological and environmental factors affected slope movements during an earthquake in the area. Engineering geology aspects concerning regional faulting and related minor structures, lithology, drainage and hydrology, geomophology, and effect of man were considered.

Field works were conducted in the whole area and among tens of slope movements, six big slopes(because of volume or resulted damages) were selected. For these slopes a detailed site investigation were performed to determine the relative significance of the factors influenced on the slope failure. To find out about the effect of weathering rate on failed slopes X.R.D analysis were conducted on the bed rocks underlying the failure zones.

This study revealed that, the area is an active area and geological structures specially faults had an important role in the ground movement. Most of the slope movements are located on or near the faults. The bed rocks underlying the area, in most of the landslides, are associated with certain kinds of the rocks such as weathered andesites, shale and marly limestones, that are indicative of the field instability. The other important factors which affected the slopes failure are existence of surface and ground water in failure areas.

Key words: Earthquake, Earthquake damage, Landslide, Ground failure, Weathering rate

INTRODUCTION

On 21 June 1990, an earthquake (Manjil earthquake) with Mb=7.3 and Ms=7.6 hit the north-northwestern Iran, just 30 minutes after midnight local time (21 hours June 20 GMT), Fig. 1. The epicenter was located about 200 km northwest of Tehran (36°50 N and 49°25 E), between Manjil and Rudbar with focal depth of about 10 km. The quake-shaken area was about 600000 km² with the maximum intensity between X and XI in modified Mercali scale at the epicenter, as reported by Moinfar and Nader Zadeh (1990).

The extent of damage to structures reported by International Institute of Earthquake Engineering and Seismology(IIEES) (1991) was between 10 to 100%. A number of accelerograph installed in the region recorded the main shock. The nearest accelograph to the epicenter of earthquake in the region is Abbar station, some 60 km to the west of epicenter, in which the recorded peak horizontal acceleration was 0.65g and in vertical direction was 0.52g(Moinfar and Nader Zadah,1990). The maximum horizontal acceleration recorded at Lahijan was 0.17g, which is 75 km far from epicenter. Several aftershocks, with magnitude larger than 5, hit the area within a few weeks after the main shook.

The results of this earthquake were vast liquefaction in the saturated sand and silty sand deposited area in northern Gilan prefecture, numerous landslides which buried villages, farms and roads and rock falls that induced damages to road and farm lands.

The purpose of this paper is to determine the geological, geomorphological, and environmental factors which affected tens of slopes failure during the earthquake in the area.

GEOMORPHOLOGY OF AREA

The region shows a clear zonation into a number of parallel, northwest–southeast trending mountain ranges and intervening valley-plains.

The geomorphological zones in this region are as follows(Fig 1):

1. The Talesh mountains

This zone, which is located in the north of the Manjil basin, forms a direct western extension of the high central range of the Alborz and exposes a broad core of old phyllites and a strongly lacunary cover a Paleozoic and Mesozoic sedimentary rocks and Eocene volcanic formations. The elevations of the crestal part reaches between 2500 and nearly 3000 m. The north sides of the mountains are thickly wooded, but forest cover decreases sharply in the crestal region and on the south flank. The wooded northern part, including the lower part of the sefidrud valley, belong to the fertile caspian region.

2. The Manjil basin is formed by the lower Qizil Uzun and Shahrud valleys, which join to form the sefid–rud valley at Manjil. The basin is a deep intermountain depression filled by hill–forming Neogene red beds and extensive river terraces.

3. The Tarom mountains, formed exclusively of Tertiary volcanic rocks and granites, have a very asymmetric profile.

REGIONAL GEOLOGY OF THE AREA

The mentioned geomorphological zones(Fig 1) form parts of the western Alborz mountain of northern Iran(NIOC,1978),an accurate system that stretches in a general east–west direction along the southern shows of the Caspian sea.

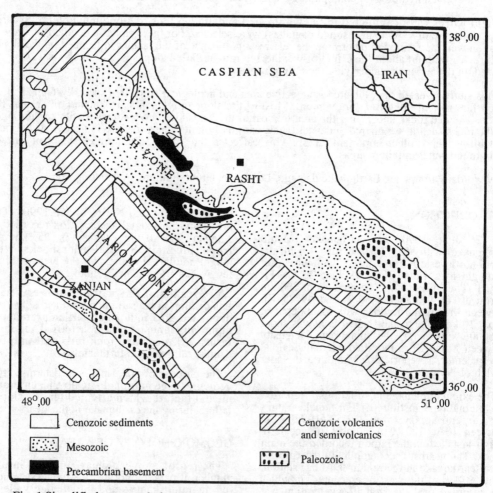

Fig. 1 Simplified geomorphology and regional geology map of Gilan area
    (after Tchalenko et al. 1974)

Developed as part of the Alpine–Himalaya orogeny in the late Cretaceous –Tertiary time, the Alborz mountains are made up of a number of parallel mountain ranges and intervening longitudinal valleys. The regional trend in the epicenteral area is WNW–ESE. The cores of the Alborz mountains consist of strongly deformed metamorphic basement complex, which are covered by the Paleozoic and Mesozoic rock formations(Stocklin and Eftekhar–nezhad,1969). The Talesh mountains are characterized by intense faulting and steep thrusting, and they contain a broad zone of phyllites of Precambrian age in the crest and northern flanks. Geological map of this study area is shown in Fig 2.

The most important geological formation which involves slidings and rock falls area are as follows :

PRECAMBRIAN BASEMENT

The rocks that attributed to the precambrian have several large outcrops in the Talesh mountains and in the Soltanieh mountains. They include three major rock complex: the Metamorphic complex, the Kahar formation, and the Doran–type granites.

The Metamorphic complex rocks of the Talesh mountains are derived mainly from argillaceous and siliceous sedimentary rocks, and the most common rock types are Slates and Phyllites, Quartz–Sericite–Chlorite Schists, and Albite–Chlorite–Muscovite Schists. The Kahar formation consists of a sequence of Argillaceous to Slightly Siliceous Slaty Shale.

JURASSIC

The Jurassic sequence of the area are Plant–bearing"Continental deposits"(Shemshak formation) in the lower part, and marine limestones in the upper part.

The plant and coal-bearing shales and sandstones constitute the Shemshak formation in its type area in the central Alborz(Assereto,1966a). Dark–green argillaceous and sandy shale, black carbonaceous shale, green–gray sandstone, and occasional thin lenticular layers of coal are the main rock types.

A local facies variation is found in the Manjil-Rudbar area of the Talesh mountains, where the typical shales and sandstones contain many intercalations of coarse–grained conglomerate, some of which attain thickness of nearly 100 m. Near Rudbar, a one meter thick coal bed immediately underlies such a conglomerate bed.

CRETACEOUS

Cretaceous rocks have outcrop in   several points in the area. The thickness of these rocks is considerable. The study by Eftekhar–nezhad, Nabavi and Valeh(1965) has been much hampered by the strongly fractured condition of the rocks and a relatively dense vegetation of that region where limestone predominate. The massive limestones which are mostly recrystallized and light–gray in color, form the high ridge of Kuh–e Kapateh. The massive limestones grade upwards into thin–bedded marl and dense marly limestone of yellow to buff color, which also include some intercalations of black classtic limestone.

TERTIARY

The Tertiary volcanic and sedimentary rocks occupy the greater part of the area, and the overlap of these sequence on older formations is not in conformity.
Karaj formation

This formation consists of a great variety of lava flows of mostly andesitic composition and of associated tuffaceous beds such as breccias, lapilli tuffs, pumice tuffs, and tuffaceous sandstones and mudstones.

PLIOCENE–QUATERNARY SEDIMENTS

In the Mangil basin, these sediments mostly are alternating sandstone, shale, and gypsiferous mudstones(Ng1); and alternation of sandstone, mudstone, and conglomerate(Ng2).

In the other area, the considerable sediments on these times are soft loamy and sandy beds and chalky limestone(plio–plistocene); several alluvial terraces consist of poorly consolidated gravel,sand and clay; and loess which distributed on the valley flanks of the lower Sefied–rud in north of rudbar and consists of light–brown to pink, poorly consolidated and poorly stratified loamy and silty materials. The loess forms a high terrace conforming with one of the older valley levels of the Sefid–rud river.

SLOPE FAILURES INDUCED BY THE EARTHQUAKE

Among various kinds of damages caused by the earthquake, one of the most significant was tens of landslides that occurred on the mountain slopes within the area   close to the activated faults.
The structural geological map of the area which also shows the location of the most catastrophic landslides is presented in Fig 3.

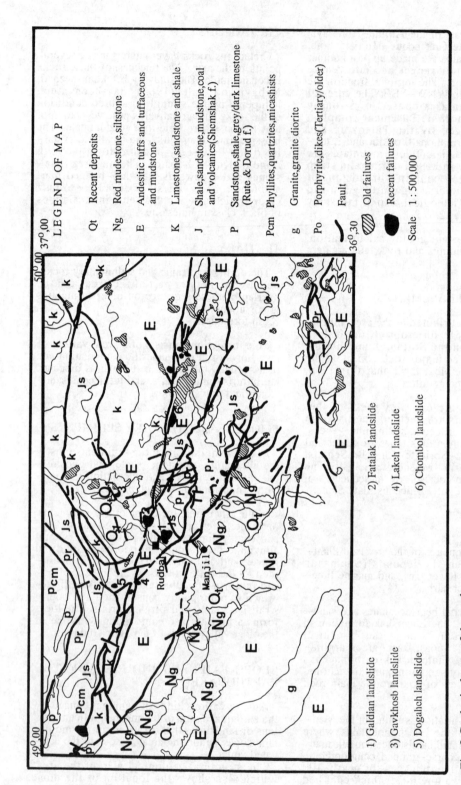

LEGEND OF MAP

Qt	Recent deposits
Ng	Red mudstone,siltstone
E	Andesitic tuffs and tuffaceous and mudstone
K	Limestone,sandstone and shale
J	Shale,sandstone,mudstone,coal and volcanics(Shemshak f.)
P	Sandstone,shale,grey/dark limestone (Rute & Dorud f.)
Pcm	Phyllites,quartzites,micashists
g	Granite,granite diorite
Po	Porphyrite,dikes(Tertiary/older)

Fault

Old failures

Recent failures

Scale   1 : 500,000

1) Galdian landslide   2) Fatalak landslide
3) Gavkhosb landslide   4) Lakeh landslide
5) Dogahch landslide   6) Chombol landslide

Fig. 2 Geological map of the study area

The rupture zone defined by site investigation by Moinfar and Naderzadeh (1990) and Berberian et al. (1991) is reported to be about 80 km long with a major trend of northwest-southeast direction (parallel to the local structural trends) and follows in the north part of the Manjil basin boundary. The maximum measured horizontal and vertical offsets of ruptured fault was reported to be about 20 and 50 cm by Moinfar Nazerzadeh (1990). The observed vertical offset by the authors is approximately 2.5-3 m in Pakdeh village which caused the settlement of road (Fig. 13).

The main aftershocks revealed the same trend as the trend of rupture zone. The trend of the failure zone is same as fault rupture zone(Fig 3).

In this work the results of site investigation on the six major failures namely Galdian landslide, Fatalak landslide, Gavkhosb landslide, Lakeh landslide, Dogaheh landslide, and Chombol landslide are reported. Among these slides, the Galdian and Fatalak landslides due to volume of carried materials and rate of damages are important.

1. GALDIAN LANDSLIDE

The Galdian landslide with about 32 million cubic meters moved materials occurred east of Rudbar city. This slide took place one day after the occurrence of main shock and caused the breaking of water and oil lines, damaged to more than 20000 big olive trees and destructed the electrical lines. The moved materials are mostly shale and sandstone of Shemshak formations, which are highly weathered. The direction of layers in the slope is same as the slope of ground surface(Fig 4).

2. FATALAK LANDSLIDE

The Fatalak landslide occurred near the Rudbar–Kelishom fault. The moved mass are mostly dune sand and alluvial deposits which overlaid on the andesite rocks and sandy tuff with considerable thickness of Karaj formation(Fig 5).

The length of failure zone is more than 500 meters which caused a 30–40 meters scarp area. The slide buried the Fatalak and Fishom villages with those populations.

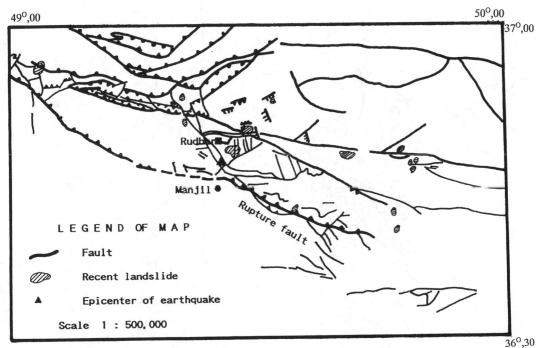

Fig. 3 Structural geology map of failure area

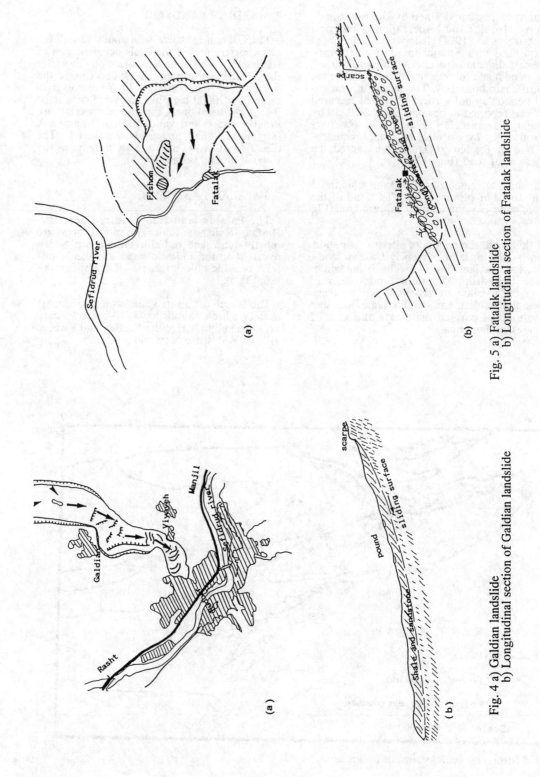

Fig. 4 a) Galdian landslide
       b) Longitudinal section of Galdian landslide

Fig. 5 a) Fatalak landslide
       b) Longitudinal section of Fatalak landslide

Fig. 6 Gavkhosb landslide

The movement area is forest zone of Alborz mountain with slope angle more than 45 degree.

3. GAVKHOSB LANDSLIDE

The Gavkhosb landslide took place 8 km away from Rudbar–Kelishom fault. The sliding materials are quaternary sediments, which are mostly silts. The mass failure had a slope angle of about 70–80 degree. The slide damaged the farmlands(Fig 6).

4. LAKEH AND DOGAHEH LANDSLIDES

In the Lakeh area, several landslides occurred. Most of their materials are berccia tuff, andesite, limestone and shaley formations. Among them, the Lakeh and Dogaheh landslides due to volume of moved materials and damage to life are considerable. Also rockfalls occurred in Kapateh village, and several long and deep cracks in Kapateh and Lakeh-Dogaheh area are remarkable.

The Lakeh landslide which is located   500 m far from Lakeh–Jeishabad fault is a part of big ground movement in the area. The main scarp of this landslide is more than 10 m and the moved materials are mostly tuff, limestone and shale(Fig 7). The surface soils mountain slope which resulted by weathering of shales and tuffus layers, traveled down the valley and buried the houses which were constructed on the slope. The failured layers belong to Karaj formation. The layers of landslide have a N80W strike and 30 E dip.

The Dogaheh landslide which is located on NW of the Rudbar city took place on a mountain slope, with a scarp about 30 m in phyllite and micaschist of precambrian(Fig 8). The sliding mass destroyed a part of village with its population, damaged   the farmlands, moved old trees, collapse in lateral area of slide and farmland. The distance of traveled old trees on the mountain slope is considerable and estimated approximately about 100 m. within the moved mass, several huge blocks of rocks are remarkable

6. CHOMBOL LANDSLIDE

The Chombol landslide took place on Kelishom–Rudbar fault. The slope failure materials, are weathered andesitic tuff rocks, of Karaj formation(Fig 9).

The landslide caused settlement in landform and destruction of a road in Chombol village. A rock failure also took place near to this landslide and the rupture of bed rock in the area are considerable.

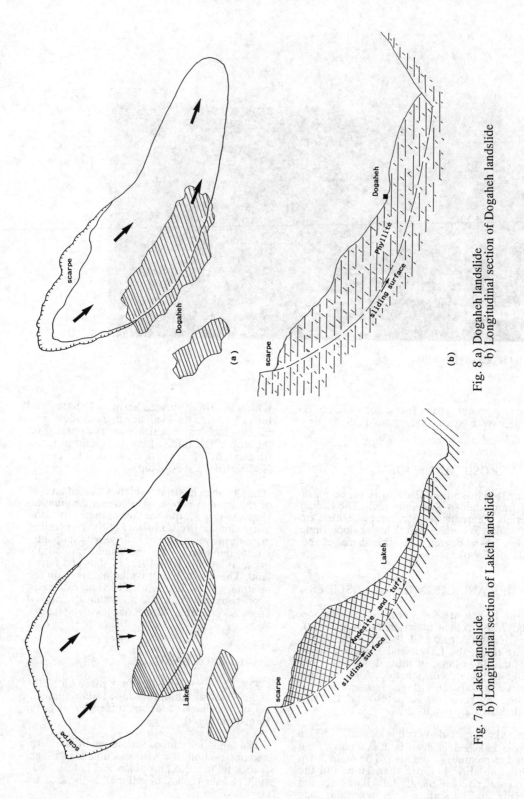

Fig. 7 a) Lakeh landslide
    b) Longitudinal section of Lakeh landslide

Fig. 8 a) Dogaheh landslide
      b) Longitudinal section of Dogaheh landslide

334

Fig. 9 Chombol landslide

DISCUSSION

In an attempt to clarify the environmental and geological factors affecting slide failures during the earthquake different parameters such as lithology, geomorphology, structural geology, land use, effect of under ground and ground surface water, and weathering rate of sediments for above mentioned slides were checked and recorded.

1. EFFECT OF LITHOLOGY

Table 1 illustrates the lithofacies of the bed rocks underlying the slides. It is obvious that more than 55% of the slides are related to the Karaj and Shemshak formations, which can be attributed to their weathering and inherent unstable characteristics of these formation. Figure 10 shows the distribution of landslides in geological formations.

2. STRUCTURAL GEOLOGY

To find out about the effect of this factor on slide failure for old and new failures the direct distance to the major faults were measured. The results are given in Fig 11, which proved the fact that most of the slides are located on or near to the faults.

3. THE WEATHERING FACTOR

To identify the effect of weathering factor on occurrence or acceleration of slides, mineralogical characteristics of landslides forming materials were examined by means of the X-ray diffraction method.

X–RAY DIFFRACTION ANALYSIS

The samples which were collected from bed rock and sliding surface of six mentioned landslides, were analyzed. The diffraction traces of these samples, revealed that, the weathering of geological formations, specially in Karaj and Shemshak formations, are considerable. The diffraction traces of Galdian and Chombol landslide samples are illustrated in Fig 12.

The results show that in the sliding zone, peak intensities of primary minerals such as feldspars decrease and those of clay minerals increase.

4. THE OTHER FACTORS

Some the other factors also affected or accelerated slides in the area. Among them geomorphological feature of the area, effect of ground water, and surface water which was due to irrigation are significant. Table 1 illustrates the different geological, geomorphological, and environmental factors affecting or accelerating the slope failures during the 1990 Iran earthquake. It is clear that most of the landslides have angle dip more than 45 degrees.

CONCLUDING REMARKS

The 1990 Iran earthquake was the most recent major earthquake, for nearly 10, to have hit Iran for the last 30 years. Most of the damages were induced by several huge landslides in mountain areas and liquefaction in plain areas. The field

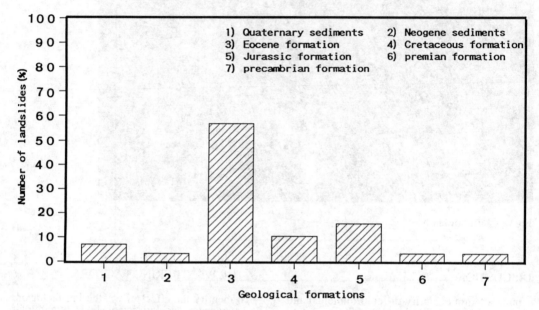

Fig. 10 Distribution of landslides in geological formations in Gilan area

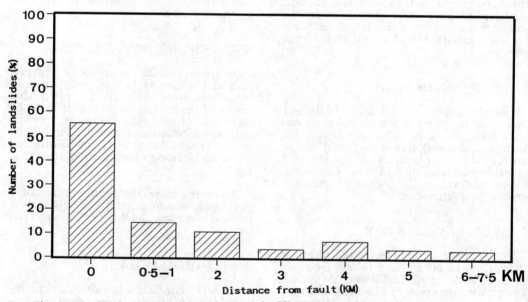

Fig. 11 Distribution of landslides from fault in Gilan area

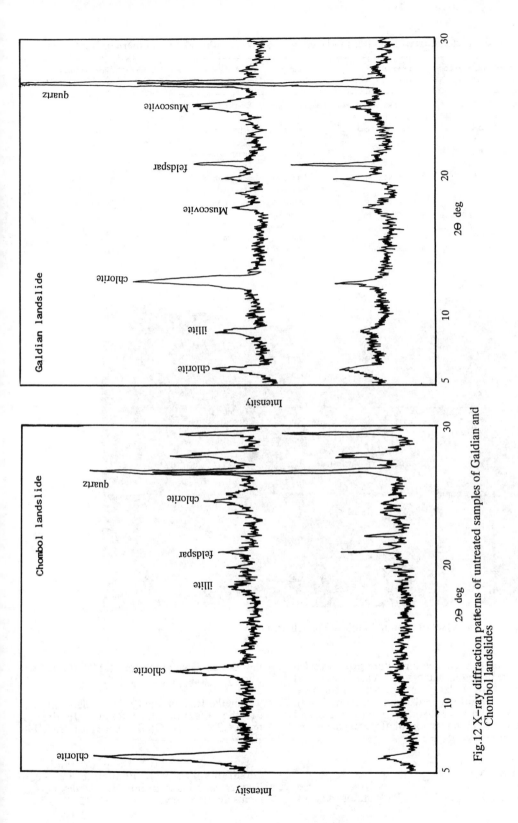

Fig.12 X-ray diffraction patterns of untreated samples of Galdian and Chombol landslides

Table 1. Characteristics of landslides induced by 1990 Iran earthquake in northern of Iran(Gilan area)

Investigated landslide	Distance from fault(Km)	Geological features	Land use	Angle of slope (degree)	Effect of ground water	Effect of surface water
Galdian	2.5	Shale, Sandstone(Shemshake formation)	Farm land	Smaller than 45	+	–
Fatalak	0.5	Conglomerate, Loss	Farm land, Forest cov.	Higher than 45	–	+
Gav Khosb	6.5	Silt	Farm land	Higher than 45	+	+
Lakeh	0.5	Tuff, Andesite, Shale and Limestone	Weakly forest cover	Higher than 45	+	–
Dogaheh	0.5	Shist, Andesite, and Limestone	Farm land, weakly forest cover	Higher than 45		+
Chombol	0.0	Sandstone, Tuff, and andesite, Aglomerate	Weakly forest cover	Higher tan 45	–	–

Fig.13 Vertical offset of ruptured fault in Pakdeh area

and laboratory studies on failure area revealed that, the important factors on occurrence of land slides in Gilan area are geological features(lithology and distance from fault), environmental factors such as(surface and ground water, and geomorphological character-istics of area).

REFERENCES

Alavi, M., et. al(1975). Explanatory text of the Zanjan Quadrangle map(1:250,000). Geol. Survey of Iran, No. D4

Annells, R.N.,et. al.(1975). Explanatory text of the Qazvin and Rasht Quadrangle map(1:250,000). Geol. Survey of Iran, No. E3,E4

Anvar S.A. ,L. Behpoor and Ghahramani,(1991). Landslide in recent Rudbar earthquake in Iran. Landslides Glisse-ments de Terrain, vol.2, 1169–1172

Assereto, R.,(1966a). The Jurassic Shemshak formation in
central Elborz(Iran). Riv. Ital. Paleont, vol.12,4. 1133–1182

Berberian,M. et all(1991). The Rudbar–Tarom earthquake of 20 June 1990 in NW Persia: Preliminary field and seismological observations, and its tectonic significance.

Clark, G.C., et. al(1969). Explanatory text of the Badar–e– Anzali Quadrangle map(1:250,000). Geol. Survey of Iran, No. D3

Eftekhar–nezhad,J. ,M. Nabavi, and N. Valeh,(1965). Geology of Tarom–Talesh area. Geol. survey of Iran, Geol. Note No. 16,24, with map(1:100000).

International Institute of Earthquake Engineering and Seismology,(1991). Preliminary report on Manjil–Rudbar earthquake of June 21,1990, Tehran,Iran.

Ishihara, K. and A. Nagao(1983). Analysis of landslides during the 1978 IZU–OHSHIMA–KINKAI earthquake. Soil and Foundations, vol. 23,1. 19–37.

Moinfar, A.A and A. Naderzadeh(1990). Preliminary report on the Manjil, Iran earthquake of 20 June 1990. Building and Housing Research Center, Iranian Ministry of Housing and Urban Development.

National Iranian Oil Company(1978). The geological map of Iran(1:1000,000).

Niazi, M. and Y. Bozorgnia(1992). The 1990 Manjil, Iran earthquake: Geology and seismology overview, PGA attenuation and observed damage. Bull. Seism. Soc. Am. 82,2 774–799.

Author index